Thomas Maitland (1885-1976). The most outstanding plant collector of Ijim Ridge. In January to June 1931 he travelled between Belo, Laikom, Mbesa and Nchain and was first to collect specimens of many of the endemic plants of the Mt Oku and Ijim Ridge area, including *Kniphofia reflexa, Habenaria maitlandii, Plectranthus punctatus* subsp. *lanatus* and *Newtonia camerunensis*. He started his career as a Kew gardener and was then recruited as a supervisor at a succession of botanic gardens in Africa, including Calabar, Entebbe and finally Victoria (Limbe).

The Plants of Mount Oku and the Ijim Ridge, Cameroon

– A Conservation Checklist –

Compiled and edited by

Martin Cheek, Jean-Michel Onana and Benedict John Pollard

assisted by Laszlo Csiba

Royal Botanic Gardens, Kew
Herbier National Camerounais

Published by the Royal Botanic Gardens, Kew

First published 2000

Printed in the European Union by Redwood Books Ltd., Trowbridge, Wiltshire, U.K.

Typeset by Maureen Bradford (Introductory Chapters) and Laszlo Csiba, Penelope Doyle and Benedict John Pollard (Main Checklist)

Cover design by Jeff Eden, Media Resources, R.B.G., Kew

ISBN 1 84246 016 1

Front cover: the endangered vine *Pentarrhinum abyssinicum* Decne subsp. *ijimense* Goyder (Asclepiadaceae) known only from three sites in the lower montane forest below Ijim Ridge. Photo of *Cheek* 9943, taken at Ntum, November 1999.

Rear Cover: Satellite images of Mt Oku and the Ijim Ridge (© BirdLife International).

CONTENTS

FIGURES

DICOTYLEDONS

MONOCOTYLEDONS

PREFACE

Christian Azenui Asanga

Manager, Kilum-Ijim Forest Project

The Kilum-Ijim Forest is important to local people, as a source of water, honey, timber, firewood, medicine and other forest products and for its role in local traditions and culture. Use of the forest is divided by ethnicity, the Oku and Nso people using the Kilum forest and the Kom people the Ijim forest. Throughout its twelve years of operation, the Kilum-Ijim Forest project has been developing a system of full community participation in forest management and from 1994 it embarked on establishing a system of community-based forest management, following the enactment of new forestry legislation. Early efforts to produce a crude working plant list for the project revealed that local names exist for most of the plants of Kilum-Ijim in the three languages: Oku, Nso and Kom, again showing the keen interest that the people have for their forests. Most of the uses listed are those given by the local people. This checklist of plants of the Kilum-Ijim Forest will go a long way to enhancing the existing knowledge and interest and mitigating the dangers of mismanaging the habitats containing these plants.

KIFP's purpose is: the biodiversity, extent and ecological processes of the Kilum-Ijim Forest are maintained, and the forest is used sustainably by local communities. The partners in this work are the communities and traditional authorities in the fondoms of Oku, Nso and Kom and government, with the project acting as a facilitator. This means a conservation partnership for community forest management. All three partners are participating fully in the process of establishing community forests. Local Forest Management Institutions (FMI's) have been formed covering the whole forest, although the various FMI's are at different stages in the process of applying for granting of community forests. Several FMI's have completed inventories of their forests and are developing management plans. As agreed in a three-fondom set of rules, each community forest will have a core conservation area with very restricted use. These areas are already showing up in the management plans. This plant checklist gives the status of the different plants and the various habitats containing them and this will certainly guide the various communities in drawing up their management plans. The Kilum-Ijim area is renowned for its traditional medicine and herbalists abound. The plants on the mountain and around Lake Oku are special to these herbalists and they will also very much use this checklist.

This community use of the plant checklist is stressed here to emphasise the special applicability of this particular work. Needless to enumerate the scientific, educational and other applications of the work. The document is the culmination of hard work that has been done over the years and it is just the beginning.

1

ACKNOWLEDGEMENTS

Firstly we wish to thank the Fons of Kom, Oku and Nso and the people of their Fondoms for the hospitality and support that they have given us during fieldwork on the plants of Mount Oku and the Ijim Ridge. It is they who protect and manage the Kilum-Ijim forest. We also thank the Ardos of Laikom and Mbesa for their hospitality and help with provision of horses. Our guides assisted us greatly, not just in showing us the paths to follow, but providing information on local names and uses. Therefore we thank Peter Wambeng, Patrick Lem, Isaac Fokom and Pa Nfome of Oku, Frederick Nkwah, Johnson Ayeah, David Ngong and Raphael Kombeh of Aboh, Kelles Anchang, Peter Yama, Amos Ndi and Andrew Talla of Laikom-Fundong. Amos Ndi is especially thanked for co-ordinating our stay at Laikom.

We are especially grateful to all at the Kilum-Ijim Forest Project (executed by BirdLife International in collaboration with MINEF – the Ministry of Environment and Forestry) for inviting us to work in the Bamenda Highlands of Cameroon, and to focus on the plants of Mount Oku and the Ijim Ridge. If it had not been for them we would never even have started this venture. They helped by arranging accommodation, helping us with logistics, participating in the fieldwork, liaison with local communities and by encouraging us with our work. In particular we would like to thank Anne Gardner and John DeMarco, Co-project Managers for the duration of most of our fieldwork and Christian Asanga, their successor, who wrote the Preface to this book. Jerry Nkengla, Site Manager at Ijim, Chiabi Lawrence and Fuchi Emmanuel played a leading part in teaching our expedition members about the Kilum-Ijim Forest Project. Njini Gideon organized our accommodation at the Palace of his Highness Fon Vincent Yuh II of Kom in 1998 and 1999. Chiambah Cletus Tuh kept our Landrover running. The 'ecomonitors' of the Kilum-Ijim Forest Project have become our close associates over the years. First and foremost, Moses Kemei, Clement Toh, Innocent Wultoff and Ernest Keming have given us much-valued help, guidance and information during the fieldwork, in which they have participated so ably and enthusiastically. We also thank their present leaders Romanus Ikfuingei and Dr Philip Forboseh Forbah, and their past leader, Dr Fiona Maisels.

Satellite images were provided by BirdLife International. BirdLife International also provided a contribution towards publication costs through a grant from GEF-World Bank under the Cameroon Biodiversity Conservation and Management Programme. Dr David Thomas of BirdLife International contributed a review of the introductory chapters of the Checklist and gave permission for reproduction of BirdLife images. Christine Alder helped with retrieval of BirdLife literature and Marc Languy with the latest report on the status of forest survival in the Bamenda Highlands.

The Darwin Initiative grant to R.B.G., Kew is the main single source of funds to our 'Conservation of the Plant Diversity of Western Cameroon' Project. We are grateful for this assistance without which the work necessary to prepare this book for publication would not have been completed at such an early date, and perhaps not at all. In particular we thank Valerie Richardson, Sylvia Smith, Helen Munro and Sarah Collins for their administration of this grant.

The Earthwatch Institute, Europe, supported by DG VIII of the European Commission, have been the main sponsors of our fieldwork in 1996, 1998 and 1999. We are grateful to all the Earthwatch fellows and volunteers that gave their time to help us in forest, grasslands and swamps of North West Province Cameroon particularly Patrick Ekpe, Gyakari John Ntim, Augusto Correira, Sam Moropane, Rephius Mashego, Kaizer Mdluli, Mandla Tembe (at Oku-Elak, 9-14 June 1996); Bernard Tabil, Shefe Siraj Bekelie, Mirutse Gebru Giday, Elizabeth Hagan, Paul Munyenyembe, Napolean Tsegaye, Sileshi Nemomissa, Henry Ndangalasi, Frank Mbago, Ben Pollard, Anne Dillen (at Oku-Elak 25 October-9 November 1996); Cecilia Maliwichi, Dickson Kamundi, Richard Duah Nsenkyire, Owusu Agyei, Sonny Moyo, Dr Manyuchi, Mr Baedermariam, Mr Gashu, Rachel Holder, Phil Callow, Ben Pollard (at Aboh, Ijim, 16-29 November 1996); Gabriel Albano, Edson Konde, Veronica Muiruri, Jane Nyafuono

2

Mukunya, Patrick Mucunguzi, Alfred Maroyi, Byamukama Biryahwaho (at Laikom 1 December-15 December 1998); Anthony Anderson Kimaro, Kambale Kioma, Kalindula Musavuli, John Amponsah, Maryam Abdul-lah (at Laikom 2-16 November 1999).

From head office in Oxford, Sally Moyes, Gill Barker and more recently Lucy Beresford-Stooke were extremely helpful in recruiting helpers for us. Julian Laird, Robert Barrington and Pamela Mackney were also of great assistance.

Earthwatch funds allowed the employment of staff in Cameroon. We thank Camp Managers Pascale Ngome and Henry Ekwoge (1996), Theophilus Ngwene (1998) and Martin Mbong and Hoffmann Ngolle Ngolle (1999). In particular we thank Martin Etuge, crack plant hunter and his assistants Freddy Epie and Edmondo Njume for their help with collecting plants and processing specimens. Local staff who helped with cleaning, cooking, carrying wood and water, and washing clothes, are listed in the expedition reports included in the introduction. We thank them all for their hard work.

In addition, the Global Environment Facility part-funded our 1996 expedition through the Cameroon Biodiversity Conservation and Management Programme.

At the National Herbarium of Cameroon, Yaoundé, we thank the present leader, Dr Achoundong, the past leader, Dr Satabié, Elvire Biye, researcher, and technicians Jean-Paul Ghoghue, Fulbert Tadjouteu and Victor Nana, all of whom have joined us in fieldwork at Kilum-Ijim, and many of whom have helped with pre-determination of the specimens on which this book is based. Boniface Tadadjeu, Laurent Kemeuzeu (drivers) and Jorobabel Moussa (general assistant) are also thanked.

Dr Louis Zapfack (and his assistant Sonwa Denis) of the University of Yaoundé I is thanked for his assistance with fieldwork, in particular for leading collecting teams which discovered species previously uncollected, and for collecting, with John DeMarco, so many orchid species that otherwise would have been overlooked.

In Douala, we are grateful to Genevieve Fauré of the British Consulate and the Brothers of the Procure Generale des missions Catholiques, for help and hospitality.

At the Royal Botanic Gardens, Kew, we thank the numerous colleagues who have assisted making determinations, Maureen Bradford for help in typesetting, Suzy Dickerson for liaison on publishing, the Publications Committee for agreeing to finance the larger part of the production costs of this book, the library staff of John Flanagan for help with references, and particularly Marilyn Ward for assistance with illustrations. Margaret Tebbs and W.E. Trevithick provided the excellent line drawings which enhance this book so much. John Harris, Mick Parker and Fiona Bradley are thanked for guidance on producing the camera-ready copy from which this book was produced, and for scanning-in images and design work.

Final determinations of the specimens gathered in the course of our fieldwork were made by ourselves and by the following world specialists. We sincerely thank them all, and their institutions, for their work, often without remuneration, towards producing the names which are used in this book. B. Sonké, J. DeMarco, L. Zapfack, D. Harris, C. Jongkind, K. Vollesen, M. Thulin, M. Balkwill, K. Lye, T. Cope, S. Phillips, D. Sebsebe, P. Edwards, A. Paton, M. Thomas, P. Cribb, W. Baker, M. Lock, B. Mackinder, B. Schrire, H. Beentje, D. Goyder, A. R.-Smith, S. Dawson, D. Bridson, R. Faden, G. Gosline, J. Kirschner, P. Goldblatt, C. Townsend, P. Wilkin, B. Verdcourt, R. Polhill, A. Leeuwenberg, R. Brummitt, Sileshi Nemomissa, U. Meve, I. Nordal, P. Green. Without the work of these specialists there would be no book.

Craig Hilton-Taylor, IUCN officer for Red Data, based at WCMC in Cambridge is thanked for reviewing the Red Data treatments.

George Gosline, Research Associate of R.B.G., Kew, is thanked for co-ordinating specimen data-basing in the field and for help in leading the expeditions in 1998 and 1999. We all value his company and support.

We are indebted to our colleagues at Kew, Eimear Nic Lughadha and Mike Lock for commenting on drafts.

BJP wholeheartedly concurs with the above acknowledgements. Here follows his list of those deserving special mention.

It is essential that I thank first and foremost two remarkable people without whose unswerving generosity and support my involvement in this project would never have come to fruition. They are, namely, the late Kathleen Tweddle and Halldora Blair.

Prior to my appointment in February 2000 as Western Cameroon Darwin Initiative Officer at R.B.G., Kew, several parties enabled me to participate in the 1996 and 1998 expeditions to Mount Oku and the Ijim ridge. Funding was generously provided in 1996 by: The Old Radleian Society, The Ernest Cook Trust, The Really Useful Group, Elstree School, Pat Fromberg and Patrick Bailey. For the 1998 fieldwork I firstly thank Valerie Finnis VMH, Honorary Secretary of The Merlin Trust and all the other trustees for their generosity in the award of such a substantial grant. Brigadier Gurdon of The Really Useful Group was again and still is a patient supporter of this work. Alice Sanders of Olympus Optical Co. (UK) Ltd. kindly loaned photographic equipment, which added greatly to the value of the trip.

It also seems fitting to detail those who have in one way or another educated me in plant matters over recent years, thus indirectly leading to my involvement in such an important field of study as this Darwin Initiative Project. They are: Dr. Paul Kessler, Dr. Jennifer Kemp, Dr. John Iles, Professor David Mabberley, Dr. Roger Hall, Dr. Peter Savill, Douglas Harris, Dr. Andrew Wallis, David Hardy, Dick Usherwood, and particularly Alan & Gabriella Pollard.

Thanks to all those who have been supportive in other equally significant ways, both in Cameroon and elsewhere: Des, Eunice & Karl Hemphill, Miriam Le Borgne, Brother Jan, Eto Wa Paule Corinne, Margot Harris, Matt Campbell, Steve Taylor and Phil Honey of Palmbrokers Ltd., Tanya Wilson, Stuart Fenton, Matt Waterfield, James Wallman, Edwin Blanchard and Fiona Lay. Finally, respect to Magnus Blair for being the 'friend in deed'.

NEW NAMES

The following are published in this volume for the first time:

FOREWORD

Mt Oku and the Ijim Ridge lie at the centre and form the highest part of the Bamenda Highlands in western Cameroon (see Fig. 1). The Bamenda Highlands in turn, are part of a line of mountains (the Cameroon Highlands) that runs from Bioko and Mt Cameroon in the south, northwards to Tchabal Mbabo and then continues eastwards as the Adamaoua area.

The vegetation of the Cameroon Highlands shares many species with those of the East African mountains. However most of the primary vegetation has been destroyed. It is estimated for example, that 96.5% of the original montane forest of the Bamenda Highlands may have been lost (see chapter on loss of forest). There is no doubt that the largest and most intact block of surviving natural vegetation in the Bamenda Highlands is at Mt Oku and the Ijim Ridge. Here is the most complete and extensive representation of the Afromontane flora west of the Congo basin. Its future is in the hands of the Fons of Kom, Oku and Nso and the people of their Fondoms. We hope that this book will help them and their advisors to manage the forest and grasslands of Kilum-Ijim so that no more plant species become extinct.

A major purpose of this book is to enable identification of the wild plant species of Mount Oku and the Ijim Ridge (Kilum-Ijim), in particular 'Red Data' taxa, i.e. those that are the highest priorities for conservation. So far as we know, of the nearly 1000 species, subspecies and varieties of plant described in this book, 52 are threatened or near threatened with extinction (see Red Data chapter). Of these, 14 are only found in the Kilum-Ijim area and nowhere else in the world. In addition to the 52 'Red Data' plants there are a further 17 that have been identified as 'Red Data Candidates'. These are probably new to science but are not yet fully worked out.

Our knowledge of the plants of Kilum-Ijim has grown enormously but is still far from complete. 91 species were known in 1987 (Thomas 1987), increasing to 269 (Tame & Asonganyi 1995) and here we increase the number to nearly 1000. With more research this figure will undoubtedly be raised further still.

We hope that this book will also be useful for visitors to the Bamenda Highlands. Those seeking to identify the correct names of plants in connection with their professional duties, whether as foresters, agriculturists, or medical practitioners, should benefit by using this book. Those taking a holiday in this beautiful terrain may also benefit by acquiring it if they wish to be able to identify the plants they meet.

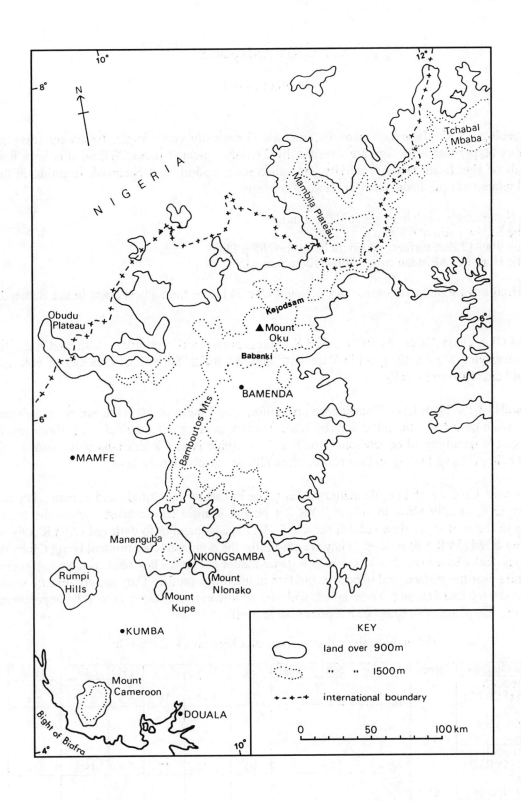

FIG. 1. THE CAMEROON HIGHLANDS
Reproduced with the permission of Birdlife International from McLeod (1987).

7

GEOLOGY AND SOILS

Martin Brunt

The geology of the Bamenda area of Cameroon is basically very simple: Basement complex granites of pre-Cambrian age are overlain by Tertiary volcanic rocks. Within this area four periods of significant tectonic activity and subsequent erosion have occurred, resulting in the development of four distinct land surfaces. These are:

- a Basement peneplain at about 300m a.m.s.l.
- the Mbaw Plain at 600m a.m.s.l.
- the 900–1200m surface, which includes the Ndop Plain
- the High Lava Plateau from 1500–2100m a.m.s.l.

The High Lava Plateau is dominated by Mount Oku 3011m – the highest point in the Bamenda area.

Mount Oku, characterised by its crater lake, is predominantly of Tertiary basaltic and trachytic lava, overlying the mainly granitic Basement complex rocks. Superficial deposits of volcanic ash and cinders occur locally.

The soils of the High Lava Plateau are very uniform, and those of Mt. Oku are no exception. They are humic ferralitic soils, derived from Tertiary lava parent material, and may contain appreciable quantities of concretionary rubble. This rubble is of a lateritic-bauxitic nature, and may occur at the surface, or in the profile, especially on steeply sloping land.

These soils have a high organic matter content, due in part to the altitude and climate. The top soil layer is usually black in colour (5YR 2/1 on the Munsel colour chart), while the lower layers in the profile are dark reddish brown (5YR 2/2–2/3) passing to dark red (2.5YR 3/6) or yellowish red (5YR 4/6) at depth. They are clay soils, the dominant clay mineral being Gibbsite, but they feel like a loam due to the high organic matter content. They tend to have a granular structure near the surface, and to be structureless in the lower profile. They are non-plastic when wet, and very free draining. These humic soils are of low nutrient status, as will be appreciated from the following table based on a representative profile.

The main analytical properties of a high lava soil (Profile)

Profile Diagram	Depth	% Clay:	Silt:	Sand	% C	% N	C/N	pH	CEC	TEB	% Sat.
Dark reddish brown (5YR2/2) "Loam". Granular very friable.	0" 4"	Not Determined			7.9	0.44	18	5.5	27.2	4.4	16
Dark reddish brown (5YR3/2) "Loam". Thickly rooted.	9"	70:	24:	6	5.1	0.30	17	5.3	19.8	1.1	6
Dark red (2.5YR3/6) "Loam" Massive. Non-sticky. Non-plastic.	18"	66:	29:	4	2.0	0.13	15	5.2	10.0	0.9	9
Dark red (2.5YR3/6) Gravelly "loam". About 50% concr. rubble.		64:	25:	8	1.6	0.12	13	5.2	8.2	1.1	13

The high figures for the cation exchange capacity (CEC) and exchangeable cations near the surface reflect the high organic matter content. The low nitrogen levels suggested by the wide C/N ratios, are the most serious deficiency in plant growth terms. The values for exchangeable calcium, magnesium and potassium are also very low; this reflects the age of the soils, and the intense leaching to which they have been subjected, due to the high rainfall and the soils extremely rapid infiltration and permeability rates.

CLIMATE

The following synopsis on the climate of Mount Oku and the Ijim Ridge is taken from Appendix A in McLeod (1987) by Tchatchoua.

The general climate of North West Province is broadly the same as that of the rest of West Africa (Hawkins and Brunt 1965) with a rainy season between May and September and a dry season between October and April.

The mountainous nature of the Bamenda Highlands alters this general pattern somewhat. Main rain-bearing winds coming from the south-west are interrupted by high land, causing very wet south-west facing slopes on the mountains and resulting in rain shadows on the north-east sides. Temperature inversions occur in valleys and depressions. The drainage of cold night air into lower areas results in lower than expected temperatures and thick mists in the foothills. During the day, warm air ascends and leads to cloud formation and misty conditions along escarpments. The intensity of the Harmattan, a dry north-easterly wind which may be heavily laden with dust, is also influential in determining weather conditions.

The combination of altitude, temperature inversion, slope orientation, harmattan, mist and cloud leads to the development of different local climatic zones (Hawkins and Brunt 1965). Nine climatic zones are defined by Hawkins and Brunt (1965) for the whole of North West Province, three of which can be used to describe the climate of Mount Oku: Zone 7 (Cool and Misty); Zone 8 (Cold, Very Cloudy and Misty); Zone 9 (Variable). The characteristics of the three zones are described below and are extracted from Hawkins and Brunt (1965).

ZONE 7 – Characteristic of most of the mountain.

Temperature. Mean maximum: 20–22°C; mean minimum: 13–14°C. November has the lowest mean minimum temperature and December the highest mean maximum. Temperature inversions at night in narrow valleys which suffer from bad air drainage leads to some ground frost, mainly in January or February.
Rainfall. Varies from 1780–2290mm per year (average annual rainfall for Jakiri is 2000mm). Most rain falls between July and September. The summit of Mount Oku leads to drier conditions to the east of the mountain.
Humidity. Generally January and February have the lowest relative humidity (average 45–52%). The monthly average exceeds 80% in July and August (Bambui Agricultural Station). During the rainy season, mist and low cloud occur frequently.

ZONE 8 – Characteristic of summit.

Temperature. Maximum temperature of 16.5–19°C, and minimum temperatures of 9–10.5°C would be expected.
Rainfall. Probably in excess of 3050mm per year.
Humidity. From the prevalence of mist and cloud, humidity would be expected to be high and the incidence of sunshine low.

ZONE 9 – Characteristic of escarpments.

Some areas of Oku mountain are also likely to fall within Zone 9. This climatic zone is variable depending on the topography and is characteristic of escarpments where strong winds are common.

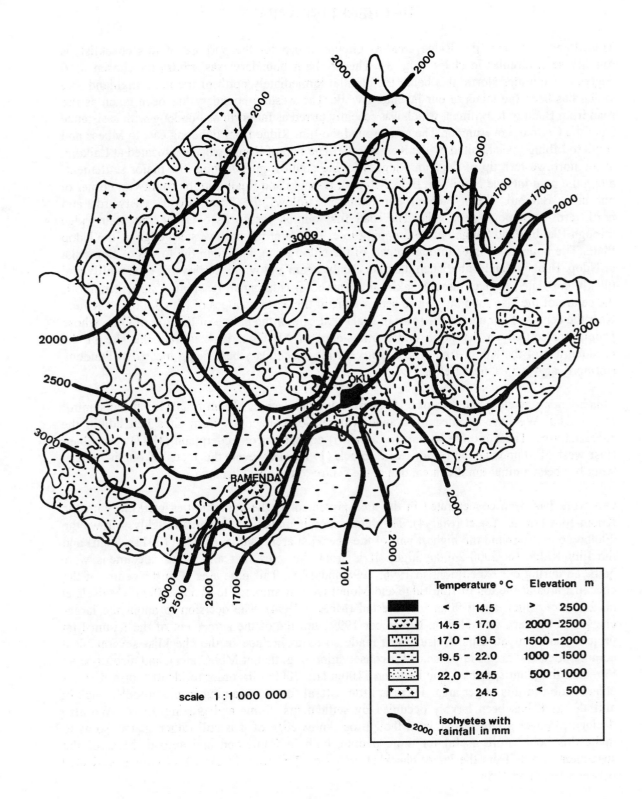

FIG. 2. RAINFALL AT MT OKU AND IJIM RIDGE
Reproduced with the permission of Birdlife International from McLeod (1987).

THE CHECKLIST AREA

The Mount Oku and Ijim Ridge area, as circumscribed for the purpose of this checklist, is roughly semi-circular in shape (Fig. 3). The northern boundary was arbitrarily chosen as 6 degrees 22 minutes North, this being the parallel immediately north of the main highland area which has been the focus of our inventory work. The western boundary has been taken as the road from Bambui to Bafmen, which was recently tarred as far north as Fundong with assistance from the German government. The area around the Ijim Ridge from Fundong east to Mbesa and south to Mbingo is inhabited by the Kom people. The Fon of Kom's palace is located at Laikom, at the northwestern tip of the Ijim Ridge. Moving south from Fundong, the major settlements along the road include Njinikom, Belo, Mbingo, Babanki and finally, in the southwest corner of our area, Bambui. From Bambui our boundary follows the "Ring Road" eastwards and northwards to a point near Ndu. The road first climbs eastwards over the mountain ridge through Bambili and Sabga before dropping to c. 1100m at the northwestern edge of the Ndop Plain. The boundary passes through a part of the Ndop Plain, including the towns of Bamunka or Ndop, Babungo and Baba. Then the road climbs again and winds northwards over a hilly and thickly populated plateau through Jakiri and Kumbo to Ndu, again reaching 6°22'N. Kumbo is the centre of the Nso fondom, whose people inhabit the eastern slopes of Mt Oku or Mt Kilum. Near the centre of our area is the town of Elak-Oku, site of the palace of the Fon of Oku, whose people are settled on the north and south slopes of Mt Oku. Elak is also the site of the headquarters of the Kilum-Ijim Forest Project, with a secondary base being located at Fundong (formerly at Anyajua, near Belo).

This boundary encloses an area of c.1550km², lying between c. 1100m to 3011m at the summit of Mt Oku. We have decided to include two sets of collections from sites that adjoin the checklist area. These are the Kufum (near Jakiri) collections of Ledermann (1907) and Mbam (just west of Fundong) collections of Zapfack (1998). However, the focus of our inventory work has been a much smaller area, that at the centre of the checklist area.

Our work has been concentrated in the area protected by the communities working with the Kilum-Ijim Forest Project (Map 4). The demarcated boundary of this area roughly follows the 2000m contour around the highest part of the checklist area, that is, Mt Oku (3011m high) and the Ijim Ridge (c. 2000-2400m high). It is from this area that almost all specimens were gathered during our expeditions in 1996, 1998 and 1999. This core area is in the centre of the checklist area at the top of Ijim ridge and Mount Oku. It amounts to about 200km² (Maisels et al. 2000) of which about 90km² is forested (McKay 1994). The decision to adopt the larger checklist boundary was taken in November 1999 with the of the agreement of the Kilum-Ijim project management. This decision was made so as to include in the checklist several plant communities that are of importance for conservation (e.g. that at Mbi Crater), and also to begin to record the plant species found between 1100m and 2000m in order to identify priorities for conservation in this larger area. There is little natural forest remaining in this middle range of altitude, as it has been largely occupied by settlements, farms and grazing areas. We also decided to cover a larger area, even though our knowledge of it is still rather sparse, so as to make this book more useful for more people, both in Cameroon and abroad. Most of the specimen records from the larger checklist area are taken from historical specimens collected between 1907 and 1986.

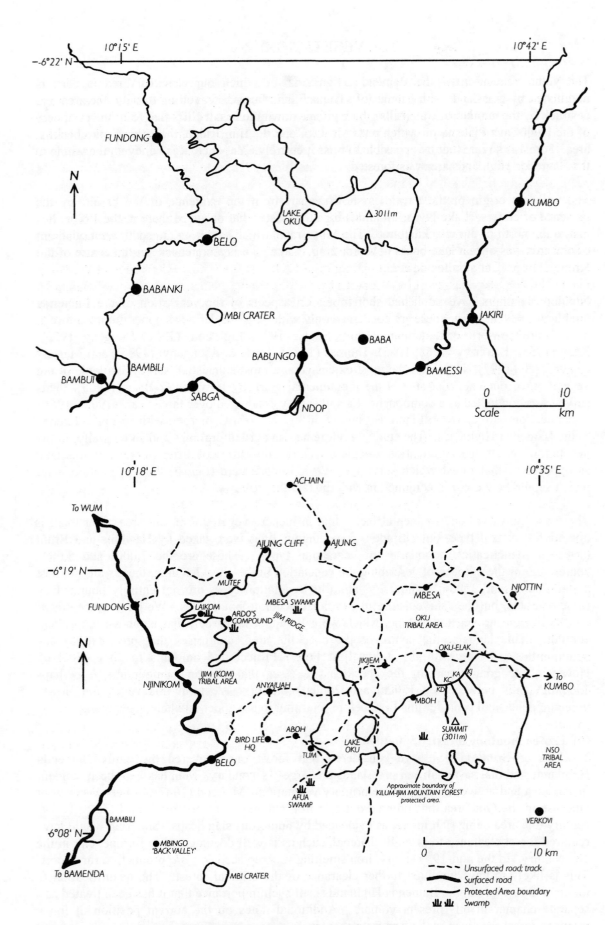

FIG. 3 (above) THE CHECKLIST AREA. Drawn by BJP.
FIG. 4 (below) SITES AT MT OKU AND THE IJIM RIDGE. Drawn by BJP.

13

VEGETATION

Today the vegetation of the Bamenda Highlands, of which our checklist area is part, is dominated by grassland, with patches of savannah, and farms. Travelling through the area one is struck by the number of waterfalls which plunge down the basalt cliffs that form the periphery of the High Lava Plateau, on which rests Mt. Oku and the Ijim Ridge, the core of our checklist area. Forest is so rare that it is possible to miss it entirely. Yet once, it is believed, the whole of the Bamenda Highlands area was forested.

The volcanic origin of the terrain is shown, apart from the evidence of the basalt, by the presence of craters. Lake Nyos, infamous for the disaster that occurred there in the 1980s, lies just to the north of our checklist area. The Mbi Crater which occurs in the south west quadrant of our area was once a lake but is now a swamp. Lake Oku is a crater-lake at the centre of the Kilum-Ijim project's protected area.

Numerous authors have published short notes on aspects of the vegetation of the Bamenda Highlands. By and large these are concerned only with what we define as upper montane forest. Notable are the works of Ledermann (1908), Engler (1925), Lightbody (1952), Boughey (1955), Keay (1955), Letouzey (1968, 1985), Thomas (1986), Tame & Asonganyi (1995), and Maisels & Forbosch (1997). However, the most extensive and most exhaustive survey available of the vegetation of our checklist area is the monumental work by Hawkins & Brunt (1965). This survey was conducted as a component of a soil survey conducted over three years (1962 - 1964) in connection with an overall land use survey of former West Cameroon with special reference to the Bamenda Highlands. The study therefore had an agricultural bias and so naturally, in the vegetation survey, special attention was given to the composition of different types of grassland as opposed to that forest which survives. We reproduce here (Fig. 5) their vegetation map, which should be viewed in conjunction with the text that follows.

The most significant environmental factor that influences vegetation in our area, after man, is not rainfall or soil type, but altitude. Two contour lines were taken by Hawkins and Brunt (1965) as demarcating boundaries of vegetation types. These are the 7,000' and 5,000' contours. For the purpose of describing the vegetation in the checklist area, these translate best as the 2000m and 1500m contours. Naturally, vegetation types are not entirely bounded by these contours, but they serve as useful and fairly accurate boundaries. We have concentrated our studies on the vegetation type, which we call 'upper montane', that is, above the 2000m contour. This land, at the centre of the checklist area, comprises that protected by the communities working with the Kilum-Ijim Forest Project and amounts to about 200km². However, the greater part of the checklist area (see that chapter), amounting to perhaps 1300km², falls between the 1500m and 2000m contour. We term this vegetation 'lower montane', and shall consider this first, before returning to the upper montane vegetation.

(a) Lower montane forest; 1500-2000m
Hawkins & Brunt (1965) have considered that forest once covered the entire Bamenda Highlands and that savannah and grasslands are largely secondary. Man has long been resident in this area and active in clearing the primary vegetation. McLeod (1987), in her overview of land use in the Oku area, has pointed out that iron smelting was common until the early 19th century, the area being rich in ore, as evidenced by numerous slag heaps. Smelting would have required forest clearance to provide charcoal, such as is well documented in Europe, e.g. in the 18th century. In the mid-19th century, iron smelting was replaced by *Cola* production for export. This is likely to have involved further clearance of the original forest. The quantification of montane forest loss in the Bamenda Highlands is of such importance that it has been treated as a separate chapter in this present volume. Additional notes on the current position of lower montane forest are given in the Red Data chapter.

It was only at the end of 1999 that we began in earnest to study this most threatened of all vegetation types in the Bamenda Highlands. Only one small area, the Ajung cliff forest, is protected by the Kilim-Ijim boundary. Our observations on the composition of the lower montane forest are highly incomplete, being based only on visits of a few hours in each of several small (each less than 1km²) and remote patches, all made with the help of John DeMarco. In addition to the Ajung Cliff forest (c. 1900m alt.), we have visited the Tum forest (c. 1800m), and 'Back Valley' forest, Mbingo (c. 1750m alt.) and the forest at the head of the Laikom Valley (c. 1900m alt.). See Fig. 4 for the location of these forests. Each of these forest patches has a slightly different composition and the species listed below are a composite of all four of these patches. In addition to those species listed, scattered individuals of upper montane forest were also found, evidence that there is not a hard and fast boundary between these types.

Canopy trees
Zanthoxylum rubescens (Rutaceae)
Cuviera longifolia (Rubiaceae)
Entandophragma angolense (Meliaceae)
Pouteria altissima (Sapotaceae)
Strombosia scheffleri (Olacaceae)
Tabernaemontana cf. ventricosa (Apocynaceae*)*
Garcinia cf. smeathmannii (Guttiferae)
Ternstroemia polypetala (Theaceae)
Symphonia globulifera (Guttiferae)
Newtonia camerunensis (Leguminosae*) Letouzey & Maitland records, extinct?*
Olea capensis (Oleaceae)
Alangium chinense (Alangiaceae*)*
Schefflera barteri (Araliaceae)
Pterygota mildbraedii (Sterculiaceae). Brunt (1963) record, lower altitudes.
Ficus vallis-choudae (Moraceae)
Ficus cf. craterostoma (Moraceae)
Ficus cf. cyathistipula (Moraceae)
Ficus chlamydocarpa (Moraceae)
Ficus natalensis (Moraceae)
Cassipourea congolana (Rhizophoraceae)
*Dalbergia lactea (*Leguminosae)

Understorey shrubs
Chassalia laikomensis (Rubiaceae)
Psychotria sp. nov. (Rubiaceae)
Deinbollia cf. pinnata (Sapindaceae)
Dovyalis sp. nov. (Flacourtiaceae)
Oncoba sp. nov. (Flacourtiaceae) extinct in the wild
Rothmannia urcelliformis (Rubiaceae)
Coffea liberica (Rubiaceae)
*Psychotria peduncularis (*Rubiaceae)
Erythrococca hispida (Euphorbiaceae)

Climbers
Thunbergia fasciculata (Acanthaceae)
Gerrardanthus paniculatus (Cucurbitaceae)
Pentarrhinum abyssinicum subsp. *ijimense* (Asclepiadaceae)
Phytolacca dodecandra (Phytolaccaceae)
Agelaea sp. (Connaraceae)
Adenia rumicifolia (Passifloraceae)
Clerodendron silvanum (Verbenaceae)
Rhpidocystis phellocalyx (Cucurbitaceae)
Millettia sp. (Leguminosae) higher altitudes
Paullinia pinnata (Sapinadaceae)

<u>Secondary and fire-resistant tree species – usually found at the forest edge</u>
Albizia gummifera (Leguminosae)
Gnidia glauca (Thymelaeaceae) higher altitudes only
Bridelia speciosa (Euphorbiaceae)
Croton macrostachyus (Euphorbiaceae) higher altitudes

So little is left of this vegetation type that it is difficult, as Hawkins & Brunt (1965) have pointed out, to assess what variants there might be or have been. However, they do point out that the 'montane woodland' of Keay (1955) is almost certainly secondary to forest and is a fire-induced feature of forest edges. The tree species present at these edges are listed above. Hawkins & Brunt also discuss a drier montane forest type that may have existed at well-drained areas furthest from the streams and damper areas to which forest is usually confined today. A remnant of this drier lower montane forest seems to exist on the approach route to Mbingo 'Back Valley'. Large areas have closed canopy, although there are many open spaces which, presumably, have been enlarged by cattle grazing and fire. The terrain is flat, and about 1600m alt.

The main canopy species here are *Eugenia gilgii, Sapium ellipticum, Antidesma sp., Croton macrostachyus* and *Phoenix reclinata. Ensete gilletii* is frequent in the gaps. *Dombeya ledermannii* is another drier forest tree species, seen by us only in the Akeh-Ajung area.

Several other species of lower montane forest are known from specimens collected in sites adjoining our checklist area. They may yet be found within the checklist area if the forest patches mentioned above are investigated more intensively (should they survive long enough for this to be possible). These species are:
Anthonotha noldae (Leguminosae) known from Chapman collections at Mambilla.
Celtis africana (Ulmaceae) known from Letouzey collections at Mbam.
Celtis durandii (Ulmaceae) known from Chapman collections at Mambilla.
Peddiea africana (Thymelaeaceae) known from several collections at Wum.

It is notable that lower montane forest has been passed over by most botanical studies of the vegetation and flora of the Bamenda Highlands, apart from Hawkins & Brunt (1965). There is an urgent need to protect and study this vegetation before it disappears for all time.

(b) Lower montane savannah and grassland; 1500-2000m
(i) Savannah
According to Hawkins & Brunt (1965), once montane forest is removed in the Bamenda Highlands, *Hyparrhenia* grassland replaces it, and forest cannot regenerate due to the regime of grassland fires that result in the dry season. Fire-resistant savannah tree species have subsequently migrated in from lower altitudes (Hawkins & Brunt, 1965). The commonest and most widespread savannah tree in our checklist area is *Entada abyssinica*. The following species are also found with it, although the composition varies from place to place. In some areas *Entada* is the only savannah tree present.
Piliostigma thonningii (Leguminosae)
Annona chrysophylla (Annonaceae)
Annona senegalensis subsp. *oulotricha* (Annonaceae)
Terminalia avicennioides (Combretaceae)
Lannaea kerstingii (Anacardiaceae)
Hymenocardia acida (Euphorbiaceae)
Bridelia ferruginea (Euphorbiaceae)
Cussonia djaolensis (Araliaceae)
Erythrina sp. (Leguminosae-Papilioniodeae)

We have not investigated savannah in any detail, having studied it for only a single day in November 1999, at a site between Babungo and Mbi Crater. Moreover, the specimens collected at that time are not yet, at the time of writing, available for study. This account is, therefore, not to be regarded as complete.

(ii) *Hyparrhenia* grassland
We have not investigated this vegetation type at all, and depend entirely on Hawkins & Brunt (1965) for the following description.

Mature *Hyparrhenia* grassland consists of tussocks up to 1m apart, growing up to 3m high. The leaves flop, and completely cover the ground between the tufts. The species composition is as follows:

Hyparrhenia bracteata (Gramineae)
Hyparrhenia diplandra (Gramineae)
Hyparrhenia filipendula (Gramineae)
Hyparrhenia rufa (Gramineae)
Melinis minutiflora (Gramineae)
Paspalum commersonii (Gramineae)
Setaria anceps (Gramineae)
Setaria sphacelata (Gramineae)
Clematis villosa (Ranunculaceae)
Commelina sp. (Commelinaceae)
Gladiolus sp. (Iridaceae)

(c) (i) Waterfalls and rapids; c.1300-2000m alt.
Such habitats are conspicuous and scattered throughout the checklist area, occurring wherever streams and small rivers flowing on the high lava plateau reach a cliff. The oxygenated water produced supports several species of Podostemaceae. Those discovered so far are: *Saxicolella marginalis, Ledermanniella keayi* and *L. cf. muscoides.*

(ii) Damp shady cliffs and banks c.1300-2000m alt.
Wetter sites, such as those near waterfalls, yield: *Utricularia striatula* and *Streptocarpus elongatus.* Drier, but still damp and shady sites such as the cliff at Anyajua (c. 1400m) yield *Impatiens* sp., *Begonia schaeferi, Pilea tetraphylla* and *Lecanthus peduncularis.* At higher altitudes e.g. the cliff at Laikom Ridge (c. 1900m), the species composition in this habitat is entirely different: *Radiola linoides, Wahlenbergia ramosissima* subsp. *ramosissima, Plectranthus* sp., *Craterostigma* sp., *Umbilicus botryoides.*

d) Grassland swamps; 1750-2800m alt.
The swamp communities of the Bamenda Highlands have not been documented before. For this reason they are discussed in detail here.

Most swamps in our area are formed when the drainage of streams becomes impeded. Thus they are often linear, as at Kinkolong Swamp (c. 2800m alt.), which is about 15m wide and 100m long, ending when the stream that feeds it falls over a cliff into forest below (Maisels *et al*, 2000). *Kniphofia reflexa, Sphagnum* spp., *Juncus dregeanus* subsp. *bachiti, Schoenoplectus corymbosus* and *Disa* sp. are characteristic of this site.

The Mbesa swamp (c. 2300m) occurs where two or three sinuous streams join. Here *Kniphofia reflexa* and *Setaria sphacelata* are amongst the largest species. A distinct community occupies the spaces between the tussocks, comprising *Laurembergia engleri, Lobelia minutula, Drosera madagascariensis, Hydrocotyle sibthorpioides, Xyris capensis* and *Utricularia scandens.* Higher up the sloping banks in the more seasonally wet areas are found swards of *Scleria distans.* Amongst this are found patches of *Ascolepis brasiliensis, Cyperus* sp. A, *Carduus nyassanus, Stachys pseudohumifusa* subsp. *saxeri, Plectranthus punctatus* subsp. *lanatus* and *Conyza clarenceana.* In areas of standing water are found *Schoenoplectus corymbosus, Cyperus niveus, Juncus oxycarpus* and *Fuirena stricta* subsp. *chlorocarpa.* The Afua swamp (c. 1950m alt.) is

17

formed by a single, linear impounded stream which is swampy for about 500m along its length, in a band c. 50m wide. A tussocky area in the downstream part bears several species not seen at Mbesa. These are: *Thelypteris confluens* (sides of tussocks), *Adenostemma caffrum* var. *asperum*, *Echinochloa crus-pavonis*, *Anagallis tenuicaulis* (between tussocks), *Cyperus elegantulus*, *Cyperus* sp. B & sp. C, *Rhynchospora corymbosa*, *Floscopa glomerata*, *Polygonum salicifolium* and *Epilobium salignum*. Species present that are shared with Mbesa are: *Xyris capensis*, *Plectranthus punctatus*, *Stachys pseudohumifusa*, *Eriocaulon bamendae*, *Hydrocotyle sibthorpioides*, *Utricularia scandens*, *Carduus nyassanus* and *Schoenoplectus corymbosus*. In close-cropped turf towards the head of the stream are found *Oldenlandia goreensis* and *Cyperus sesquiflorus*. *Heterotis angolensis* var. *bambutorum* is found at the margins of the wet area, along the flanks, and *Salix ledermannii* along the outlet stream.

Of the four swamps that we have investigated, the swamp at Mbi Crater (c. 1950m alt.), is the only one that is not derived from an impounded stream, but from a crater lake which gives rise to a sluggish outlet stream. It is several kilometres across, completely flat, lacks woody vegetation entirely, and consists of tussock-forming sedges and grasses reaching 0.8-1.5m high. *Cyperus denudatus* var. *denudatus* is the leafless, tussock-forming species that makes up about 80% of the vegetation and grows to c. 0.8m tall. *Rhytachne rottboelloides* forms the largest tussocks, these being up to 60cm wide and 30cm high. Other tussock-forming species are: *Fuirena stricta* subsp. *chlorocarpa*, *Xyris congensis*, and *X. rehmannii*. Straggling between the tussocks are: *Afrocarum imbricatum* (the only known site for the genus in Cameroon), *Andropogon lacunosus*, *Scleria achtenii*, *Leersia hexandra*, *Sacciolepis chevalieri*, *Cyperus dichrostachys*, *Oldenlandia lancifolia*, *Justicia* sp. nov.?, *Polygonum* cf. *strigosum*, *Conyza clarenceana* and *Helichrysum forskahlii*. *Xyria capensis* also occurs. No species of *Eriocaulon*, *Drosera* or *Utricularia* were located, although these genera were represented in other swamps. At the centre of the crater, colonies of *Gladiolus* sp. nov. and *Kniphofia reflexa* were found.

e) Upper montane grassland; 2000-3000m alt.
The main upper montane communities at Kilum-Ijim (Mt. Oku and the Ijim Ridge) have been mapped by McLeod (1987). We reproduce this map here (Fig. 6).
Sporobolus africanus dominates most of this vegetation type. It is an erect, clump-forming grass, growing to c. 60cm high. Clumps are spaced at intervals of 15-30cm. There are no co-dominant species and this vegetation can appear monotonous, and species-poor. However, in one area of 6m x 6m on Laikom Ridge (c. 2000m alt.) we recorded 25 species including: *Smithia elliotii*, *Anagallis djalonis*, *Crotalaria subcapitata* subsp. *oreadum*, *Trichopteryx elegantula*, *Wahlenbergia silenoides*, *W. ramosissima*, *Lobelia neumannii*, *Antherotoma naudinii*, *Antopetitia abyssinica*, *Sporobolus micranthus*, *Eragrostis camerunensis*, *Arthraxon micans*, *Oldenlandia rosulata*, *Trifolium usambaranse*, *Uebelinia abyssinica*, *Polygala tenuicaulis* subsp. *tayloriana*, *Panicum hochstetteri*, *P. pusillum* and *Otomeria cameronica*. These are all annual herbs that die as the dry season begins, leaving the ground between the clumps bare. Several perennial species with underground rootstocks that can survive the dry season aridity and fires were also found, namely *Helichrysum globosum* var. *globosum*, *Cyanotis barbata* and *Spermacoce natalensis*. *Hyparrhenia smithiana* and *Loudetia simplex* were the only other perennial grasses and these were infrequent.

Hawkins and Brunt (1965) have shown that, in the Bamenda Highlands, forest has been replaced by *Hyparrhenia* grassland and that this in turn, as a result of cattle trampling and repeated burning, has given rise to *Sporobolus africanus* grassland. Cattle were introduced into the upper montane grasslands in the 1920s. It seems possible that, before this date, *Hyparrhenia* dominated the area. There are three grassland types where *Sporobolus africanus* is absent or rare. Each occupies only a small part of the total upper montane grassland area.

(i) Basalt pavement grassland Where the surface of the basalt layer is relatively flat and unbroken, and bears no soil, an unusual vegetation grows in the thin layer of black, peaty soil that accumulates in the cracks between the heads of the hexagonal basalt columns. In the wet season such areas must be permanently wet, due to the impermeability of the basalt. In the dry season, they dry out completely. This vegetation type appears not to have been discussed before so is mentioned in some detail here. The area studied is the headland at the extreme tip of Laikom ridge, near the path down to Laikom. In all its extent is about 25m x 100m.

This 'grassland' is rendered conspicuous by the only tall (c. 60cm high) perennial grass present, *Loudetia simplex*. However this species provides only about 5% of the cover; *Scleria interrupta*, forming a dense sward c. 15cm high, provides about 80% of the cover. About 15%, in some patches 50%, of the area is bare basalt. Beneath the *Scleria* a wetland community of small annuals can be found: *Utricularia pubescens, U. scandens, Eriocaulon parvulum* and *E. asteroides*, together with *Bulbostylis densa*. A few perennial plants with underground perennating bodies are scattered here: *Cyperus* sp. 2, *Gladiolus* sp. and *?Urginea* sp.

Another patch of basalt pavement, about 1km east of the site discussed above, has a different species composition. Here the surface is tilted at about 30° from the horizontal and there is a seepage over the whole surface. *Scleria interrupta* is less in evidence. *Eriocaulon parvulum* is absent, but *Xyris cf. filiformis* and *Utricularia livida* are present.

(ii) Rocky grassland on thinner soils. This vegetation is less well demarcated than that of basalt pavement. It occurs on steep slopes and on the tops of ridges. In such areas members of the Compositae are especially numerous and diverse, particularly species of *Vernonia* and *Helichrysum*, as are species of *Crotalaria*.

(iii) Summit grassland. The community on the summit of Mt. Oku (3011m alt.) down to c. 2900m alt. contains an assemblage of species seen nowhere else in the Bamenda Highlands and with greatest affinity, perhaps, to that at similar altitude on Mt. Cameroon. Letouzey (1985) has already documented this vegetation type at Mt. Oku. A conspicuous feature is the silver carpet of the point endemic *Alchemilla fischeri* subsp. *camerunensis*, and such species as *Agrostis manniii, Anthospermum asperuloides, Delphinium dasycaulon, Veronica mannii* and *Habenaria obovata*.

(f) Upper montane forest: grassland edge; 2000-3000m alt.
The boundary between forest and grassland is one of the richest places for plant species diversity at Kilum-Ijim. Numerous species for which the forest is presumably too dark and which seem to grow poorly or not at all in the open grassland, perhaps because of trampling or wind-damage, have this ecotone as their main habitat. *Erica mannii* and *Erica tenuipilosa* are to be found here. They do not comprise a distinct 'Ericaceous belt' as is found on East African mountains, however. The fire-resistant trees that often occur at the boundary are listed under g). Sometimes they extend to form a band as much as 10m wide around the forest proper (particularly *Gnidia*), giving a 'woodland' appearance. In this habitat, Labiatae e.g. *Platostoma, Pycnostachys* and *Satureja*, are particularly common, and numerous Compositae can also be found. Shrub species that are apparently fire-resistant are: *Hypericum roeperianum, Adenocarpus mannii* and at lower altitudes *Solanecio mannii*. Forest edge shrubs not thought to be fire-resistant are mentioned in g) below.

(g) Upper montane forest; 2000-3000m alt.
This vegetation type is the most intensively and repeatedly investigated in the Bamenda Highlands. A passing reference was made to it by Ledermann (1908), and subsequently amplified, largely on the basis of Ledermann's collections, by Engler (1925). Lightbody (1952) wrote an unpublished thesis 25 pages long on forest of this type at Mba Kokeka at Bamenda. In the accounts of the vegetation of the Cameroon Highlands by Boughey (1955) and Keay (1955), those parts that deal with the Bamenda Highlands are largely occupied by upper montane forest. Hawkins & Brunt (1965) refer to this vegetation briefly, using the term 'Bamboo Forest', signalling the predominance of *Arundinaria alpina*. However, whereas this species was conspicuous in upper montane forest in Mba Kokeka, it is scarce in many parts of this forest at Kilum-Ijim, for example on the Ijim Ridge. For this reason the term is not adopted here. Letouzey (1968, 1985) and Thomas (1987) also discuss this vegetation type. The note of

Thomas (1987) is particularly succinct and accurate. Tame and Asonganyi (1995) catalogued the upper montane forest of the Ijim Ridge parcel by parcel and recognized 6 main types and several subtypes of upper montane forest, namely:

Type 1a: *Schefflera mannii/Syzygium staudtii* forest.

Type 1b: Degraded *Schefflera mannii/Syzygium staudtii* forest.

Type 1c: *Schefflera abyssinica* and monocarpic Acanthaceae.

Type 2a: *Syzygium staudtii, Nuxia congesta, Prunus africana, Podocarpus latifolius, Rapanaea melanoneura* forest.

Type 2b: Old regrowth forest.

Type 3:   *Carapa grandiflora, Syzygium staudtii, Prunus africana, Nuxia congesta, Bersama abyssinica* forest.

Type 4:   *Albizia gummifera, Croton macrostachyus, Bridelia speciosa, Schefflera abyssinica* forest.

Type 5:   *Arundinaria alpina* forest.

Type 6:   Gallery forest.

Maisels and Forboseh (1997) recognised 18 forest types in a study of upper montane forest covering the whole of the Kilum-Ijim area using TWINSPAN analysis.

Only ten species form 90% or more of the canopy of upper montane forest on Mt. Oku and the Ijim Ridge. These are: *Schefflera abyssinica, S. mannii, Prunus africana, Rapanea melanophloeos, Bersama abyssinica, Syzygium staudtii, Ixora foliosa, Carapa grandiflora, Clausena anisata* and *Arundinaria alpina.* The genus *Ficus* is notable by its absence.

Much rarer canopy species are *Podocarpus latifolius* (which can form extensive, but rare and localised patches), *Cassipourea malosana* and *Brucea antidysenterica.*

The forest in the Lake Oku crater (2200m alt.) has species not seen elsewhere within the upper montane forest of Kilum-Ijim, such as *Olea capensis, Dovyalis* sp. nov. and, around the margin of the lake, *Salix ledermannii. Croton macrostachyus* and *Neoboutonia glabrescens,* pioneer tree species, are also found at the lake forest and at the lower edges of upper montane forest elsewhere. Gallery forest along streams on Laikom Ridge (1900-2000m) falls just within the altitudinal band of the lower montane forest although for practical purposes, being dominated by upper montane species it is usually treated as such. Here can be found *Morella arborea, Ilex mitis, Agauria salicifolia, Eugenia gilgii* and *Deinbollia cf. pinnata.* These have not been found in the main blocks of upper montane forest.

Fire resistant tree species found at the forest edge are *Nuxia congesta, Hypericum revolutum, Maesa lanceolata, Cyathea dregei, Psorospermum* species and *Gnidia glauca. Nuxia congesta* and *Hypericum revolutum* can persist in the absence of fire and become components of forest. Thus when found in forest situations these species may be interpreted as indicators of forest expansion.

Duncan Thomas is credited by Tame & Asonganyi (1995) as having observed that forest succession also occurs by *Schefflera abyssinica* establishing preferentially on the forest edge species *Gnidia glauca,* and then strangling it. By contrast *Schefflera mannii* prefers *Syzygium staudtii* as host. Our field observations tend to confirm these observations.

The following liana species are present: *Maytenus buchananii, Embelia schimperi, E. mildbraedii, Clematis simense, Clematis hirsuta, Jasminum dichotomum* and *Stephania abyssinica.* Common understorey shrubs are *Rytigynia neglecta, Xymalos monospora, Allophyllus bullatus* and *Ardisia kivuensis.* Less universal are *Discopodium penninervium, Psydrax dunlapii* (lower altitudes) and *Cassine aethiopica.* At forest edges, though not thought to be fire resistant, common shrubs are: *Tarenna pavettioides, Pavetta hookeriana* and *Pittosporum viridiflorum.* Understorey herbs are dominated by the mass-flowering species discussed in the next paragraph. Other common species present are *Commelina cameroonensis* (gregarious), *Spermacoce princeae*, *Girardinia diversifolia*, *Laportea ovalifolia* and *L. alatipes*.

Synchronous mass-flowering of understorey monocarpic herbs

Our observations confirm the hitherto little documented phenomenon of the synchronous mass-flowering, followed by mass-death of four understorey herb species. These are *Mimulopsis solmsii*, *Acanthopale decempedalis* and *Oreacanthus mannii* (all Acanthaceae), *Plectranthus insignis* (Labiatae), collectively known as 'Mbum'. On our first expedition in October and November 1996, large tracts of the forest floor were bare of all but a dense carpet of seedlings c. 5cm high, dead stems 3-5m high, and thinly scattered shrublets of *Ardisia* and *Rytigynia*. Mass flowering had taken place in the previous year. A few late-flowering out-of-sequence plants of the species mentioned have been found in flower subsequently. In successive years the seedling carpet has grown in stature. There is uncertainty as to whether the cycle is seven or nine years in duration. The next mass-flowering is therefore expected in either 2003 or 2005. A similar phenomenon is known on Mt. Cameroon (see Cable & Cheek, 1998). Mass-flowering occurs on East African mountains but synchronicity, i.e. several species flowering and dying together, appears not to have been reported.

h) Lake Oku; 2200m

We investigated the Lake margin on the north side (1996) and on the south side, near the outlet stream (1998). The upper montane forest that fringes the lake grows out over it, so there is little opportunity for terrestrial 'water-margin' plants to establish. Only *Polygonum cf. salicifolium* falls into this category. The edge of the lake shelves has been gradually investigated at both sites. At 5m out from the edge it is no more than 1m deep. In this zone *Ottelia ulvifolia* grows gregariously in areas with a muddy bottom and *Isoetes biafrana* is found where there is a more solid, stony bottom. Further out, in a band of vegetation that follows the perimeter of the lake at a depth of 1-2m are: *Potomogeton schweinfurthii*, *Ceratophyllum demersum* and *Myriophyllum cf. spicatum*.

The central area of the lake has not been investigated for plants, as far as is known. No other aquatic species other than those mentioned are known to us.

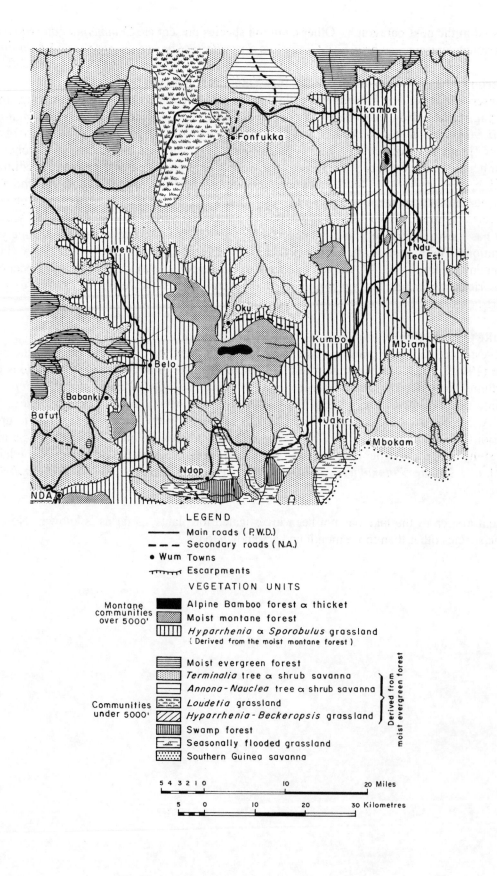

LEGEND

    ——— Main roads (P.W.D.)
    – – – Secondary roads (N.A.)
    • Wum Towns
    ⊤⊤⊤⊤ Escarpments

VEGETATION UNITS

Montane communities over 5000'

- Alpine Bamboo forest α thicket
- Moist montane forest
- *Hyparrhenia* α *Sporobulus* grassland (Derived from the moist montane forest)

Communities under 5000'

- Moist evergreen forest
- *Terminalia* tree α shrub savanna
- *Annona-Nauclea* tree α shrub savanna
- *Loudetia* grassland
- *Hyparrhenia-Beckeropsis* grassland
- Swamp forest
- Seasonally flooded grassland
- Southern Guinea savanna

Derived from moist evergreen forest

5 4 3 2 1 0   10   20 Miles

5   0   10   20   30 Kilometres

FIG. 5. VEGETATION MAP OF THE BAMENDA HIGHLANDS.
Escarpment denotes 1500m contour.
Reproduced with permission from Hawkins & Brunt (1965)

22

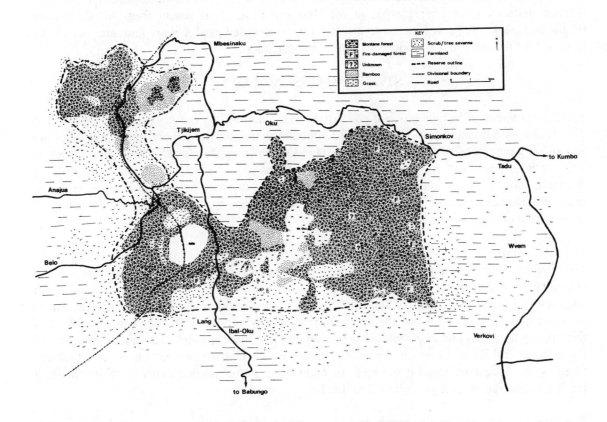

FIG. 6. VEGETATION MAP OF MT OKU AND THE IJIM RIDGE.
Reproduced with the permission of BirdLife International from McLeod (1987).

PHYTOGEOGRAPHY

Thomas (1987) produced a useful numerical breakdown and overview of the affinities of the flora of our area:

Species shared with SW Province, Cameroon	90%
Species shared with East African mountains	56%
Species shared with mountains to the west	32%
Species endemic to Cameroon Highlands	18%
Species characteristic of disturbed sites	17%

His analysis was based on only 91 upper montane forest and forest edge species. A comparison of all the 920 species in our checklist (not all habitats) shows that 505 (55%) are shared with those listed (Cable & Cheek 1998) for Mt Cameroon in S.W. Province, Cameroon. This is unexpectedly lower than suggested by the Thomas figure of 90% in the table above. The explanation is that, while the upper montane forest flora of Mt Cameroon and Mt Oku and the Ijim Ridge are extremely similar, the flora of other vegetation types in our checklist area is rather more distinctive. The lower montane forest, for example has a significant proportion of species restricted to the Bamenda Highlands and immediately adjoining areas, but not extending to Mt Cameroon. The lower montane savannah and grassland species have their closest affinities with the Sudano-Guinean savannah that traverses Africa just south of the Sahara. This savannah is absent from the Mt Cameroon area. The grassland swamp species of Mt Oku are absent from Mt Cameroon because of the lack of permanently wet sites there. Species of this habitat in our area seem to have their closest affinities with those of the mountains of E and SE Africa. Finally, the upper montane grassland flora of Mt Oku and Ijim seems to share far fewer species with Mt Cameroon than that of upper montane forest. Groups such as grassland *Dissotis* (Melastomataceae), well represented at Mt Oku and Ijim, seem absent from Mt Cameroon (Cable & Cheek 1998). We intend to present a more detailed phytogeographic analysis of the flora of Mt Oku and the Ijim Ridge in a separate publication.

STATISTICAL SUMMARY

The total number of indigenous and naturalized taxa included in the checklist that follows is 920. If the 54 cultivated taxa (listed in 'Ethnobotany') that do not appear in the checklist are added, we have a total of 974 vascular plant taxa known from the checklist area. The figure of 920 excludes imperfect specimens cited e.g. as '*Cyperus* sp.' since when better material is available these may be found to represent one of the numerous species of *Cyperus* already recorded. However, we do include taxa such as '*Strophanthus* sp.' since in this case, there is no other record of *Strophanthus* in the area.

The total number of taxa is likely to rise to above 1000 when all the specimens collected in the 1999 inventory become available. If more surveys are conducted in under-collected parts of the checklist area the figure could rise further. Whilst the 200km^2 area protected by local communities working with the Kilum-Ijim Forest Project is now relatively well collected, much of the remaining c. 1300km^2 of the checklist area as a whole remains uninventoried. Most of this area is farmland and *Hyparrhenia* grassland and not likely to yield additional species, but there are several Royal forests, and possibly some swamps that we have not surveyed which are likely to produce novelties for the area. In addition, a more thorough survey of savannah is likely to yield species not yet included in this list.

A comparison of total species numbers with other checklist areas in the Gulf of Guinea area is presented in the table below, which is expanded from that presented in Cable & Cheek (1998).

Checklist site	Area	No. Species
Bioko	2018km^2	842
Korup Project Area	2510km^2	1693
Mt Cameroon Area	2700km^2	2435
Mt Oku and Ijim Ridge Area	1550km^2	920

The relatively low species number for Mt Oku and the Ijim Ridge in comparison to Mt Cameroon reflects the fact that our area lacks the species-rich lowland forest of the Mt Cameroon foothills and also that the Mt Oku and Ijim Ridge area is only about half the size of the Mt Cameroon checklist area. Furthermore, an estimated 850 of the 920 species in our checklist are derived from the 200km^2 of the Kilum-Ijim protected area (and the forest patches adjoining). Thus in terms of species diversity per unit area, the Kilum-Ijim protected area is still extremely rich, and probably as diverse or more diverse than the land on Mt Cameroon at comparable altitude, i.e. above c. 2000m alt.

In terms of total species number, area and terrain, there is more similarity between Kilum-Ijim and the Pico das Almas of N.E. Brazil than with the Gulf of Guinea areas mentioned above. According to Stannard (1995) 1044 species are found in his checklist area of c. 170km^2 most of which is between 1000-2000m alt.

The breakdown of species in our checklist area by major group is as follows:

1	Gramineae	95
2	Orchidaceae	85
3	Compositae	83
4 =	Cyperaceae	53
4 =	Leguminosae	53
6	Labiatae	37
7	Rubiaceae	36
8 =	Acanthaceae	22
8 =	Euphorbiaceae	22
10	Umbelliferae	15

The most important genera in terms of numbers of species are as follows:

	Genus and Family	No. of taxa	Predominant Lifeform
1	*Cyperus* (Cyperaceae)	29	Terrestrial herb
2	*Vernonia* (Compositae)	14	Terrestrial herb
3 =	*Bulbophyllum* (Orchidaceae)	13	Epiphytic herb
3 =	*Polystachya* (Orchidaceae)	13	Epiphytic herb
5	*Asplenium* (Aspleniaceae)	12	Terrestrial and epiphytic herbs
6	*Hyparrhenia* (Gramineae)	10	Terrestrial herb
7 =	*Plectranthus* (Labiatae)	9	Terrestrial herb
7 =	*Helichrysum* (Compositae)	9	Terrestrial herb
7 =	*Pennisetum* (Gramineae)	9	Terrestrial herb
7 =	*Ficus* (Moraceae)	9	Tree

The predominance of the Gramineae reflects the fact that grasslands cover most of the checklist area. More unexpected is the presence of Cyperaceae amongst the top four families, and of *Cyperus* as the most speciose genus by far in the checklist. This is evidence that swamp vegetation, although occupying a relatively small portion of the checklist area, makes a disproportionately large contribution to the species diversity of the checklist.

The title of the first plant collector of Mount Oku and the Ijim ridge belongs to **Thorbecke**. He collected on the route connecting Mt Cameroon, Manenguba and Bamenda to Banyo in 1907-1908 (Letouzey 1968).

According to Engler (1925: 154), Thorbecke made a small collection of plants from Lake Oku ("Mauwe-See") itself in 1908. Those mentioned by Engler are *Agauria salicifolia, Asplenium furcatum, A. contiguum, A. sphenolobium var. usambarense, Adiantum poiretii, Dryopteris inaequalis, Thalictrum rhynchocarpum, Cardamine africana, Crassula alsinoides, Alchemilla cryptantha, Trifolium usambarense, Geranium simense, Hypericum peplidifolium* var. *diestelianum, Gnaphalium luteo-album* and *Gynura cernua.* Unfortunately his specimens are believed to have been destroyed at the Berlin herbarium in 1943.

Ledermann is famous for his botanical journey from the coast at Limbe to Lake Chad and back which he wrote up and published shortly afterwards (Ledermann 1912). At the time he was employed as a gardener at what is now Limbe Botanic Garden. In making his journey he passed along the southern boundary of our checklist area, the itinerary for his collecting tour between 30 June 1908 and 7 Dec. 1909 being "Victoria...Bamenda, Kumba, monts Oku (Banyo),..." (Letouzey 1968). Several new taxon were described in his honour from the specimens that he collected on this trip. They include *Dombeya ledermannii (Sterculiaceae), Ledermanniella (Podostemaceae)* and *Salix ledermannii (Salicaceae).* Ledermann spent Christmas of 1908 at Kumbo. The following excerpt from his travelogue is taken from the period just after Christmas when he ventured to a place named Kufum near present day Jakiri-Kumbo on the SE boundary of our area. Ledermann's 'Kufum' is probably Kovifem, the Royal Forest of the Fon of NSO. The following passage is of interest for what may be the first published description of montane forest and associated vegetation in the Bamenda Highlands:

"Nach dem Aufbruch von Kumbo beginnt für uns eine anstrengende Kletterpartie den Zentralgebirgszug des Bansso hinauf. Das Gelände ist Grassteppe mit einzelnen Raphiapalmen oder etwas Galerie in den Niederungen und an den Bächen. Vom Wege herab schaut man auf eben frischbestellte Felder von Tabak, Mais und Durra, die von Erdwällen umgeben sind. Da eins neben dem andern liegt, hat man von oben den Eindruck, als ob da unten sich riesige, mit Grün aisgefüllte Parallelgräben hinzögen. Sicher hat der Wunsch, den Kulturen jetzt in der Trockenzeit mehr Feuchtigkeit zu bieten, zu dieser eigentümlichen Bestellungsweise Veranlassung gegeben.

Bei 1900m ist das vor kurzem abgebrannte Gras viel mit Protea singwensis durchsetzt, die durch das Feuer zwar die Blätter verloren hat, aber schon wieder silbrig-weiße Sprosse treibt. Daß der Frühling einzuziehen beginnt, darauf deuten auch zahlreiche, höchstens 5 bis 20cm hohe, über und über mit Blüten bedeckte Kräuter, in ihrer Gesamtheit ein Bild schaffend, das uns an die Frühlingsflora der trockenen Abhänge des Jura gemahnt. – Kurz nachdem wir den höchsten Punkt der Straße (2000 bis 2100m) passiert haben, sehen wir vor uns einen trockenen Gebirgswald, vorwiegend aus 15 bis 20m hohen, breitkronigen Ficus-Arten, Schefflera hookeriana, Brucea antidysenterica und einer Flacourtiacee bestehend. Er hat durchaus nichts Tropisches an sich, da er licht ist und das wenige Unterholz durch krautige Acanthaceen und Balsaminen ersetz wird. Die Aste der Bäume sind von Moosen und Flechten umhüllt. Zweifellos ist es ein Sekundärwald, ähnlich wie der in Sangwa geschilderte, denn überall stößt man in ihm auf zerfallene Lehmwände von Hütten, die wohl ein Dort, die frühere Residenz, des Oberhäuptlings über das Bansso-Reich zusammengesetzt hatten. Wir übernachten im Tal, bei 1750m, in Kufum, wo es wenig Verpflegung gibt, da Bananen bur in geringer Anzahl, dafür um so mehr Mais gebaut wird."

Unfortunately, Ledermann appears to have collected only one main set of specimens, with few duplicates being distributed. His main set, being housed at the Berlin herbarium, was largely destroyed by Allied bombing in 1943 (Hepper & Neate 1971: 49). Fortunately, the identifications of many of his specimens were recorded by Engler in six pages describing the vegetation of the "Bansso-Gebirge" (Engler 1925: 162-168). "Bansso-Gebirge" is a term used by Ledermann for the high land around Jakiri-Kumbo, but it does not seem to have included Mt Oku itself, which Ledermann apparently did not visit. From Engler's text it is clear that Ledermann collected numerous specimens along the southern boundary of our checklist area. Interestingly, they are almost entirely of secondary savannah species, suggesting that, even in 1908, forest clearance was highly advanced. Many of the names used by Engler have since been reduced to synonymy. Some of Ledermann's collections, e.g. various orchids from Kufum are fully cited in FWTA, and so are included in this checklist. The remainder of the names listed by Engler (1925) for Ledermann's collections we have not used since the specimens on which they are based are not cited and the identifications have not been confirmed. Nonetheless, Engler's work (1925) is of great value for those interested in the composition of the flora in the Bamenda Highlands in 1908.

Maitland (see frontispiece), whilst not the earliest plant collector on Mount Oku and Ijim, is certainly the most notable of the early collectors, producing the most significant and comprehensive of the historical collections from the area. Between March and June 1931 he made hundreds of collections from the Ijim ridge, collecting predominantly from Laikom ("Lakom"), Nchain ("Nchan"), Mbesa ("Mbizenaku" of current maps but spelled by Maitland as "Basenako") and Belo. Maitland was the first collector of the Mount Oku and Ijim endemic *Kniphofia reflexa* (*Asphodelaceae*), which was not to be recorded again until November 1996, and of *Habenaria maitlandii* which has not been seen since he collected it at Nchain in 1931 and is still known solely from his collection. He was also the first collector of such endemic species as *Eriocaulon bamendae (Eriocaulaceae), Newtonia camerunensis* (*Leguminosae*), *Plectranthus punctatus* subsp. *lanatus* (*Labiatae*) and *Vernonia bamendae* (*Compositae*). Given the importance of Maitland's collecting work in the Bamenda Highlands, it is regrettable that his notes and collecting books have not been traced at R.B.G., Kew or elsewhere. These documents would provide useful supplementary information and clarify some mysteries. One of Maitland's localities, "Lakoni" in the Bamenda area, has not been traced using maps and gazetteers available to us now and may fall within the checklist area. However, it may simply be a misspelling of Laikom. Maitland's itinerary is still not clear. However, some of his labels indicate fragments of his routes, e.g. "Basenako to Nchan" and Lakom-Nchan" and when his collections at Kew have been databased (work is now in progress by Suzanne White) it should be possible to reconstruct his route in the Bamenda Highlands.

At the time of his visit to Ijim, Maitland was at the height of his professional career and experience as a plant collector (Anon. 1955). He retired from his post as superintendent of the Victoria (now Limbe) Botanic Garden and from 22 years in the colonial service later in 1931. He collected while travelling, often for weeks on end, on foot, on horseback or on bicycle, depending on the terrain (Anon. 1955). A horticulturalist, he had worked at R.B.G., Kew before being recruited by the Director to act as supervisor at various Botanic gardens in Tropical Africa. He was elected President of the Kew Guild in 1955-56.

During 1931 and 1932, **A.T. Johnstone** collected several hundred plant specimens in the Bamenda Highlands including some, cited here, which fall within our checklist area. According to a Kew determination list concerning his specimens, he was Assistant Conservator of Forests, Bamenda Division at this time. Lightbody (1962) cites him as author of an unpublished manuscript, dated 1947 "A list of plants collected in Bamenda Division and identified at the Herbarium, R. Bot. Gardens, Kew and at the Oxford Forestry Herbarium", which regrettably has not been traced.

Keay (Deputy Director of Forestry, Ibadan, Nigeria and editor of F.W.T.A.) and **Lightbody** (Provincial Forest Officer, Bamenda) seem to have been the next plant collectors on Mount Oku. In January 1951 Keay was on a one month's botanical reconnaissance of the Cameroons. The following is an excerpt from his travelling notes (Keay 1994):

27

FIG. 7 (above) LAKE OKU IN 1951.
Note the grassland : forest interface. Photo by Ronald Keay.
FIG. 8 (below). 'THE MOT-ORABLE ROAD' OF NIGEL HEPPER. Oku junction,
Kumbo in February 1958. Photo by Nigel Hepper.

"4 Jan. Lightbody and I intended to go to Lake Oku today, but on reaching Belo (30 miles from Bamenda on the Njinikom road) found that the messenger had not obtained carriers as instructed. Postponed this trip until the weekend, and then went via Bamenda to Banso where we stayed the night. Russell reached Bamenda yesterday and accompanied us on this trip. Saw a lot of country and collected 15 specimens.

5 Jan. Went by car from Banso to Ndu, returning slowly and botanising en route; returned to Bamenda in the late afternoon. Collected 31 specimens today, including many European genera such as Veronica, Trifolium, Rumex, Geranium and Drosera.

6 Jan. Left Bamenda early for Belo. Met carriers and set off for Lake Oku. The carriers had been collected for us very kindly by Miss Kittlitz of the Baptist Mission. The first 2^1/$_2$ hours of the trek took us up a long valley of cultivated land. Cultivation ceased only at about 6000 ft. alt. when we climbed up through Schefflera-Bersama-Clausena forest at the head of the valley. Emerged from the forest on to a grassy ridge at about 7800 ft. Walked along this ridge, down through a bit of forest, then up through a bamboo brake to another grassy ridge which overlooked Lake Oku. The lake is surrounded by forest. We descended into this forest and camped in it near the lake (alt. 7000 ft.). Collected 23 specimens today. We were glad of the fly-sheet we had brought, as there was rain for about two hours at night. Carriers were excellent.

7 Jan. (Sunday). Started the day with a bathe in the lake, then climbed up through the forest on to the grassy ridge and then up a steep knoll (alt. 7900 ft.) where we found several plants (including Blaeria) which we had not seen before. Botanised on the grassy ridge sand upper margins of the forest and collected 33 specimens. Returned down the valley to Belo. Noted Schefflera mannii commonly planted as a live hedge around the Bikom compounds. Returned to Bamenda."

Keay & Lightbody's expedition was important in providing the first specimens of the rare and near-endemic Mt Oku-Ijim species *Isoetes biafrana* and *Dipsacus narcisseanum* amongst the 57 collections made within our checklist area (in the series FHI 28463-28519: Keay *pers. comm.* 1996). On this visit, Keay also made collections at the boundary of our checklist area as follows: 4th January, Bamenda to Belo (FHI 28416-28430) and 5th January, Banso to Ndu (FHI 28431-28458). Lightbody, who accompanied Keay to Lake Oku, later wrote a thesis on the Grassland Forests of Mba Kokeka (Lightbody 1952).

Egbuta is listed by Letouzey (1968) as collecting at Kumbo (on the southeastern boundary of our area), in 1951. He is notable for an early collection of the near-endemic *Crotalaria bamendae* (see Red Data chapter).

McCulloch is listed by Letouzey (1968) as having collected at Banso (i.e. Kumbo-Jakiri) in Nov. 1953 and we have included a few McCulloch specimens in this checklist.

Hepper's collecting itinerary for 3-27 Feb. 1958 was Nkambe, Wum, Nkambe, Kumbo, Bamenda. This took him around the northern, eastern and southern boundaries and into the eastern part of our area. He made several important collections, over two days, on the road from Kumbo to Oku-Elak ("Oku"), for example the second known collection (*Hepper* 2021, 15 Feb. 1958) of the largest Oku-Ijim endemic *Eriocaulon, E. bamendae* (Phillips 2000). Here we include a fragment from Hepper's diary for those two days. It is of interest for the information contained on forest limits at that time and for observations on the state of the vegetation at the height of the dry season.

FIG. 9. THE FON OF OKU ON HIS THRONE IN 1958.
Photo by Nigel Hepper.

"Saturday 15 Feb.
Sunny.
Aimed for the Oku road we had noticed just this side of Kumbo. The notice amused us

> *The Oku road is now mot-*
> *orable a distance of 16 miles*

This goes right into the centre of the Bamenda Ring Road, but we didn't get very far along it. Our first stop was at a high point to look at the withered grasses. The place was unburnt and I could imagine a carpet of plants in the rainy season but now nothing – if only I could stay till April even when the rains begin. Smithia elliottii was in shrivelled profusion. Glad to see Celsia, like Mullein, and Hypericum lanceolatum in flower, sure indication we were in the higher parts and must have been well over 6000 ft. Stopped again a short distance further on when we were right away from the houses and down to an overgrown dip. What a wealth of things were here. Several specimens of a huge orchid 5-6 ft. high, Galium (bedstraw) and best of all Sibthopea europaea I last found in Pembrokeshire. As we beat our way through the stand of Brillantaisia and thick Melastomataceae we were treading on a mat of semi decaying sticks overlying mud and water and we hoped some venomous snakes didn't inhabit the place. Finally with huge and bulging press we went back to the car where the Forest Guard had dozed all the time. Moved on a mile till we got a good view of the road ahead and some much higher slopes ahead which must be quite 8000 ft. Here we lunched and then I spotted an Albuca by the car in the burnt grassland and this started another session for I saw plenty of the little Asclepiad I had seen one plant of it at Gembu, and other things, then another dip attracted our attention. This was quite different from the first in that it lacked shrubby growth of Hypericum lanceolatum and a Vernonia on its edge but had plenty of Xyris and a common Dissotis. What was so astonishing was to find the orchid in a literal drift. So it isn't as rare as we thought at first. Finally dragged ourselves away feeling fairly worn out and went to Kumbo for petrol, got a puncture on the way and had to change tyres. Then began the task of writing up the 30 we had collected and changing the 35 presses on the go. In six days collecting Monday – Saturday got 130 (compared with the Director's weekly estimate of 50-60 which is about right for normal areas), but I hope I have maintained the standard.

Monday 17 Feb.
Sunny.
Down the "mot-orable" Oku road; planned an early start but by the time the water was put in the car and John had fixed the spare wheel it was 8 again before we were off. Beyond where we were on Saturday where the hills roll and are covered with open short, or burnt grass and the occasional well vegetationed dip, through at mile 6 a village and scrub of old farmland. Plenty of Lobelias and Senecios 6 ft. high and a large shrubby Vernonia with the smaller orange-fruited Solanum beneath. Then at mile 9 it became clear that the ridge ahead round which we were to go was covered with real high forest and from mile 10 pretty well to Oku village at mile 16 it continued. Considering this was best part of 7,000 ft. alt. that is good going. Oku village itself is down a good deal and the bumpy road terminated in the Fon compound much like the Fon of Kumbo's. Our object really was Lake Oku further over but eventually we find it was 9 miles a good 3-4 hours trek. The locals said we should see the Fon and get a guide in any case. Taken through the entrance into the compound itself and spotted on the other side of the earthy square surrounded by huts, the chief's platform. Up two small steps of small stones and a small recess with a stone floor of radially-arranged pebbles, the roof supported on two carved posts and set in the back of the recess was a door covered over by Raphia poles, either side as door posts two smaller figures. All were painted in reddish daub and rather primitive in features and execution. We waited quite 15 mins. and wondered whether it was worth it as we wanted to get on. However, a rustling of the door as it was removed proclaimed his presence and he sat down on a chair there. There was a wicker work chair as well as an uncomfortable carved throne and foot stool which I scorned. I crossed the courtyard and he greeted me with his hand and said good morning and what a fine place he had, then that we were interested in Lake Oku and he

said he would get a guide. Although I spoke slowly he had difficulty in understanding me and called over an interpreter. More and more of the men had been coming into the compound and lounged around as we talked. While we waited for a guide I asked him for permission to take a photograph and he said only on condition I promise to send him a copy, which I agreed to do. He was quite photogenic but his platform was in the shade so I hope it turns out well. His dress was a gown of multicoloured velvet-type material, Bamenda woven cap with a long porcupine's quill stuck under it over his right eye as a barman might stick his pencil – I asked and was told only the Fon might wear this. On his feet were European sandals, he carried a staff wound round with strips of polished brass towards the top. We were still waiting when John came in to see what was happening and he also greeted the chief who when he heard he came from Ibadan said he would like to come with him. He was so serious about it John then started to say his car wasn't strong enough for his loads. Fon retorted he would take small small load. Perhaps the Fon wouldn't like the food? He would eat whatever he could get. Why did the Fon want to go to Ibadan? Then the interpreter indicated he was only joking. Then he started on me. I came from England, how far was that, first we tried miles: 3500, but it meant little; by boat it would take 14 days. Then the Fon had never seen a boat bigger than a canoe and 14 days trek wouldn't get one far, so we gave it up. The Fon wanted to see our "lorry" so out we trooped. Now either side of the outer court were these tall square huts as at Kumbo and, as I suspected, they were wives houses. Outside several were some of his wives who immediately bent double, their hands on knees and head down. As we approached suddenly one didn't see him in time and with a whimper skipped into the house as the Fon gave her a dirty look and flicked his hand at her. He peered all over the kitcar without either approval or disapproval and then led me to a vantage point from where we could look to the farthest point of a line of hills to the west and he pointed to the conical one at the end from where one descended to Lake Oku. Obviously too far, a pity as the mountains are very high about here and the lake is formed in an old volcanic crater. This area is the centre of the Bamenda Highlands and has only recently since the road was made become at all accessible and is quite remote. A few miles to the west lies Laakom where the Fon was accused of keeping 100 wives and not far away but nearer Bamenda is Bafut where Gerald Durrell collected animals and wrote the book "Bafut Beagles". With pride the Fon swept his staff over the distant hazy view "All my people, my country. Get plenty people too much". Then he said there are 1,730 on the tax roll in his area. I asked to be allowed to collect a specimen from his compound I noticed whilst waiting, a tufted Chenopodium-like plant. He was nonplussed but very willing. A silent group watched me press it back in the courtyard and even more when I started grovelling for Chondria repens between the stones at his feet – however, it's only the third record for West Africa I think so it was worth doing. A final photo of John C., Fon and self and we departed. No local guide was forthcoming so we wandered down the steep winding path towards the river and collected a lot of interesting things Acanths., mosses, Harpagocarpa and Thalictrum rhynchocarpa up to 7 ft. high but not in flower. John was feeling unenergetic and had said he would not go from the car if we attempted L. Oku but he perked up somewhat later. That was after we returned and found another front tyre flat and changed the wheel once more. On down the road to where there was a nice damp overgrown bank where the road cut through the forest. What a collection we got here. Got quantities of Stellaria mannii which has not been previously collected off Cameroon Mt. itself. Sibthorpea again, here in tangled masses! A terrible stinging thing Laportea (but it is strange this is the first stinging plant I have come across, although there are other forest species, yet in farms and cultivation at home I would come across nettles galore which are entirely absent and I have walked around with bare legs quite happily). Another nuisance was the brambles (Rubus sp.) which I collected from a huge mass. We could hardly tear ourselves away (!) but did so when we had collected what we wanted and had the long drive back. Started writing up at 5.0 when we got in and with a brief pause for chop turned in at nearly 11.0, John even helped to change all the presses, even so I didn't finish all the writing up, nor touched this log."

In the early 1950s a group of botanists from what is now the University of Ghana at Legon, led by Prof. **Boughey**, made important contributions to the study of the Cameroon uplands during a series of expeditions that in succeeding years resulted in numerous publications on the flora and vegetation of this area. **Morton, C.D. Adams** and Boughey seem largely to have confined their

32

collecting to the vicinity of Bamenda and not to have visited Oku or Ijim. However, C.D.Adams did collect around Ndop and Bambui. Adams specialised in the Compositae and published many new West African species, some of which are endemic to the Bamenda Highlands and are treated here in the Red Data chapter. Examples are *Crassocephalum bougheyanum* and most famously, the only known genus of plant endemic to the Bamenda Highlands, *Bafutia*. Morton, now based in Waterloo, Canada, did for Labiatae and Commelinaceae what C.D. Adams did for Compositae and still maintains an interest in these plants in Cameroon.

Brunt was based in the Bamenda Highlands for about three years (1962-1964) while gathering data for his book "The Soils and Ecology of Western Cameroon" (Hawkins & Brunt 1965). His itinerary for 25 Sept. 1962 to 23 Aug. 1963 was Bamenda, Wum, Nkambe, taking him around the western boundary of our area from south to northeast. His collections *Brunt* 1-1274 were focused on grassland species and in the characterisation of vegetation types. They make a substantial contribution to our knowledge of the flora of the Bamenda Highlands indeed, he is probably the most prolific collector of the plants of the Bamenda Highlands. His work is the foundation for all work on the vegetation of the Bamenda Highlands.

Letouzey can be considered the father of Cameroonian botany, having set up the National Herbarium at Yaoundé in the 1950 s and having been the driving force behind Flore Du Cameroun for about two decades. He also inspired numerous Cameroonian botanists (e.g. Ntepe-Nyame, Mbenkum, Satabié, Achoundong, Sonké) to doctoral level and collected extensively throughout most of the forest area of Cameroon, often far from vehicular roads. Many Letouzey localities have not been collected at before or since. Letouzey (1968) lists his visit of June and July 1967 to Mount Oku as his 16th expedition and his itinerary is given as "Banyo (+Nkambe et mont Oku)", during which he made collections under his number series from 8480 to 8964, i.e. 485 specimen numbers. Many of his specimens were collected on the path from the village of Verkovi (on the southeastern slopes of Mount Oku) up towards the top of the mountain. We have seen no specimens from this area of the mountain apart from Letouzey's.

Mbenkum, Chief of the National Herbarium, later Director of the Environment at the Ministry of the Environment and Forests, botanised around Mt Oku in 1973 (e.g. *Mbenkum* 349, Tadu, 11km WNW Kumbo, 26 June 1973: *Moraea schimperi*). A native of Kumbo, he went on to author a work on the ethnobotany of the plants of Mt Oku and later took a Doctorate at the University of Reading on the taxonomy of *Millettia*.

Letouzey revisited Lake Oku in November and December 1974, together with **Satabié**, later Chief of the National Herbarium of Cameroon (until 1998) and co-leader of the main 1996 expedition. On this visit Letouzey visited the Ijim (southwest) side. *Letouzey* 13451 (collected between Acha and Lake Oku on 5 Dec. 1974) is evidence of this. The specimen cited was the last to be collected of the Red Data species *Newtonia camerunensis*. Satabié's collections from this visit include some from the Kilum side, e.g. *Satabié* 99, from Tatem-Kishong, 16km NNE Kumbo (15 Nov. 1974).

Many important collections from the highlands of Cameroon were made by a collector (Meurillon is the only individual mentioned on a few labels) or more probably, collectors, under the acronym **C.N.A.D.** (Centre National Agronomique Dschang). Dschang is a University town, the main settlement in the Bamboutes Mts. C.N.A.D. collected herbaceous plant species on Mt Oku (approaching from Kumbo) in 1970. Amongst their specimens is the earliest record from West Africa of *Juncus oxycarpus* and one of the earliest of *Agrostis mannii*.

Bauer made several of the collections cited in this checklist from Bambui and Bambili in 1970. During the 1970s, several Dutch botanists from the University of Wageningen collected in Cameroon. Among them Willem **De Wilde** and his wife Brigitte were based at N'kolbisson near Yaoundé for 1-2 years from which base they collected extensively around Cameroon. During 1975 they, like so many botanists before them, traversed the southern boundary of our checklist area. However, in doing so they are notable for the earliest known collections of the endemic species *Eriocaulon asteroides* and *E. parvulum* at km 21 on the Bamenda- Jakiri road (e.g. *De Wilde* 8633, 2 Sept. 1975).

Thomas's visit to Oku in February 1985 resulted in the first published plant checklist of Mount Oku (Thomas 1986: 59-62). Thomas's "Provisional species list for Mount Oku Flora" included 91 species. His list is invaluable for being backed up by at least some specimens, although unfortunately, these are not cited in his text. However, we have traced many of them at YA and K. The most important of his specimens is perhaps that named *Oxyanthus formosus* in his checklist. His specimen is not that species but the earliest record of the narrowly endemic *Oxyanthus okuensis* (Cheek & Sonké in press).

Thomas adds suffixes to the species, indicating whether they are forest, forest edge and scrub, grassland or aquatic species. From this it is clear that he concentrated on the first two types of habitat. From references to localities in the text and from his specimens it seems that the majority were collected around Lake Oku, with others near Oku-Elak and some perhaps from Verkovi (however, we have seen no specimens of the latter). He does not seem to have visited Ijim. Besides the checklist, Thomas wrote two pages on "Vegetation of Mount Oku" (Thomas 1986: 54-56). From this he makes it clear that he did not study the grassland communities in detail. This concentration on the forest probably reflects the overwhelming interest in conserving this habitat as that most important for the two rarest bird species in the Bamenda Highlands, the Banded Wattle Eye and the Bannermann's Turaco. His work supported that of McLeod (1986) who was researching material for what is the single most important work on the biology and history of Mt Oku (McLeod 1986).

Much of what Thomas wrote in 1986 remains true today, although his observation that "only one species, *Disperis nitida*, is narrowly endemic to Mount Oku and a few nearby sites" is now known, with the benefit of more intensive research, to be far from true. Thomas's concise comments on the communities and phytogeographic links of the Mount Oku forest are still unsurpassed.

Asonganyi (1995) produced the next checklist from our area as part of a vegetation survey of the Ijim Mountain Forest conducted with Simon Tame (former BirdLife Project Manager) between 9 January and 25 February 1995. Asonganyi was Deputy Chief of the National Herbarium, Yaoundé and is a specialist in Gramineae. He gives a detailed itinerary and lists the total number of specimens collected as 186, making it plain that his six-page checklist of 151 taxa is based upon these specimens. Unfortunately he does not cite his specimen numbers. We have not traced any of his Ijim specimens. This is regrettable since, collecting at Ijim in January and February as he did, when few other botanists have, it is likely that he obtained fertile specimens of species otherwise unknown from Kilum-Ijim, and still unknown to us. For example, "Crinum sp." and "Lilia sp." may represent taxa which are not included in the present checklist. Among the 31 species of Gramineae that he lists are 15 names that are not included in our checklist. The Asonganyi specimens, if they can be located, will provide valuable material for the next checklist of our area.

Tame and Asonganyi's "Vegetation survey of the Ijim Mountain Forests, North West Province, Cameroon" (1995) contains a checklist of 269 species, i.e. 118 species more than in Asonganyi's list which seems to have been its basis. The source of these supplementary 118 species names is not clear, but they may be derived as follows.

1. Those "identified in the field" of which no herbarium specimens were gathered (Tame & Asonganyi 1995: 5). Apparently, specimens were only made of those species doubtfully identified in the field.
2. The list of Thomas (1986). For example, *Oxyanthus formosus*, a species unknown from the mountain, based on a misidentification, appears in Thomas's list and has been added to the Tame and Asonganyi list.

Tame & Asonganyi's report, in addition to its checklist is valuable for a detailed map of the Ijim area, notes of characteristic species of the main parcels at Ijim, and a classification of upper montane vegetation types for Ijim.

In the late 1990s, R.B.G. Kew, in collaboration with Herbier National Camerounais, mounted three expeditions to Mt Oku and the Ijim Ridge. These took place in 1996, 1998 and 1999. A reconnaissance EW-EU team for subsequent fieldwork was led by Cable to Oku-Elak between 8-14 June 1996. The following details are extracted from the relevant expedition reports (Cheek et al. 1997, Cheek 1999). The names of collectors who had their own series on these expeditions appear in bold.

The first expedition

KILUM INVENTORY, 25 OCT-8 NOV 1996.

PERSONNEL
Sponsored under GEF programme for National Herbarium of Cameroon, Yaoundé:
Dr **Satabié** (Chief of Herbarium), M. **Onana** (Researcher), M. Nana, M. **Tadjouteu** (Technicians), Dr **Cheek** (Technical Assistant).

Sponsored under Earthwatch Europe (supported by European Union, DG VIII) assistance to RBG, Kew:
EW managers/co-ordinators/computer data-entry: Henry Ekwoge, Paschal Ngome.
EW leading botanist: Martin **Etuge**.
EW fellows/volunteers/assistants: Dr **Sileshi Nemomissa** (Nat. Herb, Ethiopia), Mr Mirutse Giday (Nat. Herb. Ethiopia), Mr Siraj (Agric. Bureau, Ethiopia), Mr Tsegaye (Ethiopian Heritage Trust), Dr **Munyenyembe** (Chancellor College, Malawi), Mr Ndangalasi & Mr Mbago (both Botany Dept., Univ. Dar Es Salaam, Tanzania), Ms Hagan (student, Ghana), Mr Tabil (Forestry Dept, Ghana), Ms Dillen (London), Mr **Pollard** (Botany Dept., Univ. Oxford); Ms Litt (New York Botanic Garden, USA), Mr **Buzgo** (Botany Dept., Univ. Zurich, Switzerland), Dr **Zapfack** and M. Sonwa (both Univ. Yaoundé I).
Local Guides: Peter Wambeng, Patrick Lem, Isaac Fokom, Pa Nfome.
Three local people were employed for domestic duties at the main base at Manchok.

Funded by KIFP:
Kilum Mountain Forest Project Ecomonitor: Felix Bafon.

ITINERARY

Sat. 26th Oct. Travel to Oku-Elak (Manchok Guesthouse) from Douala, arrive c. 9 pm.

Sun. 27th Oct. All (except main Yaoundé team who arrived in evening) to lower part of transect KA. Collections under Cheek numbers. Heavy rain on return, virtually no rain in subsequent weeks: local dry season had begun.

Mon. 28th Oct. Divided into 3 collecting teams:
1. Transect KJ Lower forest: Buzgo, Zapfack, Sonwa, Pollard, Tabil, Hagan, Isaac Fokom, Pa Nfome.
2. Transect KA, Satabié, Onana and Cheek (assisted by Nana, Litt, Tadjouteu, Dillen), Felix Bafon & Peter Wambeng.
3. Transect KD (lower part): Munyenyembe (under his own numbers), Sileshi, Gidaye, Tsegaye, Siraj, Ndangalasi, Mbago, Lem.

Tues. 29th Oct.
1. Summit from KD: Munyenyembe, Sileshi, Tsegaye, Gidaye, Siraj, Ndangalasi, Mbago, Nana, Lem.
2. KA midforest: Satabié, Onana, Tadjouteu, Dillen, Litt, Bafon, Wambeng.
3. KJ lower forest: Buzgo, Zapfack, Sonwa, Pollard, Tabil, Hagan, Fokom, Pa Nfome.

Weds. 30th Oct.
1. unmarked trail near KC/KD towards rockface: Munyenyembe, Sileshi, Siraj, Gidaye, Tsegaye, Ndangalasi, Mbago, Lem.
2. KC lower forest: Buzgo, Zapfack, Sonwa, Pollard, Tabil, Fokom, Pa Nfome.
3. KA to grassland, Satabié, Onana, Tadjouteu, Nana, Cheek, Litt, Dillen, Wambeng, Bafon

Thurs. 31st Oct.
Satabié and Cheek, report writing.
1. Rockface near KC: Munyenyembe, Sileshi, Tsegaye, Siraj, Gidaye, Ndangalasi, Onana, Tadjouteu, Nana, Lem, returning early: Bafon, Litt & Dillen.
2. Summit up via KA, returning KD: Buzgo, Zapfack, Sonwa, Tabil, Pollard, Isaac Fokom, Pa Nfome.

Fri. 1st Nov.
1. Mboh: bamboo patch in flower, Podostemaceae: Satabié, Cheek, Onana, Litt, Dillen, Hagan, Bafon, Wambeng.
2. Rock face with waterfall near KD: Sileshi, Munyenyembe, Ndangalasi, Gidaye, Tsegaye, Siraj, Buzgo, Nana, Tadjouteu, Lem.
3. KC lower forest: Zapfack, Sonwa, Pollard, Hagan, Isaac Fokem, Pa Nfome.

Sat. 2nd Nov.
Cheek and Satabié: report writing.
Free day: market, carvings, paper and honey co-operatives.

Sun. 3rd Nov.
Cheek and Satabié report writing.
1. Shrine forest near MKFP HQ: Sileshi, Munyenyembe, Ndangalasi, Siraj, Tsegaye, Gidaye.
2. Lowland rainforest patches near Bafon's farm: Zapfak, Sonwa, Litt, Dillen, Tabil, Onana, Pollard, Tadjouteu, Bafon

Mon. 4th Nov.
1. To Lake Oku: Cheek, Satabié, Onana, Sileshi, Gidaye, Tsegaye, Siraj, Litt, Dillen, Buzgo, Ndangalasi, Wambeng, Bafon, Etuge, Pollard.
2. Lowland rainforest patches. Zapfack, Sonwa, Munyenyembe, Mbago, Tabil.

Tues. 5th Nov.
1. To Kumbo road (discovery of gentian site) Zapfack, Sonwa, Munyenyembe, Mbago.
2. At Lake Oku (as 28th Oct.)

Weds. 6th Nov.
Satabié and Cheek: terminating programme and budget for HNC/GEF 96/97.
1. To Kumbo road (new gentian site) Sileshi, Zapfack, Sonwa, Etuge, Tabil, Pa Nfome.
2. Lowland forest remnants near Bafon's farm: Buzgo, Pollard, Litt, Dillen, Onana, Tadjouteu, Nana, Bafon.
3. Cliff near Lumeto forest: Munyenyembe, Siraj, Gidaye, Tsegaye, Mbago, Ndangalasi, Lem.

Thurs. 7th Nov.
1. To Lumeto Forest: Satabié, Onana, Cheek, Tadjouteu, Nana, Zapfack, Sonwa, Pollard, Litt, Dillen, Buzgo, Bafon, Wambeng and guide from Fon of Oku's Palace.

2. Road to Mbam: Munyenyembe, Sileshi, Gidaye, Tsegaye, Siraj, Mbago, Ndangalasi, Tabil, Lem.

Fri. 8th Nov.
EW co-ordination team (Ekwoge, Ngome, Etuge) and Pollard remain behind to finish drying specimens and transfer all equipment to the next site.
All others depart for Yaoundé or Douala.

COLLECTIONS MADE, BY COLLECTOR SERIES

Buzgo	(580-707)	127
Cheek	(8426-8608)	183
Etuge	(3331-3350)	20
Munyenyembe	(701-888)	189
Onana	(444-514)	70
Pollard	(1-50)	50
Satabié		11
Zapfack		190
Total		740

IJIM INVENTORY 16-28TH NOV. 1996

PERSONNEL
Sponsored under GEF programme: National Herbarium of Cameroon, Yaoundé:
Dr **Satabié** (Chief of Herbarium), M. **Onana** (Researcher), M. Nana, M. Ghogue (Technicians), Dr **Cheek** (Technical Adviser) .

Sponsored under Earthwatch Europe assistance to RBG, Kew: Mr **Kamundi**, Ms Maliwichi (Nat. Herb. Malawi), Mr Duah Nsenkyire (Trees for Future Ghana (NGO)), Ms Owusu Agyei (Forestry Dept., Ghana), Ms Moyo (Warden, Hwangue Main National Park, Zimbabwe), Dr Manyuchi (Animal Production and Wildlife Management, Africa University, Zimbabwe), Mr Baedermariam, Co-ordinator, Ethiopia Heritage Trust, Mr Abebe (Entoto School, Addis Ababa), Mr Gashu (Biodiversity Institute, Addis Ababa), Ms Holder (Essex, U.K.), Mr Callow (U.K), Mr **Pollard** (Botany Dept., Univ. Oxford); Ms Litt (New York Botanic Garden), Mr **Buzgo** (Botany Dept., Univ. Zurich).

Guides: Frederick Nkwah, Johnson Ayeah, David Ngong, Raphael Kombeh (all from Aboh).

Kilum Mountain Forest Project Ecomonitor: Moses Kemei (of Tumuku), Clement Toh.

Cooks and Cleaners: Esther Fein, Florence Ntein.
Laundry: Charity.
Wood and water haulier: Cyprian Chi.

ITINERARY

Sat. 16th Nov. All travel from Yaoundé or Douala to Aboh (village above Ijim Project HQ).

Sun. 17th. Entire group up Gikwang Rd to *Gnidia* woodland.

Mon. 18th
1. Gikwang Rd: *Prunus* woodland HNC, Cheek, Moyo, Callow, Kemei, Nkwah
2. Tuboh Rd: *Gnidia* woodland and grassland: Kamundi, Buzgo, Abebe, Baedermariam, Gashu, Manyuchi, Johnson Ayeah.

3. Nyassoso Rd: Etuge, Litt, Maliwichi, Owusu-Agyei, Pollard, Nsenkyire, Holder, David Ngong, Raphael Kombeh.
Dr Satabié arrives from Belo.

Tues. 19th
As previous day but further:
1. To Kilum-Ijim Forest Project lodge overlooking Lake.
2. Tuboh Rd.
3. Nyassoso Rd.

Weds. 20th
As previous day.
1. To ecomonitoring transects A,B & C Bamboo forest.
2. Tuboh Rd.
3. Nyassoso Rd.

Thurs. 21st
1 Ijim Ridge above Fon's Palace, Laikom (gallery forest in grassland).
2. Tuboh Rd to track A.
3. Nyasoso Rd.

Fri. 22nd
1a. (Satabié and Cheek). Muteff: Cliff face; Laikom: sacred spring; Fundong: Touristic Hotel waterfall.
1b. (Onana and HNC). Forest below Aboh on Anyajua Rd.
2.Tuboh Rd.
3. Nyasoso Rd.

Sat. 23rd
Most people to Bamenda for break. HNC investigate forest below Aboh.

Sun. 24th
Rest day.

Mon. 25th
1a. Satabié and Cheek: Mbesa (N. side) to Jikijem junction (IMFP landrover).
1b. Onana and HNC team investigate lowland forest fragments.
2. Buzgo leads team to cliffs N. of Aboh. Kamundi repairs spectacles in Bamenda.
3. Etuge goes (E?) to stream location.

Tues. 26th
1. Cheek, Satabié, Onana, Nana, Ghogue, Pollard, Callow, Frederick Nkwah and Moses Kemei into crater from below resthouse, skirt perimeter to outlet stream.
2. Kamundi to cliffs.
3. Etuge to eastern stream location.

Weds. 27th
1. De Marco, Cheek, Satabié, Onana to transect to do 25 x 25m vegetation plot and profile in forest.
2. Kamundi to cliffs.
3. Etuge to east.

Thurs. 28th
1a. Cheek and Satabié meeting with De Marco.
1b. Onana leads team doing 25 x 25m plot and profile in *Gnidia glauca* woodland.
2. Kamundi to Tuboh rd.

3. Etuge to Afua swamp.

Fri. 29th
Most to Belo by IMFP landrovers, thence Toyota Hiace to Douala.
Co-ordinating team leave following Monday after finishing main body of drying.

COLLECTIONS MADE BY COLLECTOR SERIES

Buzgo	(?708-809)	101
Cheek	(8628-8783)	155
Etuge	(c. 3351-3703)	275
Kamundi	(602-733)	132
Onana	(515-644)	134
Pollard	(51-82)	32
Satabié	(1069-1096)	27
Total		856

The Second Expedition, 2-12th December 1998

Participants: Kilum-Ijim Project Biologists: Kemei, Wultoff, Toh, Keming and Maisels.
Peace Corps Volunteers: Cowall and Chachulski.
Forest officer, Belo: Naimbang.
Kwi Fon: Bobe Peter Yama.
Earthwatch Fellows: Byamukama, Mucunguzi, Mukenya (Uganda), Albano (Mozambique), Maroyi (Zimbabwe), Konde, Muiruri, (Kenya).
R.B.G., Kew: **Cheek, Gosline**.
Earthwatch team: **Etuge**, Njume, **Pollard**.
Herbier National Camerounais: Biye.

ITINERARY

Tues. 1st Dec. Arrived at Fon's Palace, Laikom (expedition base), from Douala.

Weds. 2nd Dec. Laikom Ridge, working on *Sporobolus* grassland and Basalt Pavement surveys.

Thurs. 3rd Dec. Laikom Ridge, general collecting on path to Ardo's compound.

Fri. 4th Dec. Anyajua. Visit to Kilum-Ijim Project Headquarters.

Sat. 5th Dec. A. Laikom Ridge, Ardo of Ijim's compound via Sappel to Mbesa Swamp. (Cheek, Gosline, Byamukama, Albano, Manu Ibrahim).
B. Akwamofu sacred/Royal Forest near Palace (Pollard, Etuge, Yama, Konde, Toh, Njume, Maroyi).

Sun. 6th Dec. Rest day.

Mon. 7th Dec. A. Afua Swamp survey. (Cheek, DeMarco, Gosline, Etuge, Kemei).
B. Swamp ½ mile W. of Ardo's compound (Pollard, Konde, Maroyi).

Tues. 8th Dec. A. Laikom Ridge, 25 x 25m plot in gallery forest. (Cheek, Byamukama, Albano, Kemei, Yama, Mukenya and Pollard).
B. Plateau forest N. of Ardo's compound. (Etuge, Konde, Maroyi).

Weds. 9th Dec. A. Anyajua (*Oncoba* site), Mbi Crater survey (Cheek, Albano, Etuge, Byamukama, Njume, Naimbang, Toh).

B. Ngengal ("Mbesa") Swamp on horseback! (Pollard, Kemei).

Thurs. 10th Dec. A. Grassland Plot demonstration by Maisels on Laikom Ridge.
B. Akwamofu Forest (Pollard & Njume).

Fri. 11th Dec. A. Forest fragment (*Oncoba*) below Palace, then up Laikom valley to savannah forest plot.(Maisels, Albano, Biye, Byamukama, Cowall, Chachuski, Wultoff, Keming, Yama, Cheek). Pollard's birthday!
B. Lower waterfall below Palace. (Pollard, Etuge, Njume, Konde).

Sat. 12th Dec. A. Fundong waterfall, Anyajua cave, Anyajua waterfall.
(Cheek, Maisels, Wultoff, Keming and Cowall).
B. Upper waterfall below Palace (Pollard & Njume).

Sun. 13th Dec. Rest day. Packing up.

Mon. 14th Dec. Leave for Douala.

COLLECTIONS MADE BY COLLECTOR SERIES

Cheek	9721-9925	205
Pollard	255-386	131
Gosline		9
Total		345

Third expedition, Nov. 1999
Objectives:
1. Survey of mid-montane forest below 2000m alt. (not previously surveyed).
2. Survey of montane savannah (not previously surveyed).
3. Resurvey of wetlands with specialist: cancelled due to absence of Dr Muasya of Nairobi.

PERSONNEL:
Herbier National
Dr Gaston Achoundong (Chef), Dr J-M **Onana** (Researcher), J-P. **Ghogue**, F. **Tadjouteu** (Technicians), Boniface Tadadjeu (Driver).

Mt Kupe-Bakossi Forest Project: Martin Etuge (first week).

RBG, Kew
Martin **Cheek**, George Gosline.

KIFP
John **DeMarco** and Anne Gardner (co-Managers), Dr Philip Forbosi (first day), Romanus Ikfuingei, Ernest Keming (last two first week), Innocent Wultof, Moses Kemei (second week).

Earthwatch fellows
Anthony Kimaro (Tanzania), Kambale Kioma, Kalindula Musavuli (both Zaire), John Amponsah and Maryam Abdul-lah (both Ghana).

Assistant camp managers
Martin Mbong & Ngole Hoffmann

Guides (Ijim area)
Amos Ndi & Andrew Talla.

Local Staff:
Sarah Nnie, Euphrasia Ngweh (cooks), Euphrasia Ajang (cleaning), Clovis Ayeah (water carrier), Lot Ayeah (night security).

ITINERARY

Tues. 2nd Nov.
HNC arrive at Laikom from Yaoundé.
EW team arrive from Douala.

Weds. 3rd Nov.
All visit Ijim ridge above Laikom.

Thurs. 4th Nov.
All visit Ntum to search for *Ternstroemia polypetala* (without success) and *Pentarrhinum* nov. (successful).

Fri. 5th Nov.
HNC team survey Palace forest at Aboh.
EW team survey Ntum area again, searching for *Ternstroemia* without success.

Sat. 6th Nov.
Martin Etuge departs for Nyasoso.
All visit forest at head of Laikom valley. Romanus and Ernest depart.

Sun. 7th Nov.
Rest day.

Mon. 8th Nov.
Whole team to Bambui, Bambili (brief survey of lowland savannah at 600m alt.), Babungo, surveying savannah at c. 1600m just before Cattle ranch at Mbi. Visit to Mbi crater abandoned due to bogging down of Landrover in mud and lack of daylight.

Tues. 9th Nov.
Visit to forest fragments at Fujua/Hausa quarter (below Laikom).
Visit to Fundong waterfall to study Podostemaceae.

Weds. 10th Nov.
Whole team to Mbingo to survey proposed hospital protected area "Back Valley".

Thurs. 11th Nov. JD, MC, GA, FT, IW & MK to Ajung, studying savannah between Mbesa, Akeh and Ajung en route.
EW team seeing KIFP work in progress. Overnight at Anyajua Day 1.

Fri. 12th Nov.
Ajung team trek to forest and return.
EW team seeing KIFP work in progress. Overnight at Anyajua Day 2.

Sat. 13th Nov.
Ajung inventory team from Ajung to Anyajua.
Anyajua to Laikom. Inventory of lower palace forest (MC, GG, IW, AN and EW fellows).
GA and FT with MK surveying at Lake Oku.

Sun. 14th Nov. Rest day. Main HNC team depart for Yaoundé. Discussions on checklist. Preparation on rapid assessment reports for Mbingo and Ajung forests.

Mon. 15th Nov.
Lake Oku: surveying rare species (*Oxyanthus sp. nov.* and *Dovyalis sp. nov.*).

Tues. 16th Nov.
Departure for Douala.

Specimens collected:

Cheek:	9,926-10,110	185
Tadjouteu:	221-304	83
Ghogue:	359-426	68
Onana:	901-924	23
DeMarco:	1-18	18
Total specimens collected		377

Maisels was Biologist to the Kilum-Ijim Project in the period 1997-1999. She collected over a hundred specimen numbers which included the first specimens of *Juncus dregeanus* subsp. *bachitii* known in West Africa. Many of the records of savannah species in this checklist (e.g. *Piliostigma thonningii*) are based on her collections. Her interest in rare plants on the summit of Mt Oku led to an article on this subject (Maisels et al. 2000).

DeMarco was co-project manager of the Kilum-Ijim Project from c. 1996 to 2000. He has a particular interest in orchids, as is evident from the number of his specimens cited in this checklist in that family.

THE EVOLUTION OF THIS CHECKLIST

This checklist began life as part of the GEF-Cameroon plant surveys and inventories programme for the National Herbarium of Cameroon (HNC) managed by R.B.G., Kew. Under this programme In 1996 a month of fieldwork on Mt Oku and at Ijim was co-funded with the Earthwatch Institute (Europe) with support from DG VIII of the European Commission. The three authors of this book (MC, JMO and BJP) first met on this occasion. Details of this and subsequent fieldwork are elaborated in 'History of Botanical Exploration on Mt Oku and Ijim Ridge'. An interim report giving details of conservation priority species discovered in the course of this work (on which the Red Data chapter in this book is based), together with other information resulting from the fieldwork was completed two months later (Cheek *et al.* 1997). A week-long reconnaissance expedition, based at Oku-Elak funded by Earthwatch, was led by Stuart Cable in June 1996.

Pre-identifications of the specimens from the main expedition were made at HNC in the months following the fieldwork, before the specimens duplicates were freighted to R.B.G., Kew for the attention of plant family specialists. The first new species to be published from these collections were two new species of *Eriocaulon* (Phillips 1998). In August 1998 JMO spent six weeks at R.B.G., Kew funded by the GEF programme, making final determinations for those large families where specialists were available at R.B.G., Kew to give advice but could not spare the time to name all the specimens themselves. These included the Leguminosae (with the assistance of Barbara Mackinder and Brian Schrire), Ferns (with the assistance of Peter Edwards) and Compositae (with the assistance of Henk Beentje and Nick Hind). A new subspecies of *Indigofera* was distinguished in the course of this work (Schrire and Onana 2000). Other major families were identified by Tom Cope (Gramineae), Kaare Lye (Cyperaceae), Kaj Vollesen (Acanthaceae) and Ben Pollard and Alan Paton (Labiatae). Rubiaceae were named by MC (shrubs and trees, with assistance from Diane Bridson) and Sally Dawson (herbs). The remainder of the families were mostly named by MC. At about this time fruiting material of a third new species of *Eriocaulon* was made available by Boo Maisels, enabling it to be described (Phillips 2000). Preparation of the family accounts began in July and August 1998 when Ray Rix and Neda Fathi (Nuffield bursary students) began work on those families already identified. The second expedition to the checklist area, focusing on wetland areas, took place in the first two weeks of December 1998. It was funded entirely by the Earthwatch Institute.

In 1999 the last of the identifications of the '1996' specimens was completed by MC and LC (Kew diploma student) in their spare evenings. A draft checklist prepared by LC was taken to the Kilum-Ijim project in November 1999 for correction and comment.

In September 1999 R.B.G., Kew was awarded a three year grant by the Darwin Initiative to fund a programme entitled 'The Conservation of Plant Diversity of Western Cameroon'. This programme sets out to catalogue the plant diversity of Western Cameroon (South West and North West Provinces) and to identify those species most in need of conservation. One of the planned outputs is to produce 'conservation checklists' for protected areas in Cameroon such as Mt Oku and Ijim Ridge. It was at this point that the decision was taken, in discussion with David Thomas of BirdLife International, to publish this book since detailed information on the rare and threatened plants at Kilum-Ijim is badly needed. Through BirdLife International and the Kilum-Ijim Forest Projects, a contribution towards publication costs was obtained from GEF Cameroon.

A third expedition was undertaken in the first two weeks of November 1999 at Ijim, inaugurating the Darwin funded project. This joint Kew-HNC expedition was jointly funded by the Darwin Initiative and Earthwatch and concentrated on lower altitude (1500–2000m) montane forest patches.

In February 2000 BJP was appointed Darwin Initiative Officer at R.B.G., Kew in connection with the Western Cameroon programme, giving fresh impetus to work on Mt Oku and Ijim. By this time the specimens from the 1998 expedition had arrived at Kew. These were divided for identification along the same lines as those arising from the 1996 expedition, except that those families previously identified by JMO were taken on by BJP. Two more new species were completed for publication at this time, a new *Oxyanthus* (Cheek & Sonké, in press) and a new *Chassalia* (Cheek & Csiba, in press). As, one by one, the family accounts produced for the draft checklist were augmented with specimens identified from the 1998 expedition, BJP added further species records by trawling F.W.T.A. Vols. 1–3, i.e. searching all families for species not yet included in our checklist, but with specimens cited from our area. The additional names were checked by MC and BJP before they were included using the Kew herbarium and Lebrun & Stork (1991-98) to ensure that they were still correct. This endeavour increased greatly the coverage of lower altitude species in our checklist.

Local names were also added to the Checklist at this stage by BJP and LC.

Finally, the descriptions were added to the species accounts by extraction from F.W.T.A. Those for Monocotyledons were compiled by BJP, those for Dicotyledons were started by Matthew Frith, but largely completed by Penelope Doyle in association with MC. Henk Beentje, BJP and Peter Edwards wrote new descriptions for the Compositae, Labiatae and Ferns respectively.

ETHNOBOTANY

Benedict John Pollard

There are two main groups of plants on the mountain that are sigificant to the local people, namely crop plants (mainly introduced) and wild, often native, species. The growing of crops and non-sustainable over-harvesting of wild species contribute threats to natural vegetation.

The data regarding plant use that were reported to us and recorded during our fieldwork are summarised with reference to standardised terms proposed by Cook (1995). This work provides a system whereby uses of plants can be described, using standardised descriptors and terms, and attached to taxonomic data sets. These descriptors operate on three levels, the third level being most detailed. Here I use just the LEVEL 1 states and their associated CODES (Cook, 1995) to provide a simple overview of different plant uses.

CODES	**LEVEL 1** states
0100	FOOD
0200	FOOD ADDITIVES
0300	ANIMAL FOOD
0400	BEE PLANTS
0500	INVERTEBRATE FOOD
0600	MATERIALS
0700	FUELS
0800	SOCIAL USES
0900	VERTEBRATE POISONS
1000	NON-VERTEBRATE POISONS
1100	MEDICINES
1200	ENVIRONMENTAL USES
1300	GENE SOURCES

1) CROP PLANTS

There are a number of crop plants that are utilised for food, in agroforestry, for timber or for other inherent properties such as their medicinal powers. Representative specimens of some of these were collected by us, and so are included in the checklist. They are, namely:

FAMILY	SCIENTIFIC NAME	ENGLISH NAME	LEVEL 1 CODE
Cruciferae	*Brassica cf. rapa*	Turnip	0100
Cupressaceae	*Cupressus lusitanicus*	Cypress	0600
Euphorbiaceae	*Ricinus communis*	Castor Oil	0300, 0600, 1100
Gramineae	*Triticum aestivum*	Wheat	0100
Myrtaceae	*Eucalyptus sp.*	Gum tree	0600
Passifloraceae	*Passiflora edulis f. edulis*	Passion fruit	0100
Solanaceae	*Cyphomandra betacea*	Tree tomato	0100
Solanaceae	*Nicotiana tabacum*	Tobacco	0800
Solanaceae	*Physalis peruviana*	Cape gooseberry	0100

Many other crop plants occur on the mountain but are not represented in the main checklist since they were not collected by us. The list below is largely derived from information on crop species provided to BJP and John DeMarco. Scientific names follow Mabberley (1998) and Purseglove (1968, 1972).

Alliaceae	***Allium cepa***	Onions
Alliaceae	***Allium sativum***	Garlic
Anacardiaceae	***Mangifera indica***	Mango
Araceae	***Colocasia esculenta***	Cocoyam
Bromeliaceae	***Ananas comosus***	Pineapple
Burseraceae	***Dacryodes edulis***	'Plum'
*Burseraceae**	***Canarium sp.***	'Bush Plum'
Caricaceae	***Carica papaya***	Papaya
Convolvulaceae	***Ipomoea batatas***	Sweet potato
Cucurbitaceae	***Citrullus lanatus***	Water melon
Cucurbitaceae	***Cucumis sativus***	Cucumber
Cucurbitaceae	***Cucurbita sp.***	Pumpkin
Cucurbitaceae	***Cucurbita pepo* 'Zucchini'**	Zucchini/courgette
Cucurbitaceae	***Cucurbita sp.***	Squash
Dioscoreaceae	***Dioscorea sp.***	Yam
Euphorbiaceae	***Jatropha curcas***	
Euphorbiaceae	***Manihot esculenta***	Cassava
Gramineae	***Brachiaria sp.***	
Gramineae	***Oryza sp.***	Rice
Gramineae	***Saccharum sp.***	Sugar cane
Gramineae	***Sorghum bicolor***	Sorghum
Gramineae	***Tripsacum fasciculatum***	Guatemala grass
Gramineae	***Vetiveria nigritana***	
Gramineae	***Vetiveria zizanoides***	Vetiver grass
Gramineae	***Zea mays***	Maize / corn
Lauraceae	***Persea americana***	Avocado
Leguminosae	***Acacia angustissima***	
Leguminosae	***Acacia mangium***	
Leguminosae	***Arachis hypogaea***	Peanut
Leguminosae	***Cajanus cajan***	Pigeon pea
Leguminosae	***Calliandra calothyrsus***	
Leguminosae	***Glycine max***	Soya bean
Leguminosae	***Leucaena leucocephala***	
Leguminosae	***Phaseolus vulgaris***	Bean
Moraceae	***Artocarpus altilis***	Breadfruit
Musaceae	***Musa cvs.***	Banana
Musaceae	***Musa* × *paradisiaca***	Plantain
Myrtaceae	***Eucalyptus spp. (3 spp.)***	Gum tree
Myrtaceae	***Psidium sp.***	Guava
Palmae	***Elaeis guineensis***	Oil Palm
Pedaliaceae	***Sesamum indicum***	Sesame
Podocarpaceae	***Podocarpus mannii***	
Proteaceae	***Grevillea robusta***	
Rubiaceae	***Coffea arabicum* 'Java'**	Coffee
Rutaceae	***Citrus* × *aurantium***	Orange
Solanaceae	***Capsicum annuum* var. *annuum* Grossum group**	Capsicum
Solanaceae	***Capsicum annuum* var. *annuum* Longum group**	Chilli pepper
Solanaceae	***Solanum intrusum***	Huckleberry
Solanaceae	***Solanum lycopersicum***	Tomato
Solanaceae	***Solanum tuberosum***	Potato
Umbelliferae	***Apium graveolens* var. *dulce***	Celeriac / celery
Umbelliferae	***Daucus carota***	Carrot
Zingiberaceae	***Zingiber officinale***	Ginger

2)   LOCAL NAMES AND USES

Many of the plants listed in our checklist are of great importance to the local people and are inextricably linked to local customs and beliefs.

In compiling this checklist of plants from Mount Oku and the Ijim ridge, it became apparent that a number of our collectors (for example Paul Munyenyembe) invested much time investigating the ethnobotanical knowledge held by the local tribes. Although this was never an intended focus of our inventory programme, a great deal of information regarding local plant names and uses has been recorded.

The local names have been included in the checklist, directly below the accepted scientific name to provide a clear link between these two knowledge systems. We have decided to exclude from the checklist details of local plant uses as the intellectual property rights to this information belong with the people of the Kom, Oku and Nso tribes, and is not to be disseminated unchecked without their express permission.

It is quite likely that there are many plants of potential economic value in our checklist area, some of which may be endemic. If these plants are to be investigated for medicinal properties, horticultural value or any other useful traits, it is essential that the correct developmental frameworks and strategies are in place. It is hoped that these issues can be addressed and perhaps discussions with stakeholders initiated so that permission to develop a more thorough and complete ethnobotanical survey may in future be granted.

Previous ethnobotanical surveys have been conducted on the mountain, e.g., by Mbenkum & Fisey (1992) and Tame & Thomas (1993). These surveys do not cite specimens, and so it is difficult to verify that the local names and uses correspond to the given scientific names. Tame & Thomas list 130 taxa with local names, 57 (44%) of which are said to be of some use to local people. Adjanohoun et al. (1996) visited Ndop, Jakiri and Kumbo on the edge of our area, and made a one day visit to Oku-Elak on the Kilum side of the mountain and gave medicinal uses of a number of species, but, regrettably, do not cite specimens.

Our data on plant uses in the Kilum-Ijim Mountain Forests are held at Kew and Yaoundé and are summarised below. A more detailed analysis is intended with the results made available to the local people and Conservation Project Managers.

LEVEL 1 states	Dicotyledons	Monocotyledons	Ferns & Gymnosperms
FOOD	19 (1)	1	
FOOD ADDITIVES	1		
ANIMAL FOOD	1		
BEE PLANTS	3		
INVERTEBRATE FOOD			
MATERIALS	27	5	1
FUELS	7		
SOCIAL USES	13	3	1
VERTEBRATE POISONS			
NON-VERTEBRATE POISONS	1	1	
MEDICINES	94(1)	4	6
ENVIRONMENTAL USES	8		
GENE SOURCES			
TOTALS	192(2)	14	8
TOTAL NUMBER OF REPORTED USES			214(2)

Numbers of uses recorded 1996–1999 for each Level 1 state, arranged by plant group.
Pre-1996 records of uses are shown in brackets.

These plants are used in a multitude of ways in Kilum-Ijim, examples of which are outlined below, according to the LEVEL 1 states.

FOOD
Leaves are eaten raw, as a vegetable, in soups. Fruits are eaten.

FOOD ADDITIVES
Spices.

ANIMAL FOOD
Monkeys eat the fruits.

BEE PLANTS
Plant flowers every 8–9 years, bees visit this, helping honey production.

MATERIALS
Carving, carving drums, axe-handles, walking sticks, weaving mats, stems used to tie firewood bundles, fibres extricated to make ropes for baskets, building beehives, timber, wooden pegs, build oil containers, paper making, dye mats.

FUELS
Firewood

SOCIAL USES
Smoked like a cigarette, used as a bait for hunting cane rats, sees off witchcraft, makes you lucky, wash a woman's stomach in order to get pregnant, gum traps birds, used for traditional dances at the Palace.

NON-VERTEBRATE POISONS
Insecticide, vermifuge.

MEDICINES
Headaches, general pains, stomach pain, curing skin ulcers, aching thighs, leprosy, relieving pain, scabies, madness, back pain, laziness, ear-ache, cough, nervous system, fever, systemic parasites, eyes, burst boils, wounds, catarrh, side pains, internal worms, joint problems, arthritis, elephantiasis, chronic gonorrhoea, prostate cancer, stops vomiting, cures snake poison, epilepsy, convulsion, cramp, filaria, navel especially in children, diarrhoea, fainting sickness, pimples in children, unsteady menstrual cycle.

ENVIRONMENTAL USES
Ornamentals, live hedges, cover crop, cultivated as 'green manure'.

DATA SUMMARY

USES RECORDED	214
TAXA WITH TWO REPORTED USES (All Dicotyledonae)	28
TAXA WITH THREE REPORTED USES	4
NUMBER OF TAXA WITH REPORTED USES	178
TOTAL NUMBER OF TAXA IN CHECKLIST	927
PROPORTION THAT ARE REPORTED TO BE USEFUL	19.2%

USEFUL TAXA (INCLUDING CROP PLANTS)	232
TOTAL TAXA (INCLUDING CROP PLANTS)	981
PROPORTION THAT ARE REPORTED TO BE USEFUL	23.6%

The data gathered to date may be valuable as a starting point for a more focused ethnobotanical survey. We intend to develop this important aspect of our study in future fieldwork if feasible.

QUANTIFICATION OF MONTANE FOREST LOSS IN THE BAMENDA HIGHLANDS

How much montane forest was there in the Bamenda Highlands, how much is left, and what are its prospects for the future?

Hawkins & Brunt (965: 208) state, concerning the location and distribution of moist montane forest, that this originally probably covered the whole of the High Lava Plateau. They give its lower altitudinal limit as about 5,000 feet (i.e. c. 1500m) and its upper limit as about 7,000 feet (i.e. c. 2100m) where it merges into the 'Bamboo' forest. They report, in 1965, extensive remnants on Mount Binka, Mba Kokeka, (including the Bafut forest reserve) and Mount Bambutos, with smaller remnants all over the plateau, mostly as gallery forest, or as remnants in the clefts and gullies.

Using Hawkins & Brunt's vegetation map, an estimate of the area of original montane forest, including 'bamboo forest' was made by using graph paper to measure the area enclosed by the 5,000' (c.1500 m) contour of the main block of the Bamenda Highlands (Fig. 5). This excludes, for convenience, outlying blocks, such as those in Western Province e.g. Mbam and Bambutos, and those in Nigeria, e.g. the Mambilla Plateau. 2820 km^2 is the total area of putative original montane forest in this block.

How much is left? McKay recently surveyed surviving forest fragments in the Bamenda Highlands in connection with assessing populations of the two forest dependent bird species *Tauraco bannermani* and *Platysteira laticincta* (McKay 1994).

McKay (1994: 8) tabulates 18 forest patches in NorthWest Cameroon. Those that fall within our block are as follows:

Forest	Area (km^2)
Bali Ngemba	c. 8
Bafut Ngemba	7.5
Bambui	7.5
Kilum	c. 30
Ijim	c. 40
Kejodsam	5
Kumbo Plateau	?
Total	c. 98

The total of 98 km^2 of surviving montane forest is only approximate. McKay may have overlooked some areas. No figure is given for the Kumbo Plateau, and more forest may survive there. Two of the figures used in the table above (those of 7.5km^2 for Bafut, Ngemba and Bambui) are converted from ranges of 5–10 km^2 which indicate the uncertainty of the data available. McKay's study concentrated on forest at 1800m and above, while our block is demarcated by the 1500m contour. However, forest at 1500m in this area, to my knowledge, is almost non-existent. Pressure for agricultural land at such low altitude seems to have been immense. Certainly, blocks as large as 1 km^2 or more are rare, and I have only seen one block below 1800m that might approach 1 km^2, and that is "Back Valley" at Mbingo at c. 1600m alt.

Taking the estimate of surviving montane forest (98 km^2) derived from McKay's work with that for the original area (2820 km^2) derived from Hawkins and Brunt, gives a figure for surviving montane forest of only 3.5%. This figures, which implies a loss of 96.5% of the original forest, is probably extreme. Apart from the caveats mentioned above in connection with the figure's derived from McKay's work, the figure for original forest cover does not take into account the fact that a significant part of the original area of the Bamenda Highlands was not forested, but montane grassland (probably restricted to rocky areas) and swamp. This is suggested by the number of species from rocky grassland and swamp habitats such as *Bafutia tenuicaulis* and *Kniphofia reflexa* respectively, that are entirely restricted to the Bamenda Highlands. Such species indicate that their habitats are far from secondary. However, even if only half the 'original area' was forested, the figure for loss of the original forest would still be as high as 93%.

The prospects for the future of montane forest in the Bamenda Highlands do not look encouraging. Based on satellite images of Mount Oku and Ijim Ridge taken in 1987 and 1995 (see rear cover), Kew's GIS specialist estimates there was a 25% decrease in forest area over this eight year period. (Justin Moat, pers. comm.).

A report based on a survey of surviving forest in the Bamenda Highlands conducted in March 2000 (Yana Njabo & Languy, 2000) makes sad reading. 99% of the canopy of Bafut Ngemba had gone and Kejodsam, which in 1994 was considered a good prospect for survival, being inaccessible and remote, now has a road through it, and a new village has appeared there. 5Ha of the original 5km^2 of forest survives, for the present.

RED DATA PLANT SPECIES OF MOUNT OKU AND IJIM RIDGE

This Red Data chapter follows the model set in Cheek et al. (1997) and followed in Cable & Cheek (1998). IUCN assignments of the level of threat to the taxa are applied using IUCN 1994 Red List Categories (IUCN 1994). These assignments are made from a global, not a local perspective. Thus species that are extremely rare and potentially threatened throughout Africa, such as *Anagallis minima* (known from only three individuals in Africa, one of which is at Mt. Oku and the Ijim Ridge) are not treated in this Red Data chapter if they are relatively common and unthreatened elsewhere. *Anagallis minima*, for example, is known from many sites throughout temperate Eurasia and extends to the New World.

The criteria for selection of taxa as threatened are as follows. Firstly we have selected taxa that are restricted to (i.e. are endemic to) Mt. Oku and the Ijim Ridge, the Bamenda Highlands or the Cameroon Uplands from Bioko northwards to Tchabel Mbabo, including the Obudu and Mambilla Plateaux in Nigeria. From these have been selected those that appear vulnerable either by their rarity or by threat from man. These judgements are made either from personal observation and/or from herbarium specimens, e.g. the fact that very few collections are in existence. Two examples will illustrate our approach: *Schefflera mannii*, is known only from upper montane forest in the Cameroon Highlands. We exclude this species from red listing here because, although upper montane forest is under great threat in the Bamenda Highlands, *S. mannii* is also a common constituent of the forest on Mt. Cameroon and it is not under pressure from clearance there. By contrast, we do include *Wahlenbergia ramosissima* subsp. *ramosissima* because, although it has an identical range, it is extremely rare within this range. Within the Bamenda Highlands it is only known from Laikom Ridge, just above the Fon of Kom's Palace, and all in all, only about six sites are recorded for the species (Cable & Cheek, 1998).

Although an attempt has been made to make this Red Data treatment complete, there is no doubt that it is not exhaustive. Undoubtedly, some Red Data species of Mt Oku and the Ijim Ridge area have been unintentionally omitted. Should you discover, or think you may have discovered, such an omission, please communicate it to the senior author of this book so that it may be included in a new edition!

HERBARIUM SPECIMENS AS A PRIMARY SOURCE

The primary source for this Red Data list is plant specimens, principally those at the Kew and Yaoundé herbaria, supplemented by personal observations of the taxa concerned, where available. All candidate species were assessed against specimens and the plant specialist for the relevant group was consulted where possible. In combination, the Royal Botanic Gardens, Kew and the National Herbarium, Yaoundé are fortunate in having the largest and most comprehensive collection of specimens in the world from western Cameroon. Most of the important collectors who have visited our checklist area have left a set of duplicates at Kew or Yaoundé. In many cases this is the top set (e.g. Letouzey at YA or Maitland, Brunt and Hepper at K), or one of the only surviving sets if the main set was lost or destroyed (e.g. Ledermann at K). Of course there are gaps in the representation at Kew, and to fill these, the literature has been checked as far as possible. Non-botanists should note that botanical literature is only a secondary source, albeit useful. Botanical literature is always founded on specimens, if it is to be of any use. Botanical literature always lags behind specimens. As soon as a checklist, Flora or monograph is published, it is the case that new collections are made which provide evidence to extend the range or reduce the rarity of the taxa included.

THE RED DATA TAXON TREATMENTS

Taxa are presented in alphabetical order, as in the checklist: species within genera, genera within families and families within the larger groupings of, respectively, Dicotyledons, Monocotyledons and ferns and fern allies.

For each taxon included, we have used the correct name (in the sense of the International Code of Botanical Nomenclature), following the most up-to-date and authoritative treatment or specialist determination. The correct name is followed by an IUCN rating which indicates the category of threat and the justification, using IUCN criteria, for giving that rating, following the 1994 Red List Categories (IUCN 1994). In brief, the categories assigned here to the species included in this red data list are: "Extinct", "Threatened" (with Extinction) or of "Lower Risk" (of Extinction). Most of the species here fall under one of the three threatened categories listed here in decreasing order of threat: "Critically Endangered" (CR), "Endangered" (EN) or "Vulnerable" (VU). Almost all of the "Lower Risk" taxa included here fall under the "Near Threatened" subdivision, and are listed as LR/nt. A note on the range of the taxon follows the IUCN category. This is based on the existence of herbarium specimens, either represented at K or YA or cited in a reputable Flora account or monograph. The main body of each taxon treatment takes the form of notes on the detailed distribution, discovery and rarity of the taxon. The detail given here is recommended for the red listing process. It is here that notes are given on whether the taxon was rediscovered, or not rediscovered, in the field surveys of 1996-99. These notes are followed by information on the habitat, threats, and management suggestions for the species. Finally, there is a detailed description (more detailed than in the main body of the checklist). The purpose of this is to enable field biologists and conservationists in Cameroon to identify the Red Data species concerned with as much certainty as possible. Diagnostic notes are added if feasible and necessary to help distinguish Red Data species from related and similar species of lesser conservation concern.

A number of the taxa included are new species, discovered in the course of the recent inventory work and as yet undescribed. These are given full treatment only in cases where work on their description and formal publication is advanced. Some other probable new species which may, after further verification, also be formally published, are mentioned only briefly in order to flag up their existence. They may appear in more detail in subsequent editions of this Red Data list if further work shows that they warrant this.

THREATS

Most plant species in the Bamenda Highlands are vulnerable to extinction not because they are specifically selected by man for destruction (*Prunus africana* might be the only exception), but because they are one of the following:

a) acutely restricted in the area they occupy (e.g. less than 100km^2) or in the number of sites (e.g. less than 5) because their habitat is scarce. Such taxa (to quote IUCN, 1994, on criterion VU D2) would thus "be prone to the effects of human activities (or stochastic events whose impact is increased by human activity) within a short period of time in an unforeseeable future, and are thus vulnerable."
b) components of a habitat that has been or is being destroyed.

The following paragraphs classify the Red Data Species recognised at the end of the chapter by vegetation type and discuss what threats they face. More detail on the species composition of the vegetation types mentioned below is given in the chapter on vegetation.

RED DATA SPECIES & VEGETATION TYPE

a) **Lower montane forest & woodland; 1500-2000m alt.**

Almost all this vegetation type has been lost to clearance for agriculture, and that which remains is disappearing rapidly (see Quantification of loss of montane forest in the Bamenda Highlands, and Yana Njabo & Languy (2000)). Several sources of information suggest an 80% reduction over the last ten years. The only hope for the survival of this vegetation type may be the Royal/Sacred forests of the traditional rulers of the Bamenda Highlands, such as those of the Fons of Kom, Oku and Nso and for those patches protected by communities such as at Ajung or institutes such as Mbingo Hospital (Back Valley). There may be as little as 10km² of this vegetation type surviving in the Bamenda Highlands. Thus any species restricted to this vegetation type in the Bamenda Highlands, including outlier areas such as the Mambilla Plateau where the same loss of habitat is occurring applies, automatically qualifies as Critically Endangered under IUCN 1994 criterion A1c. These taxa are:

Chassalia laikomensis, Psychotria sp. nov., Dovyalis sp. nov., Oxyanthus okuensis, (both up to 2200m), *Dombeya ledermannii* (drier woodland), *Ternstroemia polypetala, Pentarrhinum abyssinicum* subsp. *ijimense, Newtonia camerunensis, Oncoba sp. nov.* (now extinct in the wild), *Eugenia gilgii, Diaphananthe bueae, Genyorchis macrantha, Polystacha superposita* and *P. bicalcarata.*

b) **Lower montane savannah; 1500(-2000)m alt.**

None of the species restricted to the Mt. Oku and Ijim Ridge area or the Bamenda Highlands are known to occur in this vegetation type.

c) (i) **Waterfalls and rapids; c.1300 - 2000m alt.**

These habitats are few and scattered, occupying less than 1% of the Kilum-Ijim protected area. They contain a very few but highly specialised species which are threatened in view of their 'acutely restricted' nature, and qualify as Vulnerable under the IUCN 1994 criterion D2. In addition they are threatened by pollution. These taxa are: *Saxicolella marginalis, Ledermanniella keayi* and *L. cf. muscoides.*

(ii) **Damp shady cliffs and banks c.1300-2000m alt.**

The same observations made for (c) (i) apply here. Red Data taxa or candidates are: *Begonia schaeferi, Craterostigma sp. nov?, Wahlenbergia ramosissima* and *Plectranthus sp. nov?*

d) **Grassland swamps; 1750-2800m alt.**

This habitat transcends the 2000m contour that divides our other vegetation types. Swamps are vulnerable to trampling by cattle and potentially to drainage. *Kniphofia reflexa* disappeared from one of its three known swamp sites between 1997 and 1998. There are only four permanent sizeable swamp sites known at Mt. Oku and Ijim Ridge (several smaller, more seasonal spots are also known) and endemic species confined to this habitat automatically qualify as Vulnerable under criterion D2. The swamps are Afua (<1km²), Mbesa (<1km²), Kinkolong (0.0015km²) and Mbi Crater (c. 5km²). Red Data species confined to swamps are: *Kniphofia reflexa, Eriocaulon bamendae, Justicia sp. nov?, Stachys pseudohumifusa* subsp. *saxeri, Plectranthus punctatus* subsp. *lanatus, Cyperus* sp. A, B, C, *Cyperus niveus* var. nov., *Xyris* cf. *filiformis, X,* sp. A, X. *"welwitschii", Bulbostylis* sp. nov?, *Hyparrhenia sp. nov?,* and *Gladiolus sp. nov?*

e) **Upper montane grassland; 2000-3000m alt.**

At Kilum-Ijim, there have been concerns that this vegetation type (c. 130km²) has been increasing at the expense of forest (c. 70km²). However, most upper montane grassland in our area is dominated by *Sporobolus africanus.* Hawkins and Brunt (1965) show that the original grassland would have been dominated by *Hyparrhenia* and is replaced by *Sporobolus* only after repeated burning and trampling by cattle. Cattle were introduced to this habitat in the 1920s by Fulani people and graze here throughout the wet season. By order of the colonial government, Kikuyu grass, *Pennisetum clandestinum,* was introduced around the compounds of the graziers and now occupies significant areas having displaced, presumably, *Sporobolus.* Red Data species found in upper montane grassland are largely confined to a subset of three grassland types where *Sporobolus africanus* is absent or rare.

53

These vegetation types are relatively rare, and occupy only a small proportion of the total area.

(i) **Basalt pavement grassland** - dominated by *Loudetia simplex*. Red Data species here are *Eriocaulon parvulum* and *E. asteroides*.

(ii) **Rocky grassland on thinner soils** - characterised by an absence of *Sporobolus*. Found on the tops of ridges and on steep slopes. A highly approximate estimate of the total area of rocky grassland is 5-15km^2. Red data taxa are: *Bafutia tenuicaulis, Helichrysum cameroonense, Crotalaria bamendae, C. ledermannii, C. mentiens, Vernonia bamendae* (but habitat data inexact), *Polygala tenuicaulis* subsp. *tayloriana, Peucedanum angustisectum, P. camerunensis* and *Pimpinella c.f. praeventa*.

(iii) **Summit grassland and scrub; 2900-3011m alt.** - this highest of vegetation types in the Bamenda Highlands, though small in area, amounting perhaps to 5km^2, has a wealth of species not found elsewhere in this region. Grazing pressure from cattle, sheep and goats is a cause for concern here. The Red Data species present are: *Alchemilla fischeri* subsp. *camerunensis, Agrostis mannii, Habenaria obovata, H. microceras, Anthospermum asperuloides* and *Veronica mannii*.

f) **Upper montane forest: grassland edge; 2000-3000m alt.**
The ecotone between forest and grassland contains the densest concentration of plant species in the upper montane zone of the Cameroon highlands. It is hard to estimate the area of this habitat because of the dissected and fragmented nature of forest blocks, islands and strips at Kilum-Ijim. Red Data species are: *Lobelia columnaris, Dipsacus narcisseanum, Succisa trichotocephala, Crassocephalum bougheyanum, Pentas ledermannii* and *Tarenna pavettoides*

g) **Upper montane forest; 2000-3000m alt**.
So long as the boundaries of the Kilum-Ijim Forest are respected, and fires continue to be controlled in the dry season, upper montane forest is relatively extensive and unthreatened at Mt. Oku and the Ijim Ridge although it is highly threatened elsewhere in the Bamenda Highlands. About 70km^2 survives in our checklist area. Red Data species of upper montane forest are: *Begonia oxyanthera* (but extending to lower montane forest), *Embelia mildbraedii, Peperomia thomeana, Prunus africana* (Red Data status controversial), *Carex preussii* (in gaps).

h) **Lake; 2200m alt.**
Although Lake Oku contains aquatic plant species not found elsewhere in the Kilum-Ijim area, only one is a Red Data taxon: *Isoetes biafrana*. The main threat to this species is probably siltation of the lake margin caused by clearance or grazing on the slopes above.

Plant taxa known only from Mt. Oku and the Ijim Ridge checklist area

Taxon name	Number of Sites	Red Data Rating
Eriocaulon parvulum (Eriocaulaceae)	3	VU D2
Oxyanthus okuensis ined. (Rubiaceae)	2	CR A1C, B1+2b, C2a, D
Pentarrhinum abyssinicum subsp. ijimense (Asclepiadaceae)	3	CR A1c
Gladiolus sp. nov. (Iridaceae)	3	VU D2
Dovyalis sp. nov. (Flacourtiaceae)	2	CR D
Oncoba sp. nov. (Flacourtiaceae)	<10 trees	EW
Alchemilla fischeri subsp. camerunensis (Rosaceae)	1	CR B1+2c
Indigofera patula subsp. okuensis (Leguminosae-Papilionoideae)	2	VU D2
Bafutia tenuicaulis var. zapfackiana (Compositae)	2	VU D2
Kniphofia reflexa (Asphodelaceae)	3	EN C1

Xyris "welwitschii" (Xyridaceae)	2	Candidate
Ledermanniella keayi (Podostemaceae)	3/4	CR B1+2c
Habenaria maitlandii (Orchidaceae)	1	CR A1c, B1+2c
Crotalaria mentiens (Leguminosae-Papilionoideae)	2	EN B1+2c

Other taxa likely to be included here in future, after more research, are: *Justicia sp. nov?*, *Hyparrhenia sp. nov?*, *Bulbostylis sp. nov?*, *Cyperus niveus var. nov.*, *Cyperus A, B,* and *C, Tarenna sp. nov? Psychotria sp. nov?*, *Pimpinella sp. nov?*, *Craterostigma sp. nov?*, and *Plectranthus sp. nov?*

Extinct species
The following plants may now be globally extinct:
Newtonia camerunensis (Leguminosae). Last seen in 1974, at Achan, in lower montane forest.
Habenaria maitlandii (Orchidaceae). Last seen in 1931, at Nchain.
Ternstroemia polypetala (Theaceae) – Cameroon variant. Last seen in November 1996 at Tum, but not found in November 1999 despite intensive searches by a large team of botanists. The lower montane forest patch in which it was found had been reduced in area by two-thirds in the intervening three years.
Crotalaria mentiens (Leguminosae). Last seen in the early 1970's near Jakiri.

TAXON RED DATA TREATMENTS

DICOTYLEDONS

ACANTHACEAE

Justicia sect. *Harnieria* sp .nr. *J. heterocarpa* subsp. *praetermissa*
This taxon was identified as a possible new species (with the affinities indicated above) by M. Balkwill (pers. comm. 5 June 2000). It is thus far known only from Mbi Crater (*Etuge* 4580, 9 Dec. 1998). No other specimen has yet been found that matches this specimen. If this taxon is confirmed as a new species it is likely to figure as an addition to a future edition of this Red Data chapter.
Habitat: permanently wet crater swamp grassland; c. 1750m alt.
Description: straggling herb to 1m tall. Leaves narrowly ovate, coriaceous, c. 2.5 x 1cm, apex and base rounded. Flowers with corolla purple, c. 0.5cm long.

ASCLEPIADACEAE

Pentarrhinum abyssinicum Decne subsp. *ijimense* Goyder "NTANG" (Kom, general name).
CR A1c
Range: Cameroon, known only from Mt Oku and the Ijim Ridge.
This endemic forest climber was discovered to be a potential novelty by Dr Goyder in 1998 while identifying specimens of Asclepiadaceae from the 1996 expedition. At that time only a single specimen was known. *Etuge* 3565 was collected near Tum at Ijim on 21 November 1996. A second specimen from Kilum, collected slightly earlier (*Buzgo* 798) lacks collecting data and so had been passed over. A new site was discovered for this plant by Pollard at the medicinal forest at Laikom in December 1998 (*Pollard* 368). In November 1999 Etuge led a large party of botanists back to the Tum forest patch which had yielded his specimen in 1996. Here about a dozen plants were counted flowering and fruiting profusely on both sides of the stream that bisects the forest. On the west side of the stream this climber is locally abundant over several square metres of *Solanecio mannii*, to the extent that it is difficult to distinguish one individual from another. On the east side of the stream, three plants were seen, each separated by 10–15m. New material, including flowers in spirit, was collected (*Cheek* 9943) so that the taxon could be

formally described. Etuge (*pers. comm.*) reported that this forest patch had been reduced by about two–thirds in the three years since he was there last. A third site for this rare plant was discovered by the Ghanaian botanist Amponsah at "Back valley", Mbingo on 10 November 1999 (*Cheek* 10063). This taxon is completely unprotected.

Habitat: lower montane forest with *Garcinia smeathmannii, Solanecio mannii, Cuviera longiflora,* and *Pouteria altissima;* 1750–1900m. alt.

Threats: clearance of forest for cultivation of crops.

Management suggestions: effort should be made to protect the surviving forest patch at Tum (site of the largest known sub-population of this taxon but outside the present protected area boundary) and to establish whether other sites for this variety exist apart from the three listed.

Description: herbaceous climber to c. 6m tall, exudate clear. Stems c. 2mm diam., glabrous, internodes on flowering stems c. 15–20cm long. Leaves opposite, membranous, glabrous; blades ovate, c. 9 x 6cm, apex acuminate, base broadly cordate; petiole 4–5cm long; stipule-like subsidiary leaves ovate, c. 1 x 0.6cm. Inflorescence axillary, peduncle as long or longer than the subtending petioles, bearing a cluster of 8–15 flowers; pedicels c. 5mm long. Flowers 5-merous, c. 5mm diam., white flecked with pink. Fruits with two follicles arranged in a line, follicles cylindrical, smooth, c. 4 x 0.6cm, apex acute.

This subspecies can be distinguished from *P. abyssinicum* subsp. *angolense,* which also occurs in our area (e.g. *Thomas & McLeod* 5984 from forest above Oku village) by its smaller flowers which have a shorter corona bearing shorter teeth and have a more clearly stipitate gynostegium (Goyder *pers. comm.*).

BEGONIACEAE

Begonia oxyanthera Warb. (including *B. jussiaecarpa* Warb. of F.W.T.A. 2: 186 (1963).
LR/nt

Range: western Cameroon (including Mt Cameroon two pre-1988 collections) and Bioko, possibly S.E. Nigeria.

At Oku and Ijim *B. oxyanthera* is fairly commonly encountered as a scandent epiphyte in trees within its altitudinal range and is the commonest *Begonia* in the checklist area. However, its survival is dependent on the persistence of natural forest in the Cameroon Highlands and so it must be considered at risk. This species, of *Begonia* sect. *Tetraphila* A.DC., is closely related to *B. preussii* and can be regarded as an altitudinal vicariant (information from Marc Sosef and Hans De Wilde, Wageningen). The earliest collection known, but not the type of the accepted name, is Dusen 427 "Kamerun Gebirge 1500m". In contrast to Oku-Ijim, where there are 10 recent collections, only two recent collections of this species are known from Mt Cameroon. Thus Oku-Ijim may be the main stronghold for this species.

Habitat: submontane and montane forest 1200–2200m alt.

Threats: forest clearance for agriculture.

Management suggestions: so long as forest survives at this altitudinal band, this species is likely to survive and no intervention is needed.

Description: epiphytic herb occurring 2–5m above the ground, scandent. Stems red, 1–2m long, c. 8mm diam., beset by annular scales, glabrous. Leaves oblong (-subrhombic) c. 11 x 4cm, apex long acuminate, base shortly cordate, unequal, lateral nerves evenly spaced, 5–6 on each side of the midrib; petiole 1.5cm long; stipules lanceolate, red, c. 2.5 x 0.8cm, apex acuminate. Flowers axillary, inconspicuous, in few-flowered clusters; pedicels 0.5cm long. Female flower with perianth segments dull white, elliptic, 1cm long.; ovary 15mm long. Fruit cylindrical, red, 4.5 x 0.6cm, apex shortly rostrate.

Begonia schaeferi Engl.
VU D2

Range: Obudu, Nigeria (one collection), Manenguba (one collection), Nlonako (one collection), Bamboutos (three collections), Bamenda Highlands (one collection) and Kongoa Mts (one collection).

This yellow-flowered, non-climbing, rock-dwelling species is unlikely to be confused with any other in the Kilum-Ijim area, since no other species of the genus match this description. *Begonia schaeferi* is unusual in its section in having both male and bisexual inflorescences. It is also unusual in having branched inflorescences. The species was discovered in 1921 at Manenguba. Sosef (1994: 203–206) lists only eight collections of *Begonia schaeferi*. The terrestrial yellow-flowered species of *Begonia* are considered to be indicator species of glacial refuge forest (Sosef 1996).

Begonia schaeferi is extremely rare in the Kilum-Ijim area, being known from only two locations. One of these is at Anyajua near the Kilum-Ijim office. The site is a dripping cliff with an associated forest remnant (*Strombosia scheffleri, Eugenia gilgii*)and several other interesting species.

Habitat: on rocks and vertical rock faces; on moist to comparatively dry places, in primary submontane to montane forest, the latter sometimes with trees not higher than 6–12m; c. 1500–2300m alt (Sosef 1994).

Threats: while cliff faces generally are unlikely to be disturbed, clearance of adjoining forest for fuel and agriculture could endanger this species by removing the shade necessary for its survival.

Management suggestions: during surveys of cliff spaces in the Kilum-Ijim area, this species should be looked for and, if located, the number of plants and locality recorded.

Description: terrestrial herb 10–22cm high, scattered with minute glandular hairs, dense on petiole. Stem stout, leaf-bearing part up to 2cm long. Stipules broadly triangular-ovate, 0.3–1cm long. Leaves reddish green, peltate, blade horizontal, asymmetrical, elliptic or ovate, 3.5–14 x 2.5–7.2cm, 7–11-palmately nerved, margin finely serrate in apical part, base rounded, ciliate; petiole 4–22cm long, inserted 0.8–1.8cm from the margin of the blade. Inflorescence of two types, male and bisexual; two-thirds or as long as petioles, unbranched or branched up to 3 times, bracts in pairs at branching points and also immediately below the flowers; bracts circular to broadly elliptic, 3.9–8.8mm long, bronze-green, margin dentate, ciliate. Male flower with pedicel up to 24mm long. Perianth segments circular to broadly ovate, 7.2–14.5mm long. Fruit erect, broadly elliptic or broadly obtriangular-obovoid, 6–9mm long, strongly 3-winged. Description after Sosef (1994: 203–205).

CAMPANULACEAE

Lobelia columnaris Hook. f.
LR/nt
Range: Bioko, Mt. Cameroon, Mt Oku and the Bamenda Highlands, Chappal Wadi, Mambila Plateau.

This Giant Lobelia, with 2m spikes of blue flowers, is the only representative in continental West Africa of a group for which East African mountains are famous. It thrives at the montane forest: grassland ecotone. Although there are numerous collections from Mount Cameroon, it proved extremely rare during the 1992 survey of Etinde, when only one or two specimens were seen (Cable & Cheek 1998). In contrast, it is abundant in many localities at Mt Oku (pers. obs. 1996, 1998, 1999, where it is gregarious, 10–50 plants being commonly found at a site.

Habitat: boundary between montane forest & grassland; 2000–2200m alt.
Threats: unknown.
Management suggestions: none.
Description: herb 2m tall, unbranched. Stem 2cm diam. at base, glabrous. Leaves oblanceolate-oblong, c. 30 x 8cm at base of stem, gradually diminishing in size towards the apical inflorescence, apex acute, margin inconspicuously serrate, softly hairy. Inflorescence occupying the apical half of the plant, densely flowered, unbranched spike. Flowers with corolla pale blue, 2–3cm long.
It is difficult to confuse this species with any other.

Wahlenbergia ramosissima (Hemsley) Thulin subsp. *ramosissima*
VU B1+2bc
Range: Mt Cameroon (three collections), Bamenda Highlands, Chappal Wadi, Mambilla Plateau (one collection each).

Known in our checklist area only from the grassland on the ridge immediately above the Fon of Kom's Palace at Laikom. It is likely that this species has been overlooked elsewhere, being diminutive and seasonal. It probably germinates and grows through the wet season, flowers and fruits at the beginning of the dry season, perennating as seed in the soil.

Habitat: montane grassland: forest ecotone, between *Sporobolus africanus* grass tussocks and on shady banks, with *Radiola (pers. obs.)*; 1500–2600m alt.

Threats: unknown.

Management suggestions: none.

Description: annual herb 4–15cm tall, erect, unbranched in the lower half. Stem filiform, bearing evenly sized and shaped leaves at internodes of 5–10mm. Leaves elliptic, c. 5 x 3mm, apex rounded to acute, each side bearing 3–5 crenate teeth; petiole c. 1mm long. Inflorescence 1–10-flowered, diffuse; bracts linear, 3mm long; pedicels 7–30mm long. Flowers c. 4mm diam., petals blue, marked white at base.

COMPOSITAE

Bafutia tenuicaulis C.D. Adams
LR/nt

Range: Bamboutos Mts (two collections), Bamenda Highlands (five pre–1996 collections), Mbam (1 collection) and Mambilla Plateau (three collections).

This annual herb is the only species of a genus (named for the Bafut-Ngemba Reserve near Bamenda where it was first discovered) of a genus restricted to the Bamenda Highlands and their outliers. It seems fairly specific to rock outcrops. Within our checklist area, most of our specimens were collected at Oku-Elak during November of 1996: *Munyenyembe* 825, 878, *Zapfack* 1199, near the summit (*Zapfack* 1130), at Ewook Etele Mbae (Oku-Kumbo 10km *Zapfack* 1220, 1224), or at Shambai near Oku Elak (*Munyenyembe* 878). One of these collections was from farm fallow, showing that the species is capable of adapting to disturbance. At Ijim, *Bafutia* was recorded from an outcrop near the "Mbesa Swamp" in December 1998. According to Pollard (*Pollard* 358) 20 plants occurred at this site where it is threatened by cattle trampling and the species is otherwise extremely rare there. A large population of c. 60 plants was discovered on an inselberg at Mbingo in November 1999 (*Cheek* 10060).

Two varieties are recognized, one of which var. *zapfackiana* (*Zapfack* 1130 and *Munyenyembe* 825), described in this book, is the rarer (VU D2), being known only from Mt Oku and the Ijim Ridge.

Habitat: rock outcrops; 1800–2800m alt.

Threats: grazing by herbivores.

Management suggestions: a survey is needed to find all the sites for this species at Kilum-Ijim, and to confirm that it is indeed specific to rock outcrops. Information on its palatability to livestock would be useful.

Description: annual herb 3–15(–30)cm tall, glabrous. Stems c. 1mm diam., terete. Leaves sessile, mostly basal, linear-oblong, up to 4 x 0.4cm, apex rounded, margin subentire. Inflorescence a diffuse corymb forming the apical half of the plant, capitula 3–15; peduncles filiform, 2–3cm long, lacking bracts. Capitula 3–5mm wide; involucre with 5–8 bracts united in a cup with 5–8 acute, erect lobes, c. 1 x 1mm. Ray florets absent, disc florets 12–15, purple.

Crassocephalum bougheyanum C.D.Adams
LR/nt

Range: Bioko (three collections), Mt Cameroon (14 pre-1988 collections) and the Bamboutos Mts (one collection).

This species of *Crassocephalum* is one of many in our area. They appear to be annuals which grow through the wet season and flower and fruit at the beginning of the dry season. Intermittent forest disturbance favours these species by creating "light gaps" in which they flourish. However, they do not survive in open grassland or scrub, and so are forest-dependent species. *Crassocephalum bougheyanum* is distinguished from its congeners by the stems bearing usually just a single, very large capitulum. It was named by C.D. Adams in 1957 in honour of Prof. Boughey of Ghana, a pioneer of the study of the flora of upland Cameroon. Seven collections

A. Secondary *Hyparrhenia* grassland and forest patch, S of Aboh, c. 1900 m alt., Oct. 1996. BJP.

B. Upper montane forest with bamboo patches, Kilum forest, Oct. 1996. BJP.

C. Basalt pavement (front) & gallery forest in *Sporobolus* grassland, Laikom Ridge, Dec. 1998. MC.

D. Farm & forest boundary above Oku-Elak, c. 2200 m alt., Nov. 1996. BJP.

E. *Rothmannia urcelliformis* (*Rubiaceae*), Laikom forest, Dec. 1998. BJP.

F. *Oreacanthus mannii* (*Acanthaceae*), mass flowering, Ijim, near Lake Oku, Nov. 1996. MC.

A. *Kniphofia reflexa* (*Asphodelaceae*), Kinkolong Swamp, 1997. Fiona (Boo) Maisels.

B. *Plectranthus punctatus* subsp. *lanatus* (*Labiatae*), Afua Swamp, Dec. 1998. MC.

C. *Gladiolus sp. nov.* (*Iridaceae*), Laikom Ridge, Dec. 1998. *Pollard 334*. BJP.

D. *Ledermanniella keayi* (*Podostemaceae*), Mboh, Nov. 1996, *Cheek 8546A*. MC.

E. *Eriocaulon bamendae* (*Eriocaulaceae*), Laikom Ridge, Dec. 1998, *Cheek 9795*. MC.

F. *Alchemilla fischeri* subsp. *camerunensis* (*Rosaceae*), Mt. Oku summit, 3000 m alt., Oct. 1996. BJP.

A. *Eriocaulon asteroides* (*Eriocaulaceae*), Laikom Ridge, Dec. 1998, *Cheek* 9749. MC.

B. *Eriocaulon parvulum* (*Eriocaulaceae*), Laikom Ridge, Dec. 1998, *Cheek* 9748. MC.

C. *Dovyalis sp. nov.* (*Flacourtiaceae*), Lake Oku, Nov. 1999, *Cheek* 10113. MC.

D. *Dombeya ledermannii* (*Sterculiaceae*), Akeh-Ajung, Nov. 1999, *DeMarco s.n.* MC.

E. *Chassalia laikomensis* (*Rubiaceae*), Laikom Valley Forest, Nov. 1999, *Cheek* 10105. MC.

F. *Oncoba sp. nov.* (*Flacourtiaceae*), Anyajua, Dec. 1998, *Cheek* 9883. MC.

A. *Craterostigma sp. nov.* (*Scrophulariaceae*) at Laikom Ridge, Dec. 1998, *Pollard* 273. MC.

B. *Helichrysum cameroonense* (*Compositae*), Anyajua-Jikijem, Nov. 1999. MC.

C. *Dipsacus narcisseanum* (*Dipsacaceae*), Anyajua-Jikijem, Nov. 1996, *Cheek* 8761. MC.

D. *Bafutia tenuicaulis* (*Compositae*), Back Valley, Mbingo, Nov. 1999, *Cheek* 10060. MC.

E. *Begonia oxyanthera* (*Begoniaceae*), above Oku-Elak, Oct. 1996, *Cheek* 8441. MC.

F. *Peperomia thomeana* (*Piperaceae*), above Oku-Elak, Oct. 1996. MC.

were made in 1996 and 1998 at Kilum-Ijim (see checklist) suggesting that it is a relatively common species in our area.

Habitat: clearings and at the edge of montane forest; 1500–2500m alt.

Threats: forest clearance.

Management suggestions: none.

Description: annual herb 90–180cm tall, glabrous, lacking exudate. Stems c. 8mm wide at midheight, internodes 5–15cm, unbranched or very sparsely branched. Leaves membranous, those of the mid–stem ovate-trullate, 5–9 x 2–5.5cm, apex acute, base obtuse-decurrent, unequal, margin bidentate; petiole 2–3cm. Inflorescence of a single (rarely 2–3), terminal capitulum c. 2.8cm wide, orange-yellow.

Helichrysum cameroonense Hutch. & Dalziel

LR/nt

Range: Mt Cameroon (10 pre-1988 collections), Mt Oku (1 collection), Bafut-Ngemba (1 collection) and Chappal Waddi (3 collections). Apparently absent from Bioko.

H. cameroonense appears to be a biennial. In the first year it produces a rosette resembling a small *Dendrosenecio*. In the second year, after vertical growth, the stem produces in a dense mass up to 50 capitula, each about 2.5cm diam. These large plants are highly conspicuous and in local areas relatively common scattered in sparse grassland at about 2000m alt., e.g. on Ijim Ridge between Anyajua and Jikijem. This species appears to recolonize bare, rocky ground (see e.g. account in Cable & Cheek 1998). The fact that only two specimens of this species were gathered in the 1996 and 1998 surveys may reflect the fact that this plant is time-consuming and awkward to collect, being so bulky, rather than rarity.

Habitat: montane grassland, particularly on broken rocky ground; 1900–3000m alt.

Threats: none known.

Management suggestions: none.

Description: biennial (?) herb 1–2m tall. Stem woody at base, 2.5cm diam., unbranched, densely clothed by leaves. Leaves at mid-stem level sessile, ovate-lanceolate, c. 14 x 3cm, apex acute, margin finely serrate, aromatic, viscid white-woolly below. Inflorescence a dense, terminal, fairly flat assemblage of 20–50 capitula forming an aggregation 10–20cm diam.. Capitula 2–3cm diam., disc dull orange, ray involucral bracts glossy pale yellow.

Distinguished by being larger in most parts from the similar *H. foetidum* which has stem diam. c. 5mm, leaves c. 1.8cm wide, capitula c. 1.5cm diam.

Vernonia bamendae C.D. Adams

VU D2

Range: Cameroon, Bamenda Highlands (6 specimens) and Nigeria, Mambilla Plateau (Chappal Waddi, 1 collection).

Dennis Adams described *Vernonia bamendae* citing *Maitland* 1514 (Laikom, on grassy hill slopes, 6000', May 1931) as type specimen (Adams 1957). This remains the only specimen known from the Kilum-Ijim area. Earlier in his number series, Maitland had collected two other specimens of this species from other parts of the Bamenda Highlands. *Maitland* 1455 is labelled "Bamenda, Basenako, 5,000', on stony grassy slopes". *Maitland* 1457 is labelled "Nchan, June 1931". Keay in FHI 28410 collected the fourth specimen from 7,400' on the NW slopes of Mba Kokeka near Bamenda. *Boughey* in GC 11013, from Santa Mt. at 2500m followed on 25 Dec. 1952. Daramola in FHI 46649 was the last collection from the Bamenda area (Bamenda, 27 Jan. 1957). The first and only collection from Nigeria was made from Chappal Wadi on the border with Cameroon (*Tuley* 2041, 7,500', 19 November 1969). The most recent collection of *Vernonia bamendae* was on 1 November 1974 from Mbam, 2335m alt., 35km NW of Foumban in West Province (*Letouzey* 13100).

It is curious that despite collecting expeditions in 1996, 1998 and 1999, we have not rediscovered this species from its type locality. This species seems to flower throughout the dry season, from early November until June, during part of which (Nov. & Dec.) we have conducted fieldwork in the area. The explanation for our not rediscovering the species may be that it is not only narrowly endemic, but, within its small range, extremely rare and possibly declining.

Habitat: grassy slopes, sometimes near forest boundary or in wet hollows; 1500–2500m alt.

Threats: unknown, but incidence of grazing and fire are likely to be important factors in the survival of this species.

Management suggestions: this species still awaits rediscovery in the Kilum-Ijim area. It may be extinct within the checklist area, in which case, consideration might be given to its reintroduction, providing that a source of seed-bearing plants can be found.

Description: stout herb (1–)1.5(-2)m tall, probably perennial. Stems c. 1cm diam., densely brownish tomentose, finely ribbed. Cauline leaves triangular, c. 12 x 5cm, apex rounded, base cordate or cordate-auriculate, lower surface with raised reticulate veins, pale grey-brown pubescent. Inflorescence spreading. Capitula c. 1.5cm wide, florets mauve or purple.

DIPSACACEAE

Dipsacus narcisseanum Lawalrée
VU D2

Range: endemic to western Cameroon. Known only from the Bamenda Highlands (principally Kilum-Ijim) and the Bamboutos Mts.

Rare on Kilum-Ijim. Only 3 sites and c. 30 plants seen in 1996–1999. At Kilum, seen at two sites in 1996. 1. KA transect, in forest belt in upper sector towards grassland, on steep, thinly clad walls of ravine. Two to three plants seen. 2. KA transect, 200m west of grassland: forest boundary, at top of cliff edge below grassland facing North. About 8 plants seen.

At Ijim, seen at one extended site on roadside bank, grassland-woodland boundary amongst *Ixora* bushes c. 2m high north of Gikwang rd, between Lake and Jikijem junction, c. 2500m alt. About 15 inflorescences and more vegetative rosettes seen at this grid reference in November 1996. Monitoring in November 1999 showed this population to be virtually unchanged. This is the largest population known. Along the road from this population, towards Aboh, 3–4 other groups of 1–5 inflorescences were seen.

Dipsacus is genus of about 15 species restricted to Eurasia (Europe 8) and Tropical African mountains. This is the only species known in West-Central Africa.

Habitat: on banks and cliffs or cliff edges, in scrub or grassland near forest; c. 2000m alt.

Threats: there are no obvious threats to this species. However, it appears to be so rare that unintentional clearance (e.g. for a cattle compound, or road diversion) at any of its three known sites might endanger it.

Management suggestions: the reason for the rarity of this species is unclear and needs more study. The explanation may be that establishment of new colonies demands conditions rarely encountered. See comments below on the similar and related *Succisa*.

Description: perennial herb with one to several apparently biennial stems from a several-crowned underground rootstock. Vegetative rosettes with numerous lax leaves c. 45cm long, densely felty. Inflorescences presumably produced in second year. 1–4m tall, peduncles 1cm diam., stout, ridged, bearing 1–9 short (c. 10cm) stalks in upper part, each bearing rigid spherical capitula c. 6cm diam. Flowers whitish, c. 1cm long, not very conspicuous, in spiral rows around capitula, only one row developed at one time. Fruits not seen. Apparently flowering throughout November.

Succisa trichotocephala Baksay
LR/nt

Range: Mt Cameroon (10 pre-1988 collections), Bamboutos Mts (two collections) and Mt Oku (one pre-1996 collection).

This is the only species of the genus *Succisa* in Tropical Africa. The remaining three species are Eurasian. This species appears to be a perennial, with a subterranean perennating rootstock that withstands fire. It produces biennial rosettes that flower in their second year, though more observations are needed to confirm this. Observations above Oku-Elak in October 1996, showed that this species survived, though not abundantly, and was flowering in fired grassland that was being cattle grazed. Cattle seem to avoid eating *S. trichotocephala*

Kilum: transect KA, just above forest, about 4 plants seen thinly scattered in closely cropped *Sporobolus* grassland: apparently not grazed. Also reported common from near main peak (fide Munyenyembe).

60

Ijim: Gikwang rd: almost immediately on entering the gate into the protected area from Aboh, *Succisa* occurs intermittently as scattered individuals and sometimes in dense communities of 10–50 plants in glades in *Gnidia glauca* woodland, with light shading. This is true of a 2–3km stretch of the road at this point. At the Kilum-Ijim Forest Project resthouse further along that same road, an apparently natural grassy area occurs in which the dominant plant during our visit in November 1996 was *Succisa*. An estimated 400–500 flowering plants in about a hectare. Mt Cameroon: during Oct 1992 and 1993, MC was able to study *Succisa*, then in flower, over several days in the area south of Mann's Spring. Here it occurs only as scattered individuals in relatively ungrazed (no domestic animals) grassland, never more than about 200m from the forest. Plants were also recorded at Laikom Ridge in December 1998 (Pollard pers. comm.).

Hooker treated Mann 1309 (Jan. 1862, 10,500 ft.), the first specimen known of this species, as identical to the European 'Devil's Bit Scabious', *Succisa pratensis*. *Succisa trichotocephala* was only described as a separate species in 1952, before which the African plants were only varietally distinguished (in the 1930 s) as *Succisa pratensis* var. *kamerunensis* B.L.Burtt.

Habitat: grassland, near boundary with forest; open glades in *Gnidia glauca* woodland; 2000–2800m alt.

Threats: there are no apparent threats to this species within the Kilum-Ijim area, beyond clearance of forest for agriculture. Grazing by horses and cattle was prevalent in one area where the species was observed to be fairly common in November 1996 (Gikwang road, Ijim) but appeared to cause no adverse effects to the *Succisa*.

Management suggestions: insufficient data available to suggest management regime, but fire is possibly important at intervals in preventing shading-out by shrubs and trees. The underground rootstock is likely to provide protection in the event of dry season fire. The large, lax leaves help minimise competition from adjoining herbs by smothering them. This species appears to be in no immediate danger of extinction, being relatively common within its range. Nonetheless, if surveys at the montane forest: grassland ecotone are being conducted, the opportunity should be taken to collect data on the population size and range of this species.

Description: perennial herb, with one to several apparently biennial stems from a several-crowned underground rootstock. Vegetative rosettes with numerous thinly hairy, lax leaves c. 30cm long. Inflorescence c. 0.5m tall, lax, branching from near the base, basal branches over 30cm long. Capitula hemispherical, c. 2.5cm diam.. Flowers c. 1cm long, white or mauve. Flowering abundantly in October and November. Young fruits were seen at the end of November.

FLACOURTIACEAE

Dovyalis sp. nov.
CR D
Range: endemic to the Bamenda Highlands of western Cameroon, now known only from Kilum-Ijim.

This taxon is probably closely related to *D. spinosissima* Gilg from similar habitat (montane lake margins) in Kivu, Zaire to Malawi. It differs by the glabrous fruits (not hairy), flowers with 14–21 styles (not 8–12) and much larger flowers. This species is only known with certainty from the forest in the crater of Lake Oku (16 trees) and from a site with eight trees between Oku-Elak and Nsoh (Kemei and Innocent Wultoff pers. comm. November 1999) though was probably formerly more widespread. Trees sometimes grow at the very edge of the lake but also occur c. 50m up near the rim of the crater on the S. side. It was reported in forest above Oku-Elak, but becoming rare there *fide* Peter Wambeng (November 1996) and this has not been confirmed since. An unlocated ('Bamenda district') collection in male flower (Daramola in FHI 41575 at Kew and Yaoundé herbaria) made on 5 July 1959, almost certainly belongs to this species, but we have not yet compared male flowers. Oku was formerly part of Bamenda District.

Habitat: evergreen montane forest; c. 1900–2200m alt.

Threats: timber extraction (heavy poles c. 15cm diam. for building construction) was noted in the area where the first trees were seen in the crater-lake forest and threatens the existence of this species, as does usage as a fuel wood and clearance for agriculture.

Management suggestions: felling of trees in the lake crater should be prevented. An attempt should be made to mark, record and monitor all trees of this species. A check should be made to study seedling regeneration. Is recruitment observable? A planting programme may be advisable.

Description: dioecious, ?deciduous tree 4 to 8m tall, bearing spines to 4cm long on branches near the trunk. Leaves papery, elliptic, c. 10 x 5cm, apex acute, margin serrate. Flowers 3–4cm diam, green and white, felty. Fruits c. 6-7cm diam., glossy, ripening yellow, soft. Seeds numerous, elliptic, c. 8mm long, flattened. In flower and young fruit in early November. Mature fruits seen in late November.

Oncoba sp. nov. FEEWU A WU (Kom: Clement Toh).
EW

Range: apparently formerly endemic to Mt Oku and the Ijim Ridge, now known only in cultivation.

This species was initially confused with another potentially new species of the Flacourtiaceae, *Dovyalis sp. nov.* in November 1996 (Cheek *et al.* 1997). It is preserved as isolated trees in compounds of traditional healers (Fundong and Anyajua) or in one case, at Laikom in a grove which is believed originally to have been part of the natural forest but which is now enclosed by cultivation.

The fruits of Feewu a wu are used by practitioners of traditional medicine, when the pulp has been removed, to make rattles. Owners of trees sell fruits at a price in Anyajua (pers. obs.). Clement Toh of the Kilum-Ijim Project has been active in tracking down cultivated trees. However, no more than 10 individuals are known at present.

Habitat: montane forest c. 1900m alt.

Threats: clearance of forest for firewood and agricultural land.

Management suggestions: the trees surviving in cultivation should be mapped and recorded. Peter Yama and others report success in growing this species from seed in gardens and this activity should be encouraged. Consideration should be given to reintroducing this species to secure areas of lower montane forest in due course.

Description: tree c. 6m tall, spiny, spreading. Leaves elliptic, c. 9 x 5cm, apex slightly acuminate, base acute, margin serrate with c. 15 teeth, each c. 2mm long, on each side; petiole 5–8mm long. Inflorescence of single, axillary, pendulous, flowers; pedicel c. 3cm long. Flowers 3.5–6cm diam. Sepals c. 7mm long. Petals 6–8, white, obovate, c. 2.5 x 1.8cm. Stamens numerous, forming a dense mass c. 3cm diam. Ovary ovoid, 5–6mm wide. Fruit spherical, hard, glossy, 5–6cm diam., with c. 9 equally spaced longitudinal lines.

LABIATAE

Plectranthus punctatus L'Hér. subsp. *lanatus* J. K. Morton IJIM (Kom).
VU D2

Range: Bamboutos Mts (four specimens) and Bamenda Highlands (four pre-1996 specimens).

This subspecies was first collected by Maitland, probably either on Laikom Ridge or at 'Mbesa Swamp' (*Maitland* 1724, "Basenako-Lakom, on plateau in grassland, June 1931, 1800m"). In the Bamenda Highlands it is also known from Ndu and from the Bambili Lakes as well as Kilum-Ijim.

This plant is best distinguished from other members of the genus by the purple-blotched, white hairy stems and small stature. Its habitat is also distinctive. The largest population of this species seen was at Afua swamp in November 1999 where it lines the border of the swamp for about 100m or more. In total, five sites for the species are known at Kilum-Ijim, the numbers of individuals of which varies from 5-10 to 100-200.

Habitat: damp grassland at the edge of swamps, or banks; 1800–2600m alt.

Threats: swamp drainage or development; shading-out by growth of grasses; possibly trampling by cattle.

Management suggestions: more research is needed on the management regime needed for this plant. However plants seen in long grass at the Mbesa swamp in 1998 seemed less healthy and were far fewer than those seen in close cropped grass at Afua swamp.

Description: Herb c. 30cm tall. Stems ascending, rounded and sub-succulent when alive, nodes swollen, internodes 1.5–3 (-5)cm long, whitish green, blotched purple, clothed in long white hairs c. 2mm long. Older stem bases sometimes straggling, prostrate, rooting at nodes. Leaves of the midstem sessile, sublanceolate, c. 6 x 2cm, apex rounded, margin serrate-crenate, with c. 15 teeth per side, white hairy above and below. Inflorescence 6–15cm long, verticils 5–10, internodes c. 1.5cm long, flowers subsessile. Flower with corolla c. 12mm long, pale blue, prominently speckled with purple.

Plectranthus sp. nov. IFYENGE ZVU (Kom, Peter Yama).
Pollard 267 (2 Dec. 1998, Laikom Ridge, Ijim) and also *Pollard* 385 represent a Solenostemoid *Plectranthus* which appears to be new to science. If ongoing research at R.B.G., Kew by Pollard and Paton confirms the status of this plant as a new species, it is likely that this taxon will appear in a future edition of this Red Data chapter.

Stachys pseudohumifusa Sebsebe subsp. *saxeri* Y.B. Harv.
VU D2
Range: Bamenda Highlands (Ndu and Laikom "Lakoni", 2 pre-1996 collections), Bamboutos Mts (Dschang, 1 collection) and adjoining Mambilla Plateau (Gembu, 1 collection).
This subspecies was only recognized in 1996 (Harvey 1996). It was then known from only four localities (see above). We have located three further sites for this taxon in the checklist area, all at Ijim: the Afua Swamp (*Cheek* 9842), the Mbesa Swamp (*Cheek* 9811, *Pollard* 349) and at a swampy place in a valley between the Ardo of Ijim's compound and the Fon of Kom's Palace where about a dozen plants were seen (Cheek *pers obs.* Dec. 1999). The habitat requirements of this plant seem to be highly specific and so the number of localities that can support the species is likely to remain small.
Habitat: swamp grassland, often at the interface with montane forest or scrub; 1800–2460m alt.
Threats: unknown but possibly trampling from cattle.
Management suggestions: more data is needed on the numbers of individuals at each site, and on the type and levels of regeneration. This species is conspicuous when in flower in October–December.
Description: Erect, weak-stemmed herb to 1m tall, bright green. Stem pronouncedly square, finely prickly and sticking to clothing. Leaves erect, oblong lanceolate, up to 22mm wide, margin crenate, revolute. Inflorescence terminal, verticils 4–6-flowered, 3–28mm apart. Calyx 4–4.5mm long. Corolla bright white, corolla 10–13mm long; tube 6–7.5mm long; lower lip 5–6mm long; upper lip (2.5–)3–4mm long. Nutlets c. 1.5mm long.
Description after Harvey 1996.
Stachys aculeolata var. *aculeolata* is the only other *Stachys* in our area. It differs principally in its leaves being patent, not erect, and in them being ovate, i.e. broader in proportion to their length. *Stachys aculeolata* in our area also differs in being usually prostrate, with grey-green leaves, and is found in woodland edges, not usually in wet spots.

LEGUMINOSAE

Crotalaria bamendae Hepper
VU D2
Range: Cameroon, Bamenda Highlands (four pre-1996 collections), Nigeria, Mambilla Plateau (one collection) and Angola (two collections).
This species was published by Hepper (1956) on the basis of two specimens from the Bamenda Highlands, *Egbuta* in FHI 3763 (farmland, Bamenda, Dec. 1951) and *Tamajong* in FHI 23479, (grassland at roadside, Kumbo, Oct. 1947, type of the species). A few years later (FWTA 2: 549, 1958), he had added a third collection, *Latilo & Daramola* in FHI 34369, from Gembu in the Mambilla Plateau. Polhill (1982), in revising the genus for Africa and Madagascar, extended the range to Angola. The photograph of *Newton* 76 (Angola, Huila, Humpata, Feb. 1883) shows a specimen that differs from the Cameroonian material in having much larger and more broadly elliptic leaflets (2.9 x 1.4cm) with inflorescences partly concealed by leaves and borne on short

spur shoots. Perhaps the Angolan plants merit subspecific distinction. Polhill regards *Crotalaria bamendae* as an evolutionarily isolated species in section *Glaucae*.

More recent collections of *Crotalaria bamendae* are *Jacques-Félix* 8933 from Tchabal Mbabo and *Meurillon* in CNAD 125 from Lake Bambuluwe (November 1965). Apart from the type collection, the only specimen from our area is *Munyenyembe* 882 (Kilum, Shambai, 2500m alt.) November 1996. *Crotalaria bamendae* is evidently rare in its range. The fact that only a single collection was made during a month of collecting in its habitat, at the peak of its flowering season in 1996, by a large expedition of botanists (Cheek et al. 1997), suggests that even within its habitat it is rare and infrequent and so vulnerable to threats.

Habitat: montane grassland, sometimes wet; 1800–2500m alt.

Threats: unknown, but fire and/or grazing may effect this species adversely.

Management suggestions: the range and frequency of this species at Kilum-Ijim and factors influencing recruitment and survival need to be established.

Description: subshrub 30–60(–90)cm tall. Stems densely pubescent. Leaves shortly petiolate, 3-foliolate, uppermost often reduced, rarely 1-foliolate; leaflets narrowly elliptic-oblong to elliptic-obovate, mostly 1.5–3.5(–4.3) x 0.4–1.5cm, sparingly pilose above, silky-hairy beneath. Stipules lanceolate, 3–6mm long. Racemes subsessile, dense, many-flowered, shortly cylindrical, c. 2.5 x 2cm; bracts linear 4–5mm long; pedicels 2–5mm. Calyx 4–5mm long; densely pilose, lobes triangular-lanceolate. Standard circular, yellow, with fine darker lines, hairy along midvein outside; wings longer than keel; keel abruptly rounded, shortly beaked, 4.5–5mm long. Pod c. 5mm long, pilose, glabrescent, 2–4-seeded. Seeds 1.5mm long, yellow, glossy.

Description after Polhill 1982: 128–129.

Crotalaria ledermannii Baker f.

VU D2

Range: Cameroon, Manenguba Mts (one collection), Bamenda Highlands (six collections); Nigeria, Mambilla Plateau (one collection).

This is a rare species, having been gathered only once in the course of our three inventories at Mt Oku and Ijim. *Onana* 634 was collected near Lake Oku in November 1996. The only other specimen from our checklist area is *Hepper & Charter* 2048 from mile 11, Kumbo–Oku, 17 Feb. 1958. The collection from the Mambilla Plateau is taken from a *Eucalyptus* plantation, so *Crotalaria ledermannii* may be able to tolerate some disturbance to its habitat. This is an annual or short–lived perennial species, so a deleterious change in the habitat or poor seed set in one year could drastically reduce the population within 12 months.

Habitat: grassland and forest edge; 1200–2200m alt.

Threats: unknown but possibly conversion of land to cultivation and grazing, trampling or fires.

Management suggestions: the natural habitat of *Crotalaria ledermannii* is not well characterized . More study is needed to rectify this. A survey to find the range of this species and levels of regeneration is advisable.

Description: Annual or short-lived perennial herb 20–70cm tall, with several erect stems arising from near the base, branches appressed puberulous, the hairs dense above, glabrescent below. Leaves 3-foliolate; leaflets oblanceolate, 7–20 x 1–5mm, apex rounded or truncate, apiculate, appressed puberulous beneath; petiole 4–13mm long. Inflorescences 1.5–33cm long, comprising a terminal head and a few laxly inserted flowers below; bracts linear-subulate, 1–1.5mm long; pedicel 3–6mm long. Calyx 2.5–3.5mm long, densely brownish puberulous; upper lobes triangular, sometimes slightly acuminate, 1–1.5 times as long as the tube. Standard obovate to oblong-obovate, yellow, veined red, brownish puberulous outside, wings a little shorter than keel; keel angular, 5.5–6.5mm long. Pod shortly stipitate, ovoid globose to ovoid-ellipsoid, 4–5 x 3–3.5mm, appressed pubescent, 2–seeded. Seeds 2mm long, with prominent aril.

Description after Polhill (1976).

Crotalaria mentiens Polhill

EN B1+2c

Range: Cameroon: Bamenda Highlands (two collections).

For information on this species we depend entirely on the protologue (Polhill 1976). No specimens of *C. mentiens* are present at Kew, nor were any gathered during our inventories. The type specimen (*Brouwers* 2, BR, Jakiri) was collected on the southeastern boundary of our area and is the only specimen of the species cited by Polhill. However, the accompanying map shows two points, so it is assumed that there are two collections. Polhill states "Known only from the Bamenda area of W. Cameroon and deceptively similar to *C. ledermannii* in the same area". The collector Brouwers is not listed by Letouzey (1968) and so we surmise that he or she was a Belgian who collected a few specimens in the Bamenda Highlands in the late 1960s or early 1970s. This species is given an EN rating partly because of the fact that only two specimens are thought to exist, that it has not been re-collected during our inventories, and because it occurs in the most densely populated part of our area.

Habitat: unknown, but probably as *C. ledermannii*.

Threats: unknown, but probably as *C. ledermannii*.

Management suggestions: an attempt should be made to rediscover this species in the wild and to assess whether, as is suspected, it is threatened. The Brussels herbarium at Meise should be checked for the specimen cited, and any others of this species so that more detailed data on the locality and habitat might be obtained. This would assist in relocating *C. mentiens*.

Description: herb 20–40cm tall, differing from *C. ledermannii* as follows: young branches strigulose, not puberulous, bracts 2–3mm long, as long as pedicel, not 1–1.5mm long, much shorter than the pedicel; calyx thinly, not densely puberulous; lobes 2–3 times longer than the tube, not 1–1.5 times longer; standard glabrous outside, not puberulous.

Indigofera patula Baker subsp. *okuensis* Schrire & Onana
VU D2

Range: endemic to Mt Oku and the Ijim Ridge (2 collections known).

This subspecies came to light in August 1998 when Onana was naming the Leguminosae specimens from the 1996 Kilum-Ijim inventory at Kew under the guidance of specialists, including Schrire (Schrire & Onana 2000). The first specimen collected of this species *Munyenyembe* 814, was made between Oku-Elak and KD/KC at c. 2000m alt. on 30 October 1996 when both flowers and fruit were present. The second specimen *Etuge* 3522, was collected on the other side of Mount Oku at 2252m on 20 November 1996, when flowers and fruits were also present. This plant was found on the road from Aboh to Gikwang, towards the Nyasoso forest. Although only two specimens are known, this taxon may be undercollected because its habitat can be fallow fields, which were not the main target of plant collecting! Several other taxa of *Indigofera* occur in the area. This taxon is distinct from these in that the lower surface of the leaflets, apart from appressed biramous hairs, bear black, glandular hairs at the margins and sometimes along the midrib. These black, glandular hairs are also present on the stem.

Habitat: montane grassland & fallow areas; 2000–2300m alt.

Threats: unknown.

Management suggestions: a field survey of *Indigofera* is recommended at Kilum-Ijim in order to facilitate identification of this narrowly endemic subspecies and to gather data on its distribution and frequency. This taxon may have potential as a nitrogen-fixing, soil fertility-improving fallow-field species.

Description: perennial herb or subshrub 0.3–1m tall. Stems trailing or erect, reddish, with appressed biramous hairs and blackish gland-tipped hairs, viscid. Leaves 3–13-foliolate; leaf axils 3–39mm long including a petiole of 0.5–10mm; terminal leaflets 2–14(–19) x 1.2–7(–9)mm, slightly larger than the laterals, elliptic-oblong to lanceolate or oblanceolate, about 2–3 times as long as wide, apiculate, sparsely to densely strigose with appressed biramous hairs, margins and sometimes midribs beneath usually fringed with short, blackish gland-tipped hairs up to 0.3mm long. Racemes (4–)5–10cm long, not zigzag, equalling or up to 3 times the length of the subtending leaf, including a usually densely glandular peduncle 22–38mm long, laxly to densely c. 15–50-flowered, pedicels 1.5–2.5mm long. Calyx (2.5–)3–8mm long. Corolla 5–8mm long, pink to carmine-red. Pods 5–14 x 1.5–2mm, straight, cylindrical, spreading, sparsely hyaline strigose with hairs spreading at the tips, reddish to dark brown, densely covered with

erect gland-tipped hairs 0.1–0.5mm long, style-base forming a short beak. Seeds 2–6, 1–1.5 x 1–1.2mm, quadrangular-cylindrical to globose, yellow to brown, shiny, not pitted.
Description after Schrire & Onana (2000).

Newtonia camerunensis Villiers   ADJWA (Kom).
CR A1c
Range: endemic to the Bamenda Highlands and Bamboutos Mts. of Cameroon.
In revising the African species of the genus *Newtonia*, Villiers (1990) described a new species from Cameroon, *N. camerunensis*. Only five specimens are known of this species. Both the earliest and latest collections were made from the Ijim area, below the Ijim ridge. The earliest were on the north side at Nchian (*Maitland* 1671) and on the west side at Njinikom, also by Maitland. Two collections were then made near Bamendou in the Bamboutos Mts (*Letouzey* 58 and 61). *Letouzey* 13451 is the last specimen that we have been able to trace. It was collected south of Lake Oku on the Ijim side, between the lake and the village of Acha at 1700–1800m alt. Despite collecting plants in this area over several years (though admittedly, we have not concentrated on the altitudinal band of this species), we have never seen this tree. All collecting locations are in densely populated areas. These facts, taken with the few collections in existence, indicate that this species is extremely rare, even within its small range.
Habitat: montane forest with *Albizia gummifera*, *Carapa grandiflora*, *Syzygium staudtii*, *Prunus africana*; 1600–1800m alt.
Threats: this species appears not to fall in any protected area. The sites at Kilum-Ijim fall outside the boundary of protection. It is likely that this substantial tree is used as a timber and it may well have been over-exploited. This species may already be extinct.
Management suggestions: every effort should be made to rediscover and protect any trees that might be left of this species, if it is not already extinct. Forest destruction in the Bamboutos Mts is as extensive, perhaps more so, than in the Bamenda Highlands (*pers. obs*). Trees tentatively identified as *Newtonia* in the Ajung area in November 1999 (Kemei, DeMarco, Cheek, Achoundong pers. obs.) should be reinvestigated: they may prove to be this species.
Description: Tree, trunk 50cm diam. with flared flutes at the base. Leaves c. 25cm long, petiole 1.2–1.5cm long; rhachis with 8–10 pairs of pinnae, a large gland between each pair; pinnae with 25 pairs of leaflets, margin ciliate; leaflets oblong, 15 x 3.5mm, apex rounded, base rounded to slightly retuse, upper surface glossy with midrib prominent. Flowers unknown. Fruits elliptic-oblong, 19–30 x 1.8–2.3cm, straight or gently curved, apex rounded-apiculate, base attenuate-stipitate, nervation oblique, not prominent in young pods. Seeds winged, elliptic to ovate, 4–8 x 1.5–2.2cm.
This species is distinguished from *Newtonia buchananii* (Baker f.) Gilbert & Boutique by the higher altitudinal range (1600–1800m, not 900–1500m alt.), the presence of a prominent midrib, the larger leaflets and the more conspicuous rhachis glands.

LORANTHACEAE

Tapinanthus letouzeyi (Balle) Polhill & Wiens   EYMBENMBEN-EMLAN (Oku).
VU A1c, D2
Range: Cameroon, endemic to the Bamenda Highlands (five pre-1996 collections).
On our 1996 expedition, this species was found at only two locations: in the forest just above Oku-Elak (*Zapfack* 739) and at Lake Oku near the Baptist Rest House (*Cheek* 8586). Two of the pre-1996 collections were made at Bafut-Ngemba. A recent report (Yana Njabo & Languy 2000) states that 99% of the natural canopy of this forest reserve has been destroyed. This indicates the level of threat to forest trees in the Bamenda Highlands. Kilum-Ijim may already be the last refuge for this species.
Habitat: parasitic on trees, particularly *Gnidia glauca* on the edge of montane forest; 1300–2500m alt.
Threats: some harvesting of host-parasite haustorial connections occurs in this area (pers. obs.) and may have a slight effect in threatening this species, but this is unlikely to be important. Felling of host trees for firewood, and to clear ground for agriculture is the main threat.

Management suggestions: more detailed data should be gathered on the size of the population and its area in Kilum-Ijim.

Description: Parasitic shrub to 1m tall, plant glabrous. Leaves opposite or nearly so; petiole 3–15mm long; lamina thinly coriaceous, green, paler beneath, reddish flushed when young, elliptic, apex acuminate, base rounded to cuneate, 6–12 x 1.5–6cm, with 4–8 pairs of lateral nerves. Umbels axillary and clustered at older nodes, 4–6-flowered; peduncle 3–5mm long; pedicels 0.5–1mm long from distinct sockets; bract cupular. Receptacle 1–1.5mm; calyx tubular, (2–)3–4mm. Corolla tube 2.7–3cm, red, paler above or greyish between red lines; bud heads white turning blackish, ellipsoid, obtuse, slightly angled to ribbed, 3.5–4 x 2–2.5mm; basal swelling obovoid, 5–6 x 3–3.5mm, the tube narrowed to 3–4mm above; lobes 6–7mm, reflexed. Stamens white, anthers red. Style green.

Related to *T. globiferus* which mostly occurs at lower altitudes. Distinguished as follows:

Calyx in mature bud and at anthesis being saucer-shaped, 0.5–1.2mm long, shorter than the barrel-shaped receptacle; umbels (4–)6–12 flower.......................................*T. globiferus*
Calyx cup-shaped to tubular, 1–4mm long, generally longer than obconic receptacle................
.... *T. letouzeyii*

Description and couplet after Polhill & Wiens (1998: 199–200).

MYRSINACEAE

Embelia mildbraedii Gilg & Schellenb. (*E. sp. aff. welwitschii* (Hiern) K. Schum., F.W.T.A. 2: 32). NTOH (Oku).
LR/nt
Range: Mt Cameroon (six collections) and the Bamenda Highlands (one pre-1996 collection).
Eight collections of *E. mildbraedii* were made in the 1996 and 1998 inventories, showing it to be relatively abundant at Mt Oku and the Ijim Ridge. Most of these collections were made above Oku-Elak in 1996. Although *E. mildbraedii* is highly threatened in the Bamenda Highlands by forest clearance, it is given a non-threatened rating here because of the fact that it also occurs on Mt Cameroon, where there is little pressure on its survival in the higher altitude forest. This plant is used in beehive construction at Oku (*Munyenyembe* 830).
Habitat: montane forest; 2100–3200m alt.
Threats: clearance of forest for agriculture.
Management suggestions: continued protection of surviving forest.
Description: liana, about 4m high; twigs dark brown, covered with white lenticels; leaves elliptic, 4–8 x 2.5–4.5cm, apex acuminate, base obtuse-acute, margin serrulate, abaxial surface silvery green, secondary veins 12–18, venation brochidodromous, glabrous; petiole about 1cm long; inflorescence racemose, axillary; peduncle about 2–2.5cm long; pedicel 0.2–0.4cm long; flowers small, pinkish; fruit spherical, red, 4mm diam., stigma persistent.

MYRTACEAE

Eugenia gilgii Engl. & Brehm.
CR A1c
Range: Cameroon, Bamboutos Mts (three collections), Bamenda Highlands (seven pre-1996 collections), Ngaoundere (one collection) and Nigeria, Mambilla Plateau (13 collections).
Described in 1917 from *Ledermann* 2131 (NW Cameroon, Tapare to Riban, gallery forest, 1300m, Jan. 1909). Although many herbarium collections of *Eugenia gilgii* exist, it seems highly threatened by the fact that a) its natural habitat has been almost completely destroyed, b) what is left is disappearing rapidly and c) by the fact that, apart from a few trees on Laikom Ridge at Ijim, no individuals are in protected areas. Nonetheless, this was probably once a relatively common species in the Bamenda Highlands. Apart from two widely separated trees in the Anyajua area and a small population at Laikom, the only trees of *Eugenia gilgii* that I have seen are in an extensive sub-population of at least 50 trees discovered in 1999 on the path to Mbingo "Back Valley". This may be the main refuge for the species in the Bamenda Highlands although it has no formal protection. The relatively numerous collections from the Mambilla

Plateau were all made over 23 years ago. It seems reasonable to infer that the situation with regard to habitat destruction in Nigeria over this time has been no better than it has been in the Bamenda Highlands.

Habitat: lower montane forest, often at edges; (1200–)1500–2000m alt.

Threats: clearance of forest for wood and land for agriculture.

Management suggestions: the possibility of protecting the sub-population at Mbingo should be investigated.

Description: Tree 3–10(–15)m tall, bole pale brown, spiny when juvenile, exudate absent. Young stems brown-puberulent, glabrescent. Leaves opposite, elliptic, c. 9 x 5cm, apex subacuminate, rounded, base obtuse, secondary nerves 5–6 on each side of the raised midrib, margin revolute at leaf-base; petiole 0.5–1cm long. Inflorescence fasciculate, 5–10-flowered, below the leaves; pedicels c. 7mm long, with a minute pair of bracts in the upper half. Flowers 6–8mm diam., petals pink or pinkish white. Fruit red, shortly ellipsoid, c. 1 x 0.7cm.

PIPERACEAE

Peperomia thomeana C.DC. (including *P. vaccinifolia* C.DC., F.W.T.A. 1: 82 (1954)).
LR/nt

Range: São Tomé (three collections), Bioko (one collection), Mt Cameroon (five pre-1988 collections) and the Bamenda Highlands (three collections).

This species appears to be relatively common (seven collections in 1996) in the forest above Oku-Elak although it has not been recollected in recent surveys on Mt. Cameroon suggesting that it is much rarer at the latter. However, at Kilum-Ijim, this species is at the top of its altitudinal range. Were forest at e.g. Oku-Elak to be cleared to a slightly higher altitude than it is (c. 2200m) then *P. thomeana* might become locally extinct.

Habitat: epiphytic in montane forest; 1460–2300m alt.

Threats: forest clearance for agriculture and firewood.

Management suggestions: none are needed, so long as the existing Kilum-Ijim boundaries are respected.

Description: epiphytic herb, stems trailing and rooting when young. Flowering stems erect, 10–15cm tall, c. 2mm diam., glabrous. Leaves alternate, deep green and glossy above, pale green below, obovate or oblanceolate, 2.5–3 x 1–1.5cm, apex notch c. 1mm deep, base cuneate, glabrous; petiole 1–7mm long. Inflorescences terminal, 1–5, pale green, erect spikes c. 6cm long.

P. thomeana can be distinguished from other species of *Peperomia* on the mountain by the combination of obovate to oblanceolate, alternate, glabrous leaves with a distinctly notched apex.

PODOSTEMACEAE

Ledermanniella keayi (G. Taylor) C. Cusset (syn. *Inversodicrea keayi* G. Taylor, F.W.T.A. 1: 127 (1954)).
CR B1+2c

Range: Cameroon, known only from Mt Oku and the Ijim Ridge area.

Discovered (*Cheek* 8546A, 1 November 1996) on boulders in a small stream, 2–3m wide, running through intensively cultivated land below forest, fed from a high, unnamed waterfall east of Mboh (above Jikijem on N. side of range). This species is otherwise known only from two other collections, both also from the checklist area. These are *Keay* in FHI 28457 from Banso and *Adams* 11073 from Sagbo near Ndop. At Mboh, *L. keayi* was seen only on 3 or 4 boulders, but each had numerous plants. It was difficult to determine where one plant ended and another began. In November 1999, what was probably this species was observed in the stream running from Ajung cliff.

Habitat: basalt boulders in clear, turbulent running water in full sunlight; 1500–2000m alt.

Threats: possibly Endangered if this stream was diverted or lowered by irrigation or coffee-processing demands which are fairly common in cultivated areas around mountain. Pollution at this site is unlikely as it is above any habitation. However, if cultivation upstream of the site is extended up to the edge of stream, erosion may introduce sufficient silt to eliminate the population.

Management suggestions: alerting local farmers to the importance of this plant for conservation and advising them of what the threats are might help this species survive.

Description: rheophyte, thallus inconspicuous, leafless aerial stems naked, stout, erect, black, c. 2–4cm long, bearing at apex numerous short (c. 1.5cm long), slender leaf-bearing branches. Leaves ovate, c. 2mm long, in two ranks, numerous, appressed to stems and concealing them.

Ledermanniella cf. musciformis (G. Taylor) C. Cusset

The Podostemaceae at the Anyajua waterfall (*Cheek* 9920, 12 Dec. 1998) shows several differences from typical *L. musciformis*. Its specific status has not been resolved. It is likely that it will be included in the next edition of this Red Data chapter, whether it proves to be *L. musciformis* (only otherwise known from Mba Kokeka at Bamenda) or a new taxon.

Saxicolella marginalis (G. Taylor) Cheek (*Butumia marginalis* G. Taylor, F.W.T.A. 1: 124 (1954).)
CR B1+2c
Range: Nigeria, Butum falls (one collection) and Cameroon, Fundong falls.

This very rare rheophyte was known previously only from one collection at the Butum falls, Obudu Plateau in adjoining Nigeria before it was newly recorded by us from Cameroon (Cheek et al. 1997). In Cameroon it is known only from the waterfall at Fundong near the Touristic Hotel (*Cheek* 8740, 22 November 1996). What may prove to be a second site for this species in Cameroon was seen at Ajung in November 1999 (pers. obs.) but a specimen is needed to confirm the identification. All streams and rivers between Fundong and Belo were examined for Podostemaceae at road crossing points, but only this site was discovered. The waterfalls of the Bamenda Highlands have been fairly well investigated for Podostemaceae thanks to the work of Ledermann and Keay. *Saxicolella marginalis* is locally common. At least several hundred plants were seen here in 1996, and probably many more were present. Observation of this species is difficult in its habitat due to spray, with turbulent water and slippery boulders. Monitoring at this site in December 1998 and November 1999 suggested that this species was present in approximately similar numbers to 1996.

Habitat: on exposed basalt boulders below a c. 30m high waterfall; c. 1400m alt.

Threats: probably requiring clean, well-oxygenated water as usual in this family. Silt in water known to be inimical to survival of Podostemaceae. Therefore this population is possibly Endangered at this site from pollution from laundry operations at the town of Fundong just upstream: much debris was seen at side of this pool.

Management suggestions: monitoring of the population is suggested from year to year to establish if there is any diminution of numbers of individuals. Cultivation of Podostemaceae known to be extremely difficult and so *ex situ* conservation is not an option.

Description: rheophyte, probably annual, thallus dark green and inconspicuous when alive, white when dried, adhering firmly to the rock, c. 0.75cm wide, 5–6cm long, dichotomously branching like a liverwort. Margin of thallus bearing at intervals sessile tufts of linear leaves 2–3mm long, surrounding a single subsessile flower. Flower about as long as leaves, with sessile, olive green ovary bearing two purple stigmas. Stamen single, about as long as ovary. Fruit 6-ribbed. Flowering and fruiting in late November. Illustrated in F.W.T.A. 1; 125 (1954).

POLYGALACEAE

Polygala tenuicaulis Hook.f. subsp. *tayloriana* J.Paiva
LR/nt
Range: Cameroon, Bamboutos Mts (two collections), Bamenda Highlands (four pre-1996 collections) and Ngaoundere (two collections); in Nigeria, Mambilla Plateau, (two collections) and Vogel Peak (two collections).

This plant seems relatively common at Kilum-Ijim, in view of the fact that seven collections were made in 1996 and 1998. However, it tends to be highly localized. It appears to be an annual that can compete with tall-growing grasses. Usually it is gregarious, 6–20 plants being seen at one site. Above Oku-Elak in fallow fields, some plants of this species persisted at one place, suggesting that it might survive as a weed if its natural habitat is destroyed.

Habitat: rocky grassland; 1900–2400m alt.

Threats: unknown.

Management suggestions: none.

Description: Erect annual herb 30–45cm tall. Stem unbranched in basal half, c. 2mm diam., internodes 0.5–1cm long, wiry, glabrous. Leaves alternate, sessile, linear–lanceolate, 25(–35) x 1–2mm, acute, margin revolute, glabrous. Inflorescence a cluster of (1–)2–5(–10) terminal one-sided racemes, each bearing c. 20 flowers; bracts linear, 1.75mm long; pedicels 1–2mm long. Flowers pale pink, obovate in side profile, c. 7mm long.

This taxon is only likely to be confused in the field with the type subspecies, not yet recorded from our checklist area but known to occur in the Bamenda Highlands. The second is much more slender and shorter in habit, with smaller, purple flowers.

ROSACEAE

Alchemilla fischeri Engl. subsp. *camerunensis* Letouzey   MBAKLUM (Oku).
CR B1+2c

Range: Cameroon, summit of Mt Oku (three pre-1996 collections).

This attractive, silvery-leaved herb is relatively abundant in the summit area of Mt Oku (four collections were made in the 1996 expedition), but is unknown elsewhere on Mt Oku and Ijim Ridge. Letouzey (1978), in his protologue for the taxon, reports in detail on its habitat and relationships. Pollard (pers. comm.) confirms Letouzey's (1978) observations that *A. fischeri* subsp. *camerunensis* forms a continuous silvery carpet near the summit of Mt Oku. According to author BJP, the carpet that he saw in November 1996 measured c. 5 x 20–30m in area. He reports seeing only one such stand of the taxon, but concludes that another or others might exist in the summit area.

Reproduction by stolons in this taxon may be more important than seed.

Habitat: rocky grassland and scrub; 2800–3000m alt.

Threats: fires and grazing by cattle, goats and sheep and by insect larvae (Pollard pers. comm.).

Management suggestions: monitoring of the population is needed to determine whether this taxon is declining or not.

Description: robust, carpet-forming, erect subshrub; main stems c. 10–15(–25)cm high, 6–8mm diam., stolons over 50cm long; leaves up to 8 x 10cm, 7–9-lobed, median lobe 25–31-toothed, bidentate, thickly silvery hairy. Petiole up to 12cm long. Stipules oblanceolate, apex acute, 6-toothed. Inflorescences paniculate cymes, borne on stolons, c. 20cm tall; peduncle c. 8cm long; bracts resembling stipules. Flowers green, 2.5mm long.

Description after Letouzey (1978).

Prunus africana (Hook f.) Kalkman
LR/nt

Range: montane Africa (to East and South Africa) and Madagascar.

This species is one of about ten Pan-African montane tree species (including e.g. *Agauria salicifolia, Ilex mitis* and *Morella arborea)* and is not remotely in danger of extinction, so long as some montane forest survives somewhere within its enormous range. Locally it can be very common. However, on Mt Oku and Ijim Ridge, as with other areas within the range of this species, many trees have died as a result of girdling caused by bark removal. The bark from the trees is transported to the Plantecam factory at Mutengene where it is extracted to produce a powder for export to France. However, mature trees still survive in the wild. A great number of individuals of *Prunus africana* have been planted as the boundary of the protected area on the Kilum side. They are now 6–8m high and appear to be growing well. A great deal of attention, and funding has been paid by international conservation organizations to investigate and address this harvest and, perhaps for this reason, the species has received a 'threatened' conservation rating. In the opinion of this author, it merits a "Lower Risk, near threatened" at most.

Habitat: montane forest, usually at about 1800–2200m alt.

Threats: harvesting of bark for the European medicinal market.

Management suggestions: none are needed.

Description: evergreen tree 20(–30)m high, 0.4(–1)m diam. Breast height, sometimes bearing buttresses; bark brown, slash pink, soon oxidizing to brown. Leaves alternate, elliptic, 6–15 x 3–6cm, acute or acuminate, base rounded, margin serrate; petiole red, glandular at apex, 1–2cm long; stipules soon falling. Inflorescence axillary racemes c. 10cm long, several inserted at the base of young leafy shoots. Flowers white, c. 5mm diam. Drupes 1-seeded, red, c. 1cm diam.

RUBIACEAE

Anthospermum asperuloides Hook.f. (syn. *A. cameroonense* Hutch. & Dalziel, F.W.T.A. 2: 223 (1963)).

LR/nt

Range: Bioko (one collection), Mt Cameroon (seven collections), Bamboutos Mts (one collection) and the Bamenda Highlands.

Most of the c. 40 species of *Anthospermum* are found in Southern Africa, *A. asperuloides* being the only species to occur west of the Congo basin. Although the genus has been revised in detail by Puff (1986), little is known about our species. Some of the southern African species are known to be adapted to fire by having underground rootstocks, others are known to withstand grazing by goats and sheep, but information of this sort on *A. asperuloides* is lacking and we cannot assume that this species also has these traits. Concerning frequency, we have only the note on the sole collection from the Bamboutos Mts: "Not common" (*Sanford* 5590, 24 November 1968). However the fact that this species, so far as we know, has only been collected once from the Bamenda Highlands, from the summit of Mt Oku (*Munyenyembe* 835, 31 Oct. 1996), suggests that it is rare and restricted to this location. Given the potential threats (see below) all of which are known to occur at the summit, the survival of this species in the Bamenda Highlands must be regarded as a matter of concern.

Habitat: upper montane grassland and scrub; 2400–3000m alt.

Threats: unknown, but grazing, fire and trampling, known in this habitat, may have adverse effects on this species.

Management suggestions: a survey of this species on Mt Oku is needed. The information required includes numbers of plants, their distribution, regeneration levels and threats.

Description: shrub or subshrub 0.2–1m tall. Stems c. 5mm diam. at the base, branches sparse, persistently leafy, sinuous. Leaves subsessile, leathery, opposite or in whorls of three; blades 5–10(–15) x 0.8–1.5(–2)mm oblanceolate to linear-lanceolate, apex subspinulose, and base slightly attenuate, margin flat or revolute, inconspicuously white-hairy; stipules sheathing, cup-like, 1–1.5mm long, with (1–)3–5(–6) gland-tipped setae 1.4–2.4mm long, the lateral ones shortest. Inflorescences inconspicuous, of axillary, subsessile clusters of flowers. Flowers subsessile. Corolla yellowish (rarely purplish) green, hairy or glabrous, tube in bisexual flowers narrowly funnel-shaped, 0.5–0.7mm long, lobes lanceolate, recurved, 1.4–2 x 0.6–0.7mm long. Stamens 4, filaments 1–1.4mm long; anthers 0.9–1.7mm long. Fruits reddish brown, breaking into two fruitlets when ripe. Fruitlets oblong, 1.9–2.7 x 1–1.4mm, outer side convex, with white hairs; calyx lobes minute or absent in fruit.

Description based partly on that in Puff (1986: 294–295).

Chassalia laikomensis Cheek ined.

CR A1

Range: Cameroon, Bamenda Highlands (seven pre-1996 collections) and Nigeria, Mambilla Plateau (one collection).

This forest understorey shrub was part of a mixture referred to in the Flora of West Tropical Africa as *Chassalia sp. nr. umbraticola* (Cheek & Csiba in press). Collections are known from various parts of the Bamenda Highlands and it may once have been fairly common in this area. However, since forest in the Bamenda Highlands has seen such a dramatic reduction in the last few decades, it is considered that the best hope for the survival of this species is probably at Mt Oku and the Ijim ridge. Since the forest included within the Kilum-Ijim boundary is mostly above the 2000m contour and *Chassalia laikomensis* occurs in forest mostly below this altitude, the main populations of the species occur at the periphery of the protected area.

The greatest number of individuals of this species that have been seen are at Laikom (hence the specific epithet). In the undisturbed part of the Laikom Palace forest, *Chassalia laikomensis* is the dominant understorey species: 25–30 plants were counted at this site in November 1999 (*Cheek* 10105). At the head of the Laikom valley, the species is also common in forest (at least four plants were seen in November 1999) where the dominant understorey species is *Psychotria peduncularis*. The third most important site for this species is at "Back Valley" Mbingo (two plants seen in November 1999, *Cheek* 10053).

Habitat: understorey of montane evergreen forest; 1650–2000(–2400)m alt.

Threats: firewood; forest clearance for cultivation.

Management suggestions: more information is needed on the number of individuals of *Chassalia laikomensis* at the known sites, and on levels of regeneration.

If surveys are conducted in other areas of forest in the Bamenda Highlands, the opportunity should be taken to observe and record this species.

Description: Shrub, rarely small tree, 2–3.5m tall, glabrous. Leaves narrowly elliptic, 4–12 x 1.5–4cm, acuminate, base cuneate, domatia absent; petiole 2–12mm long; stipule shortly sheathing, sheath 2–3mm long, becoming chaffy, limb broadly triangular, 2–3 x 5.5mm, apex mucronate, outer surface conspicuously scattered with minute, yellow, rod-shaped raphides, each c. 0.1mm long. Flowers in loosely branched terminal inflorescences of 8–30 flowers, often with accessory lateral branches from the axils of the first pair of leaves, overall 5–8.5 x 2.8–5.8cm. Flowers 5(–6)-merous; calyx lobes triangular, c. 0.5 x 0.5mm; corolla white, slightly curved in bud; tube 6.4–10mm long; lobes ligulate-oblong, lacking appendages; stamens included or excluded; style exserted by 2–5mm. Fruit a black drupe held obliquely on white peduncles, ovoid or spherical, 6–9mm long; pyrenes usually single, with pre-formed germination slit.

Description after Cheek & Csiba (in press).

Oxyanthus okuensis Cheek & Sonké ined.

CR A1c, B1+2b, C2a, D.

Range: Known only from Mt Oku and the Ijim ridge in Cameroon.

This species was first collected, in fruit, by Duncan Thomas from Lake Oku in Feb. 1985 (*Thomas* 4377) and mistakenly identified as *Oxyanthus formosus* (Thomas in McLeod 1986: 62). It was suspected of being new to science once it was recollected in the 1996 survey (Cheek et al. 1997). It was thought to be restricted to the forest immediately around Lake Oku (where seven plants are known, Kemei *pers. comm.*), until c. four sterile treelets were seen in the reconnaissance to the Ajung Cliff (*Cheek* 10103) led by DeMarco in November 1999. It is notable that surveys in other areas of forest on Mt Oku and the Ijim ridge have not located this species. However, according to Kemei (*pers. comm.*), there is a third site (a single plant) for this species near Ntum at Ijim. A full description and notes on this species can be found in Cheek & Sonké, in press.

Both the Lake Oku and Ajung cliff sites are inside the Kilum-Ijim boundary, and so are protected, so long as the boundaries are respected. However, the forest at Ajung was only recently included inside the boundary and there is still evidence that the more accessible part of this forest was still being cleared, possibly as late as early 1999. The forest at Ntum is not protected at the time of writing. *Oxyanthus okuensis* is probably most closely related to *Oxyanthus montanus* Sonké which is endemic to Bioko, Mt Cameroon and the Bakossi Mts.

Habitat: understorey of montane evergreen forest; 1800–2200m.

Threats: possibly cut for firewood; forest clearance for agriculture

Management suggestions: more information is needed on the number of individuals of *Oxyanthus okuensis* present at the two known sites, and upon levels of regeneration. The plant at the potential third site (Ntum) should be vouchered. Fertile material is still needed from the Ajung site in order to confirm the specific identity. This species should be looked for in other areas of forest in the Bamenda Highlands. Confusion with other species of *Oxyanthus* is unlikely since this is the only known member of the genus at Kilum-Ijim.

Description: Shrub or small tree 3–8m tall. Leaf blade elliptic or elliptic oblong, 9–13 x 2.5–6cm, apex acuminate, nerves 8–9 on each side of the midrib, domatia hairy; petiole 5–8mm long. Stipule oblong in the lower 2/3, triangular in upper 1/3, 12–19 x 3–9mm, apex aristate, 2–

4mm long. Inflorescences c. 3 per branch, alternate on consecutive nodes, erect, 20–50-flowered, condensed panicles c. 2–3.5cm diam. Flowers white, calyx cup-shaped, lobes 1.5mm long, Corolla tube 3.3cm long, lobes 5, lanceolate, 7.5 x 2mm. Fruit with pedicel 12–14mm long; fruit body ellipsoid, 35–45 x 13–16mm, including an apical rostrum. Seeds c. 10 per fruit, faceted, c. 6–8mm diam.
Description after Cheek and Sonké (in press).

Pentas ledermannii K.Krause emend. Verdc.

This shrub of upper montane forest edges only came to light in the last stages of writing this chapter and there was insufficient time to evaluate its conservation status fully or to prepare a description. It is known only from the Bamboutos and Bamenda Highlands apart from one collection from Mt Kupe.

Psychotria sp. nov.?

There are several collections of an understorey shrub from forest patches below c. 2000m around Mt Oku and the Ijim Ridge which appear to be a new species of *Psychotria* These are *Munyenyembe* 873, *Etuge* 3548 and 4518 gathered in 1996 and 1998. In 1999 more distributional data was collected on this taxon. It is now known from Upkim Forest at Kilum, and in the Ijim area from the head of the Laikom valley, the Tum forest and from Mbingo "Back Valley". At the second and fourth sites only one or two plants were found. No evidence on frequency is available for the other sites. Vegetatively this *Psychotria* is similar to and easily confused with *Chassalia laikomensis* (see above). However, it can be distinguished by the lack of chaffy stipule bases, the dull red nerves of the lower leaf surface and the domatia, which are glabrous, circular lacunae c. 0.5mm diam. in the web linking the secondary nerves with the midrib. More research is needed to confirm that this is an undescribed endemic of the Bamenda Highlands and not an East African montane straying species. If the former hypothesis is supported it is likely that this species will appear in a future edition of this Red Data chapter. . In F.W.T.A. 2:202 (1968) this taxon is, I think mistakenly, treated as *Psychotria chalconeura* (K. Schum.) Petit, based on *Maitland* 1744.

Tarenna pavettoides sensu F.W.T.A., non (Harv.) Sim.

This is a fairly common shrub of forest above 2000m on Mt Oku and the Ijim Ridge. Five collections are listed in the checklist that follows. Bridson (pers. comm.) who has treated the East African species, has pointed out that this taxon is not *T. pavettioides* of that area as it is identified in F.W.T.A., but a separate species. More work is needed to determine its specific identity, but, when this has been done, it is probable that it will appear in a new edition of this Red Data chapter. Superficially, this plant resembles *Pavetta hookeriana*, a common shrub in the same habitat. However, the *Tarenna* can be distinguished by the glossy (not dull) stipules that dry black (not grey) and by the puberulent (not glabrous) stems.

SCROPHULARIACEAE

Craterostigma sp. nov?

Pollard 273 (2nd Dec. 1998, Laikom Ridge, Ijim) may represent a new species in this genus. More work is needed to confirm its specific status. If this is confirmed, and it proves to be restricted to Kilum-Ijim, it is likely to be included in the next edition of this Red Data chapter.

Veronica mannii Hook. f.

LR/nt

Range: Bioko (four collections), Mt. Cameroon (24 pre-1988 collections), Mt Oku (two recent collections).

Formerly considered endemic to Mt Cameroon and to Bioko, the range of *V. mannii* was extended to the Bamenda Highlands by the discoveries of two of our colleagues on the 1996 expedition (*Munyenyembe* 820 and *Zapfack* 1133, both found above Oku-Elak on 31 Oct. 1996). First collected in December 1860: "on the very top of Clarence Peak, Fernando Po" (Mann 604), this species has often been collected from Mt Cameroon. However, nearly half of those collections were made by two people in two days collecting which has inflated considerably the number of specimens! On Mt Cameroon, most of the collections have been made above Buea, between Hut 2 and Hut 3, but Morton records it from a location 3 miles from Mann's Spring. Descriptions of flower colour vary from "vivid royal blue" to "violet" and it has generally been collected in flower from December to April.

Habitat: montane grassland; (2700–)3000–4000m alt.

Threats: grazing pressure and/or fires may be deleterious for this species.

Management suggestions: more information is needed on the location and numbers of individuals of *V. mannii* on the mountain, and on the question of the threats posed by fire and grazing on this species.

Description: Multistemmed perennial herb to 20cm tall. Stems erect, woody at base, 3–4mm diam., glabrous. Leaves opposite, subsessile, ovate or lanceolate, 1.5(–3.5) x 0.6–0.7(–1.3)cm, apex obtuse, margin finely serrate. Flowers solitary in axils of uppermost 3–5cm of stem, purple-blue, c. 1cm diam.

Veronica abyssinica, the only other species of the genus to occur in the Bamenda Highlands, is not likely to be confused with *V. mannii*. It has prostrate, herbaceous (not erect and woody) stems, and is widespread.

STERCULIACEAE

Dombeya ledermannii Engl.
CR A1c

Range: Cameroon, Bamenda Highlands (5 pre-1996 collections) and Nigeria, Mambilla Plateau (4 collections); Jos Plateau (2 collections).

This tree is known only in the checklist area from collections from one tree below the Akeh-Ajung road (*DeMarco* in Maisels 113, Apr. 1998). It is easily counted since, in April, trees are clearly visible from a distance on account of their white flowers. Flowering can occur in November (pers. obs.). The habitat of this species is highly threatened in all its known localities. Seyani (1982) reports that *D. ledermannii* is characteristic of forest edges and the early stages of forest succession, but that on exposed rocky slopes is a normal component of stunted, more open forest. In the Bamboutos Mts, it is propagated by cuttings and planted to form hedges (*Letouzey* 201, cited by Seyani 1982).

Habitat: woodland; (700–)1220–1980m alt.

Threats: clearance for agricultural land, over-exploitation for bast fibre.

Management suggestions: more information is needed on the extent, distribution and threats to *D. ledermanniii* within the Kilum-Ijim area.

Description: tree 3.5–15m tall, usually more than 7m tall. Leaf-blade suborbicular to ovate, slightly 5-lobed, 11–19 x 5.3–15cm, shortly acuminate, cordate, margin slightly denticulate; petiole 4–7cm long. Inflorescence 6.8–12cm long, axillary cyme; peduncle 3–7cm long, 3–4-branched; pedicels 1–2cm long. Epicalyx bracts 3, 0.2–0.35cm, obtuse. Calyx lobes lanceolate, 0.5cm long. Petals 5, white, 0.8–1.3cm long; staminodes c. 1cm long. Fruit capsule depressed globose, 0.3–0.5cm diam.

Description after Seyani, J.H. (1982).

THEACEAE

Ternstroemia polypetala Melchior
CR A1c

Range: Bamboutos Mts (1 collection in 1974), Ijim (1 collection in 1996) Cameroon; Eastern Arc Mts of Tanzania (c. 5 collections).

This canopy tree was thought to be restricted to the Eastern Arc Mts of Tanzania until discovered by Letouzey (*Letouzey* 13380, 29 November 1974, Baranka to Chefferie Fossimondi, c. 25km NNW Dschang) in the Bamboutos Mts of Cameroon. Letouzey (1977) pointed out that there are slight differences from the Tanzanian material and that the Cameroonian plant might be distinct at the subspecific level. Its presence at Kilum-Ijim was revealed in 1997 when one of the specimens gathered in November 1996, *Etuge* 3537 collected at Tum, was identified at Kew as belonging to this species. Despite a search of the Tum area (just outside the K-I boundary) by a large group of botanists including Etuge over two days in November 1999, no trace of this species was found. Etuge (pers. comm.) noted that in the intervening three years, the patch of forest at Tum had been reduced by about two-thirds its 1996 area. *Ternstroemia polypetala* may now be extinct in the Bamenda Highlands. Since the site of this species in the Bamboutos Mts was noted at the time of its discovery to be Endangered (Letouzey 1977), it is quite possible that *Ternstroemia polypetala* is now extinct in Cameroon and West-Central Africa, and survives only in the Eastern Arc Mts of Tanzania.

Habitat: montane forest 1500–2100m alt.

Threats: forest clearance.

Management suggestions: future forest survey workers should keep an eye out for this species, in case it survives. If trees are located, they should be given the highest priority for conservation and serious consideration should be given to propagating them for planting in protected areas.

Description: tree c. 18m tall, c. 40cm diam. at breast height. Flowering branches stout, c. 4mm thick, showing *Terminalia*-type branching. Leaves clustered at apex of branches, leathery, oblanceolate, 6–7.5 x 2–2.5cm, apex rounded or slightly notched, base attenuate decurrent, margin revolute, nerves c. eight pairs on each side of the midrib, glabrous. Inflorescence of single flowers, up to 4–5 per stem, held immediately below the leaves; pedicels 12mm long; bracts two, at apex of pedicel, 1.5 x 2mm. Flowers c. 10mm wide. Petals c. 8, white. Stamens 40–50. Ovary superior, ovoid, style single. Fruit enclosed at base by persistent sepals, ovoid, c. 16 x 12mm, fleshy, orange-yellow. Seeds 1–3, papillose, up to 8 x 6 x 4mm.

Description based in part on Letouzey (1977).

UMBELLIFERAE

Peucedanum angustisectum (Engl.)Norman
LR/nt

Range: Cameroon highlands from Mt Cameroon (9 collections) to Bamboutos Mts (1 collection), Tchabal Mbabo (1 collection), extending to the Mambilla Plateau (Chappal Waddi, Nigeria, 2 collections). Dubiously identified specimens from Congo (*Kassner*) and Kaduna, Nigeria (*Sharland*).

Peucedanum angustisectum is known from the length of western Cameroon, but was not collected from the Bamenda Highlands until located at Ijim in November 1996 (*Cheek* 8698, grassland, 1900m alt., path from Fon's Palace to Mboraro settlement). This remains the only collection of the species known from the area. Described by Engler as *Lefebvrea angustisecta* in 1921, this species was transferred by Norman to *Peucedanum* in 1934 (Jacques-Félix 1970: 92).

Habitat: rocky montane grassland; 1900–3200m alt.

Threats: as for *Peucedanum camerunensis* (q.v.).

Management suggestions: as for *Peucedanum camerunensis* (q.v.).

Description: annual herb up to 1.2m tall, slightly branched, glabrous. Stems smooth or slightly striate. Leaves biternate to bipinnate, gradually reducing in length towards the apex of the stem, the largest 25 x 20cm, 2–3-jugate, petiole 10cm long, basal sheath c. 3cm long, slightly canaliculate; lowermost pinnae with petiolules c. 2cm long, blade often subpalmatisect; pinnules linear-lanceolate to narrowly lanceolate, the terminal the longest, up to 8 x 1.5cm, coarsely and unequally acutely serrate to incised serrate. Umbels dispersed in a slender inflorescence; involucre absent; rays 6–8, 1.5–1.8cm long; involucel reduced to a few short straight bracts c. 5 x 1.5mm; pedicels subequal, 4mm long. Fruits narrowly elliptic, 6–7 x 3–3.5mm, 2.2mm broad, margin straight, dorsal face with slightly raised ridges; commissural face with four bands, the inner broad, the outer pair submarginal and sometimes evanescent.

Description after Jacques-Félix (1970: 92–93).

Peucedanum camerunensis Jacq.-Fél.
EN B1+2c

Range: Cameroon, Bamenda Highlands (four pre-1996 collections).

This species was described by Jacques-Félix in 1970, from collections at two locations straddling the Bamenda Highlands (Kounden, Mt. Nko Gham, 2400m, Oct. 1964, *Koechlin*

7541 (type) and Bamboutos, 2300m, (sterile, and therefore uncertainly identified), *Jacques-Félix* 5482). Since then, two further collections have been identified: just south of our checklist area (Mbam, 2335m alt., 35km NW Foumban, *Letouzey* 13099, November 1974 and Lake Bambili, Aug. 1951, *Ujor* in FHI 29968).

Three collections were made in the course of the 1996 inventory. They were at locations each c. 10km or so from each other: cliff-face, Oku-Elak, *Munyenyembe* 819 (31 Oct. 1996); Ewook Etele Mbae, Oku-Kumbo, 10km, *Zapfack* 1229 (6 November 1996) and on the southern (Ijim) side of Lake Oku, *Onana* 639 (26 November 1996). No collections were made on Laikom/Ijim ridge during the expedition there in Dec. 1998 probably because, although the habitat there might be suitable, it was then later in the year and the flowering season for that species (which seems to be restricted to Oct. and November) was past.

There is little information on the population density of this species available from the herbarium specimens seen. However, given that 2–3 duplicates of each of our specimens were made, it seems likely that plants of this species are fairly thinly scattered, although not completely isolated (e.g. by a kilometre or more) from each other.

Peucedanum camerunensis appears to be an annual species (Jacques-Félix 1970), given the relatively diminutive size of the plant and absence of a substantial rootstock. If this fact is confirmed, it makes this species more vulnerable to adverse environmental effects than if it were a perennial, and makes seedling recruitment of utmost importance for this species.

Habitat: rocky montane grassland; 2300–2800m alt.

Threats: not known, but grassland fires and grazing by domesticated animals are likely to affect the survival and establishment of this species.

Management suggestions: more detailed data on frequency and range of this species is needed. October and November, when the plant is conspicuous since in flower, is probably the best time to conduct this work. Direct observations of threats to this species are desirable.

Description: annual (?) herb, 30–40cm tall, erect, glabrous. Stems cylindrical, smooth, pale green, firm to subwoody. Basal leaves 18–20cm long, petiole 5–6cm, sheath 0.5cm, rhachis 2–(–3-)jugate, petiolules 2.5cm long, the pinnae ternate, rarely pinnate, leaflets sessile, linear, 7–8 x 0.2–0.3cm, variable in size and number of leaflets, from 3 to 1. Umbels usually terminal, on relatively numerous branches; involucre nil; rays 5–7, straight, 1.5cm long; umbellules poorly developed; involucel of several filiform bracts 1–1.5mm long; flowers 8–12; pedicels 1cm long. Fruits elliptic, 5 x 3mm, moderately flattened on the back, wing straight, 0.5–0.6mm wide, commissural bands 4, the inner thick and conspicuous, the outer pair filiform, situated in the wing.

Description after Jacques-Félix (1970: 93–94).

Pimpinella cf. praeventa Norman

Numerous collections of what seems to be a new species closely related to *P. praeventa* (which occurs in the Tchabal Mbabo area) were made during our inventories of 1996–1999 (see checklist). More work is needed to confirm its specific status. If it is confirmed as a new species that is endemic to Kilum-Ijim it is likely to appear in an update to this Red Data chapter.

MONOCOTYLEDONS

ASPHODELACEAE

Kniphofia reflexa Hutch. ex Codd
EN C1

Long known only from a single collection from a single locality above Laikom (`Lakom', April 1931, *Maitland* 1624) where it was then recorded as "scattered in considerable numbers in the grass on the plateau", *Kniphofia reflexa* was then lost to science for over 60 years until Martin Etuge's team rediscovered it in the Afua swamp while searching for material of *Eriocaulon bamendae* in November 1996 (Cheek et al. 1997). Etuge observed about seven plants at Afua. Maisels confirmed plants at this site in 1997. However, when this site was revisited in November 1998, no plants were found and the species may now be extinct at Afua. The reason for the loss is not clear. In May 1997 Boo Maisels found a new site for *Kniphofia reflexa* at

Kinkolong, near the summit of Mt Oku where several hundred plants were recorded, and two more locations have since been found. One, the "Mbesa Swamp" on Ijim Ridge, is probably the site at which Maitland collected his specimen in 1931. It has about 200 individuals. Mbi crater is the third site. Maisels estimates that this also has about 200 individuals (Maisels *et al.* 2000).

It is remarkable that such a conspicuous species has remained unobserved by scientists for so many decades. Before our work the genus was not represented in the Cameroon National Herbarium at Yaoundé.

Kniphofia reflexa is the only species of this temperate, horticulturally important genus known from West-Central Africa. The other 64 species of the genus are known from Arabia, through the mountains of East Africa, to South Africa and Madagascar.

Habitat: swamp or stream edges; 1750–2900m.

Threats: Asonganyi (1995) remarks upon the Afua area as suffering from extensive recent habitat destruction and so we consider this extremely rare species highly Endangered.

Management suggestions: annual monitoring of the three sites for this species is needed. Numbers of individuals should be counted on these occasions, signs of damage recorded and regeneration noted. Afua should be revisited to confirm that this species has indeed disappeared. Similar sites to those already known should be searched to discover potential new locations. Some evidence of fire scorching on the Afua specimens was noted, so some burning at the correct season may be required in management of habitat for this species or important for promoting flowering. Propagation by seed and replanting in e.g. Afua Swamp might be considered.

Description: perennial herb from yellow-staining rootstock 3–4cm thick. Leaf-rosettes one to few from each rootstock, about 1m wide, acaulescent. Leaves long-linear, 14–16mm wide, soft. Inflorescence a stout spike 0.6–1.6m tall (to 2m tall in fruit) bearing in the upper part numerous tubular flowers; bracts oblong, boat-shaped, sharply reflexed, 5–6mm long, 2–3mm wide, margin finely papillose-fimbriate. Flowers yellow, 8–8.5mm long, tube 5–5.5mm long, campanulate; lobes 3–3.5mm long, the outer 3mm broad at the base, the inner 2.5mm, ovate-triangular, rounded. Filaments 7–8mm long; anthers c. 2.5mm long. Style 5mm long. Fruiting in November. Flowers in April?

Description partly after Marais (1973).

CYPERACEAE

Bulbostylis sp. nov.

A collection from Laikom Ridge (*Pollard* 293, Dec. 1998) represents a possible new species of this genus according to Prof. Lye (NLH). If this is confirmed it is likely that this taxon will be included in a new edition of this Red Data chapter.

Carex preussii K.Schum.

LR/nt

Range: Mt Cameroon (two pre-1988 collections), Bamenda Highlands, particularly Mt Oku (nine pre-1996 collections) & Mambila Plateau (one collection).

Although restricted in distribution on Mt Cameroon from where it was first described (Cable & Cheek 1998), *Carex preussii* appears widespread, perhaps common, in its habitat in the Bamenda Highlands and particularly at Mt Oku and the Ijim Ridge.

Habitat: in gaps in montane forest; 2100–2400m alt.

Threats: clearance of montane forest in the Bamenda Highlands is a major threat.

Management suggestions: none.

Description: perennial tufted herb 40–80cm high, lateral shoots arising at base in a line. Leaves slightly folded, smooth, c. 80 x 0.3–0.4cm wide. Inflorescences with 3–7 nodding spikes; peduncle 3-angular; c. 45cm long; partial-peduncles up to 6cm long; spikes c. 4 x 0.5cm; male flowers below female, glumes brown.

Cyperus niveus Retz. var. nov.

This bulbous herb, with bright white capitula, has been determined as a new variety by Prof. Lye (NLH) but awaits formal description, so is not treated in full as a Red Data species here. It is so far known only from the Mbesa Swamp (*Cheek* 9810, *Pollard* 354, both November 1998*)*.

Cyperus sp. A, B. & C.

These three taxa (see main checklist), which lack names, are possibly new and/or rare species. Further work is needed to resolve their identity. It is likely that, when this is done, they will feature in a new edition of this Red Data chapter.

ERIOCAULACEAE

Eriocaulon asteroides S.M.Phillips
VU D2

Range: Cameroon, Bamenda Highlands, Mount Oku and the Ijim Ridge area, and Nigeria, Mambilla Plateau (Chappal Waddi, 1 collection).

This minute annual often co-occurs (in three of the six known sites) with the superficially similar *Eriocaulon parvulum*. The second species is not known to occur without the first. Indeed, both species were originally described (Phillips 1998) from an unwittingly mixed collection of the two species (Zapfack 1205, November 1996) from the Kumbo-Oku road (Iwooketele Mbai). At the time of its description, *Eriocaulon asteroides* was only known from that site and from Chappal Wadi, the highest mountain in Nigeria, situated on the border with Cameroon. Subsequently, in December 1998, a large colony of the two species was found on a headland of the Laikom spur of Ijim Ridge. This site was studied in some detail (see vegetation chapter: basalt pavement). 1–2km along the path from this site to the Ardo of Ijim's compound another site for this species was found. The basalt pavement area here was wetter than on the headland, perhaps accounting for the absence of *E. parvulum*. Finally at about 1700m alt., between Laikom and Fundong, below the Fulani settlement, another colony of *E. asteroides* was encountered by Tadjouteu of our party, in November 1999. Again, the species was associated with an outcrop of wet basalt pavement, and again *E. parvulum* was absent, although about ten of us spent an hour searching for it. Finally, in listing the *Eriocaulon* holdings of YA for the monographer of the African species of this genus, Dr Phillips, another mixed collection of the two species was encountered. This was collected in September 1975 (*De Wilde* 8633) from km 21 on the Bamenda-Jakiri road at the southern boundary of our area.

Habitat: basalt pavement, i.e. thin, peaty, seasonally waterlogged soil in the cracks between blocks of basalt; with *Utricularia scandens*, *Loudetia simplex* and *Scleria interrupta*; c. 1700–2500m alt.

Threats: unknown. However, too much trampling by cattle might cause damage to these small annual *Eriocaulon* plants by dislodging from the basalt substrate the thin layer of peaty soil in which they grow. Conversely, lack of grazing or of intermittent grassland fires might permit the build up of enough soil on the pavement to allow a *Sporobolus*-based community to encroach upon the basalt pavement and smother or compete with the *Eriocaulon*.

Management suggestions:

1. A survey of basalt pavement should be made in the Kilum-Ijim area. When areas are located, these species should be searched for and vouchered if found. A rough estimate of the area of occupation and total number of plants should be made. This will allow more complete mapping of the species and a more comprehensive understanding of their population size.

2. Consideration should be given to using experimental means to examine the effects of the possible threats mentioned above on these two rare annual species. Several square metres of one population could be fenced off and protected from fire and grazing. The effect of this could be monitored on an annual basis. Another area could be subjected to cattle or horse trampling to look at the effect of this on the soil that hosts these species. Results of this experimentation could then be used to guide management of the habitat of these species.

Description: annual with leaf rosette c. 2–3cm diam. Leaves linear–subulate, 0.8–1.5cm long, c. 1mm wide, acute. Scapes up to 10, 1–2.5cm high, 0.2–0.3mm thick, 3–4-ribbed. Capitulum 5–7mm wide, few-flowered, star-like; involucral bracts radiating, much exceeding the floral disc, mostly in one series, membranous, narrowly lanceolate, long acuminate, 3.0–3.8 x 0.7–1.2mm, whitish buff or flushed grey. Seeds ellipsoid, 0.5x 0.3mm, brown, smooth.

Eriocaulon bamendae S.M. Phillips
VU D2
Range: Cameroon, Mount Oku, the Ijim Ridge and the Bamenda Highlands (four pre-1996 collections); Nigeria, Mayo Daga (1 collection).
This perennial herb was only recently described as new (Phillips 2000). Previously it had been treated as a south-central African species, *E. zambesiense* Ruhland. It was first collected in June 1931 at Laikom "in a pond" (*Maitland* 1400). The second locality that was discovered, given as Kumbo-Oku 5km (*Hepper* 2021, 15 Feb. 1958) has not been refound by us. The site of the third collection (*Brunt* 1092, 11 April 1963), is given as "near Pinyin" (south of Bali-Ngemba reserve). The fourth (*Bauer* 35, 28 Feb. 1970) is Bambili Lakes. A fifth collection, from nearby Nigeria (*Hall* 1748) is slightly morphologically anomalous.
We first located this species in the field in November 1996 at the Afua Swamp (*Etuge* s.n.). Revisiting this site in December 1998, an assessment of the population size produced the figure of 2000–3000 plants. The Afua Swamp is thus one of the two most important localities for this species, in terms of population size. In December 1998, in travelling along the Laikom ridge from Laikom towards Mbesa, two more localities were discovered. The first of these, just east of the Ardo of Ijim's compound, resembles a seasonal pond, and may be Maitland's locality of 1931. Here only five plants were seen. Further east still, at the "Mbesa Swamp", 1000–2000 plants were found (*Cheek* 9819) between the tussocks in the bed of a seasonal stream and this is the second most important population of the species that we have located. In the meantime, on the Kilum side of Mount Oku, Maisels had located two more populations. The first of these was at the Kinkolong swamp near the summit of Mount Oku (*Maisels* 115, 26 June 1998), the second at Tadu stream (*Maisels* 146, 11 July 1998). From the first of these localities Maisels collected the seed which enabled Phillips to confirm her hypothesis of the specific distinctness of this species. The most recently discovered site for this species is at Mbingo. Here, at the swamp which must be crossed to reach "Back Valley", a dozen plants were seen on 10 November 1999 (Cheek *pers.obs.*). In summary, seven of the nine sites of *E. bamendae* are at Kilum-Ijim.
Eriocaulon bamendae flowers between April and July, i.e. at the end of the dry season and the beginning of the wet season, setting seed in the wettest months of the year July, and presumably August. At this time its leaf rosettes are up to 40cm diam. By the end of the wet season and the earlier part of the dry season (Oct.-Dec.) there is no evidence of flowering or fruiting and the leaf rosette is only c. 10cm diam.
Habitat: swampy grassland, often at the seasonally inundated margins of streams or ponds, often between tussocks; 2000–2900m alt.
Threats: *Eriocaulon bamendae* is vulnerable above all, to changes of the watertable. Drainage of swamps or conversely, flooding for use as reservoirs would threaten this species with extinction. The species is also vulnerable to trampling by cattle.
Management suggestions: a population census of the known sites should be completed, and the sites monitored for changes in numbers of individuals annually. Information on the demography of this species is deficient. Work is needed to discover the relative importance and levels of recruitment from division of clumps, seed and parthenogenesis from viviparous capitula.
Description: perennial, leaf rosette c. 10cm diam. (when sterile in dry season) to 30–40cm diam. (when fertile in wet season). Leaves strap-shaped, 10–22cm long in fertile specimens, 6–16mm wide, bright green, apex rounded. Scapes up to 15, 30cm high, 1.3–1.5mm diam., 5–7-ribbed. Capitulum 4–7mm wide, globose, black and white, often viviparous; involucral bracts as wide as the capitulum, straw coloured, tinged grey, leathery, 1.7–2.3mm long, oblong with broadly rounded tip, reflexing at maturity. Seeds ellipsoid, 0.7mm long, pale yellow-brown.

Eriocaulon parvulum S.M. Phillips
VU D2
Range: Cameroon, endemic to Mount Oku and the Ijim Ridge.

The discovery of *Eriocaulon parvulum* runs in tandem with that of *Eriocaulon asteroides*. Both are annual species restricted, so far as is known, to basalt pavement. However, *E. parvulum* is much rarer than *E. asteroides*, being known from only three, rather than six sites (see *E. asteroides*). Having studied three sites at which *E. asteroides* occurs, I have noted that *E. parvulum* occurs at only the driest and flattest of these and I conclude that, compared with *E. asteroides*, it may require slightly drier conditions.

The two species are distinguished by the fact that *E. parvulum* (the epithet means small) has slightly smaller capitula than *E. asteroides*, and by the lack of radiating, pointed bracts from the margin of the capitulum which give to the capitulum of *E. asteroides* the resemblance to a little star from which it takes its name.

Habitat: see *Eriocaulon asteroides*.

Threats: see *Eriocaulon asteroides*.

Management suggestions: see *Eriocaulon asteroides*.

Description: annual, leaf-rosette 1.5–3cm diam. Leaves linear subulate, 0.8–1.5cm long, c. 1mm wide, acute. Scapes c. 7, 1.5–3cm high, 0.5cm thick, 4–5-ribbed. Capitulum subglobose, 4–4.5mm wide, dirty white, more or less glabrous, bracts loose and untidy. Seeds ellipsoid, light brown, almost smooth.

GRAMINEAE

Agrostis mannii (Hook.f.) Stapf subsp. *mannii*
LR/nt
Range: Bioko (six collections), Mt Cameroon (15 pre-1988 collections), Bamboutos Mts (1 collection) and Mt Oku (2 pre-1996 collections).

Letouzey 1729 (8 Dec. 1974, Verkovi to summit, 2850m alt.) and CNAD 1729 (Aug. 1970, Prairie, 2800m) are the only two collections known of *Agrostis mannii* from our checklist area. This plant was not relocated during our 1996 or 1998 expeditions. However, this may reflect a lack of collecting of Gramineae at the summit. The only two collections known are clearly from the summit grassland area.

Habitat: montane grassland, sometimes at forest edge; 1850–3800m alt.

Threats: unknown.

Management suggestions: none.

Description: perennial, tufted herb 0.15-1m high. Leaves variable, convoluted, up to 20cm long; ligule hyaline, 2-3mm long. Inflorescence paniculate, lax, open, c. 25cm long, branches flexuose, filiform, to 4cm long; spikelets at branch extremities. Spikelets 3.5-5.5mm long, purple; rhachillae sometimes elongated. Glumes longer than the lemma. Lemma 3-3.5mm long, pilose, carrying 3 setae, two lateral and one median. Palaea as long as the lemma.
Description based on van der Zon (1992).

Hyparrhenia sp. nov. ?
Asonganyi 1327 and 1328 from Elak and *Cheek* 9792 from a swamp on Laikom ridge have been identified as a possible new species of *Hyparrhenia* by grass specialist Dr Cope. Further work is needed to elucidate these collections. If specific status for these collections is confirmed, this taxon may warrant inclusion in a future edition of this Red Data account.

IRIDACEAE

Gladiolus sp. nov.
VU D2
Range: Cameroon, Ijim Ridge and Mbi Crater.

Laikom ridge west of Ardo of Ijim's compound (c. 30 individuals), Ngengal stream at "Mbesa Swamp" (c. 20 individuals seen), Mbi crater (c. 10 individuals seen, but area not searched exhaustively).

Habitat: edge of swamps; 1750–2400m alt.

Threats: trampling and grazing by cattle (sites of this species can be used as watering holes for livestock), changes of watertable (e.g. draining).

Management suggestions: raising the awareness of the Fulani population to the existence and significance of this species may help conserve it. An investigation to assess the effect of cattle on sites of this species is recommended. It is unclear whether they are beneficial or detrimental. Seed is set abundantly (Pollard pers. obs.) and might be used to multiply the number of individuals known.

Description: perennial, cormous herb 0.6–1.2m tall. Spikes with 1–2(–3) flowers. Flowers bright orange, c. 5cm long, petals forming a hood, not splayed.

ORCHIDACEAE

Diaphananthe bueae (Schltr.) Schltr.
EN A1c+2c
Range: endemic to Mt Cameroon (two pre-1988 collections) and Mt Oku, Bamenda Highlands (two collections).
As the specific epithet suggests, this species was first collected at Buea on Mt Cameroon (*Deistel* s.n., type, collected 24 July 1905, "auf der altern Rinde hoher Baumen in gesellscaft anderer Orchideen. Walden in d. Ungebung Buea"). It was rediscovered there over 40 years later (*Gregory* 153). More recently, in the 1970 s, it has been found near Mt Oku (*Letouzey* 8889 and *Mbenkum* 354). However, we were not able to rediscover this species at Mt Oku and the Ijim Ridge during 1996 or 1998. A possible record of this species from Ivory Coast (*Perez-Vera* 725, in fruit November 1974) needs support from flowering material before it is confirmed.
Habitat: submontane forest as an epiphyte; 1000–?1800m alt.
Threats: all localities known are believed to be under pressure for forest clearance for agricultural, firewood collection and (Buea, Mt Cameroon) urban expansion.
Management suggestions: more data on the distribution and frequency of this species at Kilum-Ijim is needed.
Description: epiphytic herb; leaves distichous, ligulate-oblong, 4–15 x 0.8–2cm long, acute; inflorescence 6–15cm long, 6–10-flowered. Flowers white and green. Sepals 6–8.5 x 2.7–2.8mm. Petals oblong, 5–5.6 x 1.3–1.6mm. Labellum narrowly or lanceolate ovate, 4–8.7 x 3.3–4.4mm, apex acute, tooth longitudinally placed on the central vein; spur 12–15mm long, apex swollen, emarginate. Stipites fused to apex, margins fimbriate; viscidium 0.5mm long.

Disperis nitida Summerh.
EN B1+2b
Range: Cameroon Highlands from Manenguba (one collection) to the Bamenda Highlands (3 pre-1996 collections: Mt Neshele, Lake Bambuluwe and Lake Oku).
This species flowers during the second half of the wet season (Aug.-Sept.), fruits and then, by October, starts to die back to the resting tubercle. Plants occur as isolated individuals, but are also frequently gregarious. During the dry season this species is below ground and difficult to locate.
The earliest collection from Kilum-Ijim of this species, which is also the type, is *Savory* in UCI 451 from Lake Oku. This species seems relatively abundant in the upper part of the montane forest, near the grassland, above Oku-Elak (Kilum), permanent transect KA (Cheek *pers. obs.* Oct. 1996).
Habitat: montane forest, on lower branches or leaning trunks of trees in densely canopied areas, rarely terrestrial; 1800–2800m alt.
Threats: forest clearance for agriculture and, possibly, grazing pressure from cattle and goats.
Management suggestions: more surveys are needed of this species at Kilum-Ijim. This species seems relatively secure so long as closed canopy upper montane forest persists. However, it may be vulnerable to the current practice of dry season grazing by livestock of forest herbs (practice reported by Tame and Asonganyi, 1995) and whole tree felling for firewood (seen extensively by us in forest above Oku-Elak in November 1996).
Description: terrestrial or low epiphytic herb 10–15(–50)cm tall, glabrous. Tubercle single, ovoid to ellipsoid, 15 x 6mm. Leaves 2, opposite, situated about the middle of the stem, petiole 5mm long, ovate 1.5–7 x 1.2–4.5mm, acute, base cordate, dark green with surface like satin.

Inflorescence lax, 2–2.5cm long, 2–(3-)flowered. Bracts elliptic, apex acute, leaf-like, 13mm long. Flowers white. Hood conic-saccate, 6mm deep, finely three-toothed. Dorsal sepal linear-lanceolate, 1–2.5 x 1.8mm, acute. Petals B-shaped, 12.5 x 4mm, the lobes triangular, subacute. Lateral sepals, semi-ovate, 11 x 6mm, acute to acuminate, canaliculate at the base, with little conical pockets 1mm deep near the margin. Labellum clawed, claw linear, 7mm long, curved in an S-shape, limb 1mm long, reflexed, linear, acute, papillose at the base with two weakly bilobed, papillate appendages 2 x 2mm.

Disperis parvifolia Schltr.
EN B1+2a-e
Range: Malawi (two collections), Tanzania (one collection) and Bamenda Highlands, Cameroon (one collection).
This diminutive terrestrial orchid is only known from a single collection west of the Congo basin, at Ijim between Mbesa and Nchain (*Maitland* s.n., "Basenako to Nchan, 5000', 1931"). It has not been seen in Cameroon for 69 years and may now be extinct there. Maitland did not record the habitat, so the notes below are taken from a Malawian specimen. The reduced leaves suggest that *D. parvifolia* may be a saprophyte, in which case individuals may remain underground for several years before flowering, making locating the species particularly difficult.
Habitat: grassy banks near river crossing; c. 1660m alt.
Threats: unknown, but potentially, trampling by cattle.
Management suggestions: this species, if it is not extinct, should be refound and protected.
Description: herb 4–11cm tall. Tubercle globular, hairy, 0.7–1cm diam. Leaves 1–2, oval, circular or elliptic, obtuse to shortly acuminate, base rounded, inserted at the base of the stem, dark green. Inflorescence 1-flowered. Bracts leaf-like, enclosing the lower part of the ovary, 6–8mm long, shortly acuminate, Flower pink, orange or green, apex of the spur red. Ovary 10mm long. Dorsal sepal 6–6.5mm long, spur conical. Petals obliquely oblong, 3.5–4 x 2mm, a little lobed on the free margin, united to the dorsal sepal at the throat of the spur. Lateral sepals free, obliquely elliptic to rhombic, concave, 2.5–4.5 x 1.3–2.2mm,with barely visible pockets. Lip 4–4.5mm, basal claw linear, limb short, reflexed, c. 0.5mm long, appendage situated on the blade at the point of reflexing, 1.5mm long, linear, enlarged and hairy at the apex.
Description after Szlachetko & Olszewski (1988).

Genyorchis macrantha Summerh.
VU D2
Range: Mt Cameroon (two collections) and Mt Oku and the Ijim Ridge.
A specimen believed to be this species was collected near Mbesa in March 2000 (*DeMarco* 56). If confirmed, this will be the first and only record of *Genyorchis macrantha* away from Mt Cameroon. First discovered in 1948 at Mann's Spring on Mt Cameroon, it was not rediscovered until March 1988 on the NE slope of that mountain. "It has the largest flowers so far recorded in the genus" (Summerhayes 1957). The fact that, despite its relatively large flowers, it has so rarely been collected suggests that it is extremely scarce and thus especially Endangered.
Habitat: epiphytic on tree trunks; c. 1800m alt.
Threats: see *P. bicalcarata*.
Management suggestions: see *P. bicalcarata*.
Description: epiphyte to 7cm high, rhizome creeping and branching, slender, c. 1mm diam. Pseudobulbs 0.5–2cm apart, erect, ovoid, ellipsoid or conical-ovoid, obtusely 5-angular, 1–2 x 0.4–1cm, green, 2-leaved, slightly flattened. Leaves divergent, rigid, shining, oblong-ligulate, 1–3 x 0.3–0.7cm, apex rounded, slightly notched. Inflorescence from the base of the pseudo-bulbs, erect, racemose, to 7-flowered; peduncle to 3.5cm long. Flowers patent, not resupinate, with sepals white or pale pink; labellum pink with yellow apex; ovary sessile, c. 1mm long. Dorsal sepal incurved, oblong-lanceolate, 4.5 x 2.5mm, acute, lateral sepals obliquely triangular-ovate, 5.2 x 4mm. Petals minute, transversely oblong, 0.5 x 1mm, obtuse apiculate. Labellum oblong-elliptic, U-shaped in section, incurved, 5mm long, apex trilobed, median lobe strongly recurved, lateral lobes short and round.
Description after Summerhayes (1957).

Habenaria maitlandii Summerh.

CR A1c, B1+2c

Range: known only from the type collection at Ijim (*Maitland* 1386) at Nchian, Ijim Ridge, Cameroon.

This species has not been seen since it was first discovered, in 1931. The fact that it has not been recollected in the last 69 years suggests that it may be highly localised geographically, and possibly extinct.

Habitat: rocky grassland; 1860m alt.

Threats: unknown, but probably habitat destruction for agriculture.

Management suggestions: Nchain should be revisited and this species relocated, if not extinct, so that it may be protected.

Description: perennial herb c. 30cm tall, probably with a single tubercle. Leaves 3–6, the basal ones narrowly lanceolate, 7–10 x 0.1–0.14cm, apex acute; upper 1–3 leaves bract-like, lanceolate, acute, appressed to the stem. Inflorescence c. 7cm long, with several flowers. Flowers small, white or green, resupinate; floral bracts ovate-lanceolate, 14–15mm long, acuminate; pedicel-ovary 16–20mm, twisted, glabrous; ovary slightly swollen, slightly reflexed. Dorsal sepal 4.5–5.5 x 2–3mm, triangular-oval, acute, hooded, glabrous. Petals divided to the base, both parts papillose; anterior lobe lanceolate-elliptic 6.8–8.5 x 1–1.5mm, acute. Lateral sepals 6.5–7.5 x 4.5mm, obliquely obovate, concave, with a lateral apicule. Labellum tripartite, basal part 1.5–2mm long, glabrous; median lobe linear, 9–10 x 0.5mm, acute; lateral lobes filiform, 6–8mm long, acute. Spur 10–11mm, reflexed in the upper part, pendulous, auricles shortly stipitate.

Description based on that of Summerhayes ms, K.

Habenaria microceras Hook.f.

LR/nt

Range: Bioko (one collection), Mt Cameroon (five collections) and the Bamenda Highlands (two pre-1996 collections).

Discovered on Mt Oku in the last few days of October 1996, by Zapfack and Munyenyembe who made four collections above Oku-Elak in the summit grassland and at the forest edge. Apparently not known elsewhere at Kilum-Ijim.

The three collections on Mt Cameroon are *Mann* 2116 (type), *Johnston* 31, 32 ("8,000 ft. 12/86"), and *Preuss* 967, ("Buea 2500m, 24/9/91").The two collections in the Bamenda Highlands are *Savory* UCI 462, Bambuluwe and *Letouzey* 14271, Mt Nseshele. Rediscovered in October 1992 on Mt. Cameroon, when two collections were made near Bokwangwo Hut 3: *Thomas* 9381 at 2400m alt. and *Cheek & Tchouto* 3641, the latter at a study of the forest: grassland ecotone at about the same altitude.

Habitat: montane grassland, sometimes at the ecotone with forest (rarely epiphytic); 2000–3050m alt.

Threats: unknown, but grazing and trampling by cattle, sheep and goats are a cause for concern in the summit area.

Management suggestions: a survey to estimate the range, density and threats (if any) to the population at Kilum-Ijim seems worthwhile.

Description: Terrestrial, rarely epiphytic herb 30–75cm tall, tubers 1–2, 1–3 x 1cm. Leaves cauline, 7–8, sessile, lanceolate to oblanceolate, 4–11 x 1.4–4cm, acute. Inflorescence 5–13cm long, bearing 20–30 flowers arranged spirally; bracts oval, 8–9mm long, acuminate; pedicel-ovary 6–7mm long, twisted, cylindrical. Flowers resupinate, minute. Dorsal sepal elliptic to obovate, 2–2.8 x 1–1.5mm, obtuse. Petals obliquely elliptic, 2–2.6 x 0.7–1.2mm, obtuse. Lateral sepals obliquely oblong-ovate, 2–2.6 x 1–1.5mm, obtuse. Labellum strongly trilobed, 1.5 x 4mm, central lobe broadly oval, 1.1–1.5 x 1–1.4mm, with a central ridge, lateral lobes almost as long as the median lobe, 2 x 0.6–0.8mm, oblong-triangular, subacute, gradually diverging. Spur cylindric-swollen, 0.7–1.5mm long, gradually falcate, obtuse.

Description based on that of Summerhayes ms, K.

Habenaria obovata Summerh.

VU D2

Range: Mt Cameroon (seven collections), Mt Oku.

Dr Zapfack appears to be the first to have collected this species away from Mt Cameroon (*Zapfack* 1140, 31 October, 1996, summit of Mt Oku). Until 1992, *Habenaria obovata* was known only from five historical collections: *Johnston* 29 (type, "comm. 12/86"), *Mary Kingsley* s.n. (c. 1890 "comm. iii.96"), *Preuss* 980 (24 November 1891), *Maitland* 804 ("very frequent" December 1929) and *Boughey* s.n. (September 1954). It was recently rediscovered (*Sidwell* 39) in grassland at 2820m alt. in early October 1992 above Mann's Spring "reasonably common in grassland, seen down to around 2400m alt." The most recent collection there (*Banks* 104) was made in early October 1995 between Huts 2 & 3 above Buea.

"A remarkable species without any near relative known to me." (Summerhayes, in the protologue of *Habenaria obovata*). Curiously *Habenaria obovata* has been omitted from the Flore Du Cameroun treatment (Szlachetko & Olszewski, 1988) that covers the terrestrial orchid genera.

Habitat: terrestrial in montane grassland; 2150–3050m alt.

Threats: fire, grazing and trampling are potential threats to this species in its habitat at the summit of Mt Oku.

Management suggestions: a survey to estimate the size of the population on the summit of Mt Oku would be useful. An assessment of the significance of the threats mentioned, and annual monitoring of the population is advisable.

Description: terrestrial herb 15–45cm tall; leaves 3–4, towards the stem base, lanceolate or oblong-lanceolate, to 9 x 1.7cm, acute, base vaginate, upper part of stem with bract-like leaves. Raceme fairly dense, narrowly cylindrical, 5–11 x 2cm; bracts lanceolate, acuminate, 7–17mm long. Flowers ascending, green. Dorsal sepals oblong-ovate, subacute, 5 x 2.5–3.5mm; lateral sepal obliquely semi-ovate, acute or obtuse, acuminate, 5–6 x 2.5–3mm, reflexed. Petals obliquely obovate, 3.5–5 x 3–4mm, obtuse, anterior margin dilated, distinctly nerved. Labellum base 2.5–3mm long, tripartite, central part linear, 5–6mm long, 0.8–1mm wide, lateral part linear, more or less falcate, recurved, 5.5–6.5 x 0.6–0.8mm, spur dependent, apex strongly inflated on one side, 10–13mm long.

Description based on that of Summerhayes ms, K.

Polystachya bicalcarata Kraenzl.

EN A1c+2c

Range: Bioko (one doubtful collection), Mt Cameroon (five pre-1988 collections), Bamboutos Mts (one collection) and Mt Oku.

This epiphyte had not been recorded from the Bamenda Highlands (see range above) until Dr Zapfack obtained a specimen from near the Project Guest House at Oku-Elak (*Zapfack* 1187, 4 November 1996). Of the other six Cameroonian collections, only one (*Jacques-Félix* 5439, Bamboutos Mts), is not from Mt Cameroon. The five collections known on Mt Cameroon are: *Deistel* 62 C (type), 79; *Maitland* 730, *Dundas* in FHI 15303 and *Sanford* 488/65, all with Buea given as locality, where known. *Thomas* 9178 (1992) is the only recent collection of the species from that mountain.

Habitat: submontane and montane forest, epiphytic; 950–2000m alt.

Threats: forest clearance for agriculture and firewood.

Management suggestions: a survey is needed to find the range and frequency of *P. bicalcarata* at Mt Oku and the Ijim Ridge.

Description: epiphytic herb, pseudobulbs sessile, articulated in the middle, 50–70 x 0.5–1.5mm. Leaves single, linear-ligulate, 6–15 x 0.3cm. Racemes few-flowered, capitate. Calyx purple, petals white with median purple stripe, scentless. Dorsal sepal 3.5mm long; lateral sepals free, not forming a mentum, 7mm long; spur 2mm long. Labellum widened from a wedge-shaped base, 3-lobed; lateral lobes linear, rounded; middle lobes cuneate, 4.5 x 3mm, retuse.

Description from material at K.

Polystachya superposita Rchb.f.

EN A1c+2c

Range: Bioko (one collection), Mt Cameroon (two collections) and Mt Oku.

First recorded in the Bamenda Highlands from the collections of Dr Zapfack at Kilum (*Zapfack* 824, 9 June 1996, Elak and 1169, 3 November 1996, Kwifon forest). *Polystachya superposita* was first discovered on Mt Cameroon (*Mann* 2125, type, "Nov. 1862, 5,000 ft.").

Habitat: submontane and montane forest, epiphytic; c. 900–2000m. alt.

Threats: see *P. bicalcarata*.

Management suggestions: see *P. bicalcarata*.

Description: epiphyte, stems erect, slender, 7–15cm long, articulated; pseudobulbs 4-angular, red, 1.5 x 0.7cm. Leaves 4–8(–12) x 0.6–0.7cm, ligulate, apex acute, base cuneate. Inflorescence branched, 2–3(–7)cm long, 5–89–40) flowered. Flowers red, dorsal sepal 1.9 x 1.3mm; lateral sepal 2.1 x 2.6, mentum 2mm wide. Petals 1.5 x 0.7mm. Labellum trilobed, c. 2 x 2mm, central lobe broadly elliptic, 1.25 x 1.2mm, lateral lobes ovate, subfalcate.

Description from material at K.

XYRIDACEAE

Xyris cf. filiformis Lam., *Xyris "welwitschii"* sensu Lewis non Rendle and *Xyris sp. A*

The above three taxa (see main checklist) probably all represent new taxa and, when worked out by the specialist Dr Lock, are likely to appear in a new edition of this Red Data chapter.

ZINGIBERACEAE

This family, particularly *Aframomum*, the most species-rich genus, is being actively revised in Africa by several workers. Dr Harris (Edinburgh) is dealing with Cameroon species. Until he has concluded his work it is not possible to apply species names or to make a conservation assessment for the two (possibly three) taxa of this genus present in our area, although they may well be threatened.

PALMAE

Raphia mambillensis Otedoh

LR/cd

Range: western Cameroon mountains from Bamboutos Mts to Adamoua (10 collections), extending into the Mambilla range of Nigeria (one collection), and through C.A.R. (two collections) to Sudan (one collection).

This species, possibly both the smallest and the most cold-tolerant (high altitudinal range) in the genus, was established in 1982 (Otedoh 1982: 163). *Raphia mambillensis* has a subterranean or prostrate stem, suckers freely, and is only about 7m tall. Fortunately it is so useful to man, being tapped for Palm wine, that populations are carefully maintained along streams in the Kilum and Ijim areas.

Habitat: along water-course, 1250–1600m alt.

Threats: over-harvesting is a potential threat for this species.

Management suggestions: mapping the localities of this species in the Kilum-Ijim area, and recording numbers of plants in one or two sites would provide baseline data for future monitoring.

Description: "trunkless" palm, stem prostrate, subterranean, suckering freely. Fronds erect, 5–8m high; petioles 1.5–2m long, 5cm in diam., terete to triangular in cross-section; leaf-sheath with bract-like external, marginal fibres 20 x 3–4cm on each side at ground level, the fibres leathery, red brown. Leaflets 60–90 on each side, light green to yellowish brown, alternate or in 4 indistinct ranks with relatively few, weak, brown spines on the upper surface of the midveins. Main inflorescence axis up to 1.5m long, 1.2cm diam., slightly curved, primary branches 3–5, 25–120cm long, pendulous, largely wrapped in sheathing second order bracts, tertiary branches short, crowded with flowers in 2 ranks. Male flower with calyx 1/3 as long as corolla. Fruit

ellipsoid, 3.5–7.5 x 2.5–3.5cm, with a sharp apical beak 3–4mm long; scales slightly furrowed, in 10–11 vertical rows.
Description after Otedoh (1982).

FERN ALLIES

Isoetes biafrana Alston
VU D2
Range: Bioko (Lake Moka) and Cameroon, Lake Oku.
Known only from the two montane crater-lakes above. Two collections at YA from low altitude seasonal pools in the north of Cameroon have been mistakenly labelled as this species. *Isoetes biafrana* could become vulnerable if the factors mentioned below become more extensive. Several hundred plants were recorded below the Baptist Rest House (N side of lake) in November 1996 (Cheek et al. 1997). The largest plants are always in the deepest water. The rosettes do not form a thick mat, but are scattered, the leaf tips of adjoining rosettes rarely touching.
Habitat: margin of lake, locally common where slowly shelving shallow gravel bottom, not seen in mud bottom areas; 2,200m alt.
Threats: Increased goat grazing pressure or forest clearance in the Lake's crater could increase surface erosion, resulting in more silt entering the lake edge and covering the gravel bottom which seems to be the only habitat of *Isoetes biafrana*.
Management suggestions: aforementioned activities should not be expanded and preferably should be stopped. A survey of this species could be made using an inflatable boat to travel around the perimeter of the lake.
Description: submerged (45–150cm deep in November) aquatic, stemless and bottom rooting. Leaf rosettes 7–20cm diam. and tall of c. 15–20 dark green linear leaves, the outermost each with a swollen white base containing numerous black megaspores (reproductive bodies). Fertile in November.

BIBLIOGRAPHY

Anon. (1955). Thomas Douglas Maitland. J. Kew Guild. 7: 176-177.

Adams, C. D.(1957). Compositae. J. West Afr. Sci. Assoc. 3(1): 116.

Adjanohoun, J.E. et al. (1996). Traditional medicine and pharmacopoeia, contribution to ethnobotanical and floristic studies in Cameroon. Organisation of African Unity. 641 pp.

Asonganyi, J. N. (1995). A report on the vegetation survey of Ijim mountain forest carried out by Mr Simon Tame and J. N. Asonganyi. Herbier National Yaoundé. 10 pp. Cyclostyled.

Barthlott, W., Lauer, W. & Placke, A. (1996). Global distribution of species diversity in vascular plants: towards a world map of phytodiversity. Erkunde Band 50: 317–328 (with supplement and figure).

Brummitt, R.K. (1992). Vascular Plant Families & Genera. Royal Botanic Gardens, Kew.

Brummitt, R.K., & Powell, C.E. (1992). Authors of plant names. Royal Botanic Gardens, Kew.

Cable, S. & Cheek, M. (1998). The Plants of Mount Cameroon, a Conservation Checklist. Royal Botanic Gardens, Kew.

Cheek, M. (1999). R.B.G., Kew-Herbier National Camerounais (HNC) expedition to Ijim 2 Nov.-15 Nov. 1999. 3 pp. Cyclostyled. (to which is appended two extra pp: The plants of Back Valley, Mbingo, N.W.P. Cameroon and the forest below Ajung Cliff.

Cheek, M. & Cable, S. (1997). Plant inventory for conservation management: the Kew-Earthwatch programme in Western Cameroon, 1993–96. pp. 29–38 in Doolan, S. (ed.) African Rainforest and the Conservation of Biodiversity. Proceedings of the Limbe Conference. Earthwatch Europe, Oxford. pp 170.

Cheek, M. & Csiba, L.(in press).A new species and new combination in *Chassalia (Rubiaceae)* of western Cameroon. Kew Bull.

Cheek, M., Satabie, B. & J.-M. Onana (1997). Interim report on botanical survey and inventory for Kilum and Ijim Mountain Forest Projects by the National Herbarium Cameroon and R.B.G., Kew Oct./Nov. 96. Cyclostyled. 16 pp.

Cheek, M. & Sonké, B. (in press). A new species of *Oxyanthus (Rubiaceae-Gardeniinae)* from western Cameroon. Kew Bull.

Collar, N.J. & Stuart, S.N. (1988). Mount Oku (Cameroon) pp. 29–30 in Key Forests for Threatened Birds in Africa. ICBP Monograph No. 3. IUCN.

Cook, F.M. (1995). Economic Botany Data Collection Standard. R.B.G., Kew. 146 pp.

Engler, A. (1925). In Engler, A. & Drude, O. Die Vegetation der Erde 9, Die Pflanzenwelt Afrikas 5, 1. Engelmann, Leipzig. 341 pp.

Harvey, Y.B. (1996). The *Stachys aculeolata/aethiopica* complex in Tropical Africa. Kew Bull. 51: 433–454.

Hawkins, P. & Brunt, M. (1965). The Soils and Ecology of West Cameroon. 2 vols. FAO, Rome. 516 pp, numerous plates and maps.

Hepper, F.N. (1956). New taxa of *Papilionaceae* from West Tropical Africa. Kew Bull. 11: 113–134.

Hepper, F.N. (1958). Diary. Unpublished mss.

Hepper, F.N. & Neate, F. (1971). Plant collectors in West Africa. Regnum Veg. 74. Utrecht. pp. 96.

Holmgren, P.K., Holmgren, N.H. & Barnett, L.C. (1990). Index Herbariorum. Eighth ed. New York Botanical Garden. 693 pp.

IUCN (1994). IUCN Red List Categories. IUCN, Gland, Switzerland.

Jacques-Félix, H. (1970). Apiaceae. Flore Du Cameroun.

Keay, R.W.J. (1994). Visit to Cameroon in 1951. Unpublished notes and photos.

Keay, R. W. J. & Hepper, F. N., eds. (1954–1972). Flora of West Tropical Africa, 2nd ed., 3 vols. Crown Agents, London.

Lebrun, J-P. & Stork, A.L. (1991–1997). Enumeration des Plantes a Fleurs d'Afrique Tropicale. 4 vols. Ville de Geneve, Geneva.

Ledermann, C. (1912). Eine botanische Wanderung nach Deutsch-Adamaua, Mitt. Deutsch Schutzgebieten 1: 20–55, map 3.

Letouzey, R. (1968). Les Botanistes au Cameroun. Flore du Cameroun 7.

------------- (1977). Présence de *Ternstroemia polypetala* (*Theaceae*) dans les montagnes Camerounaises. Adansonia ser. 2 17(1): 5–10.

------------- (1978). *Rosaceae* pp. 195–244 in Flore Du Cameroun 20.

-------------(1885). Notice de la carte phytogéographique du Cameroun au 1: 500 000. IRA, Yaoundé, Cameroon.

Lightbody, J.S. (1952). The mountain grassland forests of Bamenda (Cameroons U.B.T.): some factors influencing their structure and composition. 42 pp. Unpubl. Cyclostyled.

Mabberley, D.J. (1998). The Plant-Book. 2nd ed. CUP. 858 pp.

Mackay, C. R. (1994). Survey of Important Bird Areas for Bannerman's Turaco *Tauraco bannermani* and Banded Wattle-eye *Platysteira laticincta* in North West Cameroon, 1994. Interim report. BirdLife Secretariat.

Maisels, F. & Forboseh, P. (1997). Vegetation Survey. Kilum-Ijim Forest Project. 51 pp. Cyclostyled.

Maisels, F.M., Cheek, M. & Wild, C. (2000). Rare plants on Mt Oku summit, Cameroon. Oryx 34 (2): 136–140.

Maley, J. (1997). Middle to late Holocene changes in Tropical Africa and other continents: Palaeomonsoon and sea surface temperature variations. Palaeoenvironnments & Palynologie (CNRS/ISEM & ORSTOM). NATO ASI series, Vol. 149: 611–640

Maley, J. & Brenac, P. (1998). Vegetation dynamics, palaeoenvironments and climatic changes in the forests of western Cameroon during the last 28,000 years B.P. Review of Paleobotany and Palynology. 99: 157–187.

Marais, W. (1973). A revision of the tropical species of *Kniphofia* (*Liliaceae*). Kew Bull. 28: 465–483.

Mbenkum, T.F. & Fisey, C.F. (1992). Ethnobotanical survey of Kilum mountain forest. WWF.

McLeod, H. L. (1986). The Conservation of Oku Mountain Forest, Cameroon. Study Report No. 15. 90 pp. ICBP.

Morton, J.K. (1961). The upland flora of West Africa - their composition, distribution and significance in relation to climate changes. Extrait des comptes rendus de la IVe réunion plénière de l'AETFAT (Lisbonne et Coïmbre, 1–23 Sept. 1960).

Morton, J.K. (1972). Phytogeography of the West Africa Mountains. in Valentine, D.H. (ed.). Taxonomy, Phytogeography and Evolution. Academic Press, London. pp 221–236.

Otedoh, M.O. (1982). A revision of the genus *Raphia* Beauv. (Palmae). J. Nigerian Inst. For Oil Palm Research. 6 (22): 145–189.

Phillips, S.M. (1998). Two new species of *Eriocaulon* from West Africa. Kew Bull. 53: 943–948.

---------------(2000). Two more new species of *Eriocaulon* from West Africa. Kew Bull. 55: 195–202.

Polhill, R. M. (1982). *Crotalaria* in Africa and Madagascar. Balkema, Rotterdam. 389 pp.

Polhill, R. & Wiens, D. (1998). Mistletoes of Africa. R.B.G., Kew. 370 pp.

Puff, C. (1986). A Biosystematic Study of the African and Madagascan Rubiaceae-Anthospermeae. Springer-Verlag, Wien. 535 pp.

Purseglove, J.W. (1968). Tropical Crops. Dicotyledons. Longman. 719 pp.

Purseglove, J.W. (1972). Tropical Crops. Monocotyledons. Longman. 607 pp.

Schrire, B. & Onana, J.-M. (2000). A new subspecies of *Indigofera patula* Baker (*Leguminosae-Papilionoideae*) and a new record for the species in West Africa. Kew Bull. 55: 219–223.

Seyani, J.H. (1982). A Taxonomic Study of *Dombeya* Cav. (Sterculiaceae) in Africa. D.Phil. thesis, Univ. Oxford.

Sosef, M.S.M. (1994). Refuge Begonias. PhD thesis, Wageningen.

Stannard, B. L. (1995). Flora of the Pico das Almas. Royal Botanic Gardens, Kew. 853 pp.

Szlachetko, L. & Olszewski, S. (1998). Orchidaceae. Vol. 1. in Flore Du Cameroun. 34.

Summerhayes, V.S. (1957). African Orchids XXIV. Kew Bull. 11: 107–126.

Tame, S. & Asonganyi, J. (1995). Ijim Mountain Forest Project Plant List. Appendix 2. 16 pp.

Tame, S.P. & Thomas, D.W. (1993). Kilum mountain forest project plant list. KMFP

Thomas, D.W. (1986). Provisional Species List for Mount Oku Flora, pp. 59–62 in McLeod, H. L. (1986). The Conservation of Oku Mountain Forest, Cameroon. Study Report No. 15. 90 pp. ICBP.

Thomas, D.W. (1986). Vegetation in the Montane Forests of Cameroon. pp. 20–27 in Stuart, S.N. (Ed.) Conservation of Cameroon Montane Forests. International Council of Bird Preservation. Cambridge, 263 pp.

Verdcourt, B.V. (1950). A revision of the genus *Otiophora* Zucc. (Rubiaceae). Bot. J. Linn. Soc. 53: 383–412.

------------------ (1962). Theaceae in Flora of Tropical East Africa.

Villiers, J.F. (1990). Contribution à l'étude du genre *Newtonia* Baillon (*Leguminosae-Mimosoideae)* en Afrique. Bull. Jard. Bot. Nat. Belg. 60: 123–124, fig.1.

White, F. (1983). Long-Distance Dispersal and the Origins of the Afromontane Flora. Sonderbd. Naturwiss. Ver. Hamburg. 7: 87–116.

Yana Njabo, K. & Languy, M. (2000). Surveys of selected montane and submontane areas of the Bamenda Highlands in March 2000. Cameroon Ornithological Club, Yaoundé. Cyclostyled.

Van der Zon, A.P.M. (1992). Graminées du Cameroun. Vol. 2. Wageningen Agricultural University papers 92-1. Wageningen.

READ THIS FIRST!

Before using this checklist, the following explanatory notes to the conventions and format used should be read.

All scientific names in the checklist (not those in the introductory chapters) are indexed at the end of the book. Names of taxa are organized alphabetically: species within genus, genus within family, and family within the groups Dicotyledons, Monocotyledons, Gymnosperms, Fern allies and Ferns, respectively. The families and genera accepted here follow Vascular Plant Families and Genera (Brummitt 1992) except in Cyperaceae, where we follow Lye in sinking *Pycreus* Pal. and *Kyllinga* Rottb. into *Cyperus* L.

The species names adopted follow the most recently published research that is available and acceptable: The Énumération des Plantes à Fleurs D'Afrique Tropicale (Lebrun & Stork 1991-1997). In many cases the specimens cited, after preliminary naming in Cameroon (see acknowledgements) had their names confirmed or determined by a family specialist. This person (or persons) is credited at the head of the family account. The remaining families were named by us, the authors of this book. Uncredited families were named by MC. Below the accepted scientific name we give the local name, if known and indicate if the plant is Red Data listed or a Red Data candidate. There follows a short description, usually drawn from information given in Flora of West Tropical Africa, but completely new in the case of Ferns, Compositae and Labiatae. Following the plant description is a brief description of the habitat. Altitudinal range can be determined by reference to the specimens cited.

Most of the species referred to in this checklist can be found in F.W.T.A. (The Flora of West Tropical Africa, Keay & Hepper 1954-1972). However, in some instances names used in F.W.T.A. have been superseded and are thus synonyms. If this is the case the name as in F.W.T.A. is listed *below* the description. Lastly, the specimens (to which the name used refers) are cited chronologically, under the following system. Firstly, specimens are cited from the Kilum area, that is those near the villages E and SE of Mt Oku, from Jikijem in the N to Verkovi in the S. Secondly, specimens are cited from Ijim, that is, those from Ajung, Nchain and Mbesa in the NE to Fundong, Njinikom and Belo in the W. Finally, specimens collected outside the central Kilum-Ijim area but inside the checklist area are cited starting with locations in the NW corner, moving anticlockwise around the peripheral road that marks the boundary. All specimens cited are located at the herbaria of the Royal Botanic Gardens, Kew, U.K. or the National Herbarium of Cameroon. Those specimens gathered during the 1996-1999 inventory work on which this book is based have the top set deposited in Cameroon, the first duplicate at Kew and further duplicates in the process of being distributed to other herbaria. The latitude and longitude given for the specimens cited are usually based on GPS readings. 6.1112N, 10.1952E is equivalent to: 6°11'12" North, 10°19'52" East. Readings such as 6.15N, 10.26E, represent a best estimate of position. Specimens cited e.g. as OKU 84 are plot vouchers collected whilst conducting 25m x 25m vegetation plots.

VALIDATION OF NEW NAMES

A NEW PENTARRHINUM (ASCLEPIADACEAE) FROM CAMEROON
David Goyder

The following new subspecies of *Pentarrhinum abyssinicum* is known only from three sites in lower montane forest (c. 1700–1900 m alt.) on the western side of the Ijim Ridge in the Bamenda Highlands of western Cameroon, from which the subspecific epithet derives. This vegetation type is highly threatened by clearance for agriculture, posing a grave threat to the survival of this taxon (see Red Data chapter).

Pentarrhinum abyssinicum *Decne*. subsp. **ijimense** *Goyder* **subsp. nov.** a subsp. *abyssinico* atquea subsp. *angolensi* gynostegio conspicue stipitato nec subsessili, lobis coronae erectis nec patentibus, lobis etiam brevioribus apicem tantum gynostegii attingens, et dente intro directo minus conspicuo differt. Typus: Cameroon, N.W. Province, Boyo Division, Ijim, forest at Ntum, 6°10.87'N, 10°24.67'E. Alt. 1810 m., fl., fr. 4 November 1999, *Cheek* 9943 (holotypus K; isotypi BR, EA, MO, MSTR, P, PRE, WAG, YA).

The predominant form of *P. abyssinicum* Decne. in Cameroon is subsp. *angolense* (N.E. Br.) Liede & Nicholas. The subspecies described here has smaller flowers with less ciliate corolla lobe margins and the gynostegium is more strongly stipitate. The corona lobes are of a similar form but the prominent inward-pointing tooth characteristic of the species is poorly developed. The corona is also smaller in size, reaching only to the top of the gynostegium, and the lobes are held erect rather than spreading.

Known only from Mt Oku and the Ijim Ridge in the Bamenda Highlands of Cameroon. See the checklist at the end of this book for a full listing of the collections, and the Red Data chapter for a discussion of the discovery, the habitat and the conservation status of the taxon. The author is grateful to Martin Etuge for the first collection of this species and to Melanie Thomas for the Latin diagnosis.

A NEW BAFUTIA (COMPOSITAE) FROM MOUNT OKU, CAMEROON
Henk Beentje & Benedict John Pollard

Bafutia C.D. Adams has remained a monotypic genus since its discovery nearly 50 years ago. It is restricted to the Bamenda Highlands of western Cameroon. In the course of identifying specimens of Compositae from botanical inventory work on Mt Oku and the Ijim Ridge (see elsewhere in this book for details), the authors realised that, amongst the collections of *Bafutia*, more than one entity was involved. Accordingly, a new variety is described below. The varietal epithet commemorates Dr Louis Zapfack of the University of Yaoundé I who collected the type material of this plant. Dr Zapfack is a noted specialist of epiphytic plants, especially those of the Orchidaceae. However, he is also a general collector of plants, and in leading an Earthwatch botanical collecting team at Mt Oku in 1996 collected many specimens from which several new taxa have been described.

Bafutia tenuicaulis *C.D. Adams* var. **zapfackiana** *Beentje & B.J. Pollard* **var. nov.** a varietate typica caule foliisque et praesertim pedicello involucroque piloso vel pubescenti neque glabro differt; fundus involucri multo maior (2.5–6.5mm) est, lobi involucri 8–14 sunt (8 in varietate typica), zonis findentibus sub sinu pallidioribus (azonus in v.t.) atque setae pappi 0.5–0.8mm longae sunt (0.1–0.5 in v.t.). Typus: Cameroon, N.W. Province, near the summit of Mt Oku, 2800 m., fl. fr., 31 October 1996, *Zapfack* 1130 (holotypus K, isotypi WAG, YA).

An English description of this taxon is given in the main checklist part of the book. Notes on the conservation status are given in the Red Data chapter. The authors are grateful to Alan Radcliffe-Smith for translating the diagnosis into Latin.

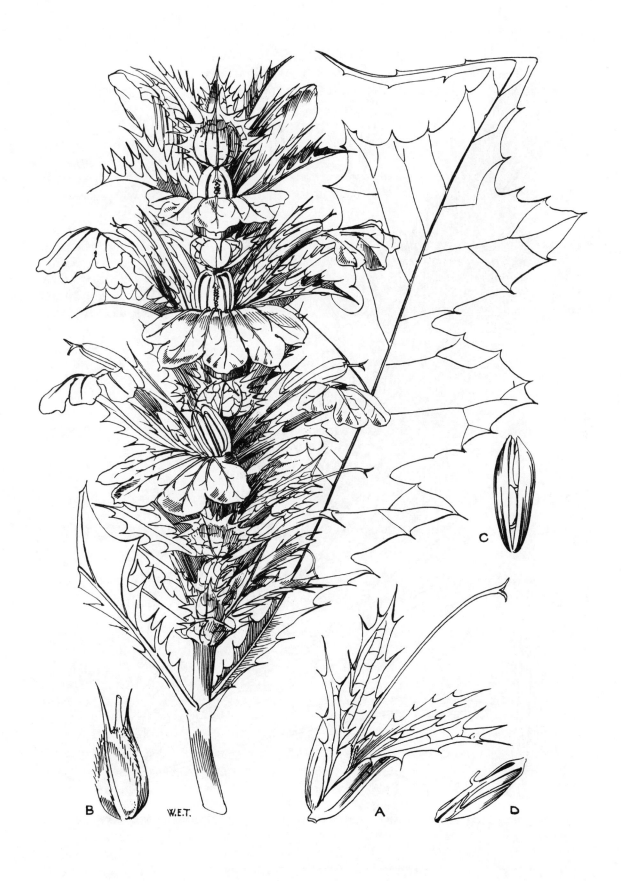

Fig. 10. *Acanthus montanus* (Acanthaceae)
Drawn by W.E. Trevithick

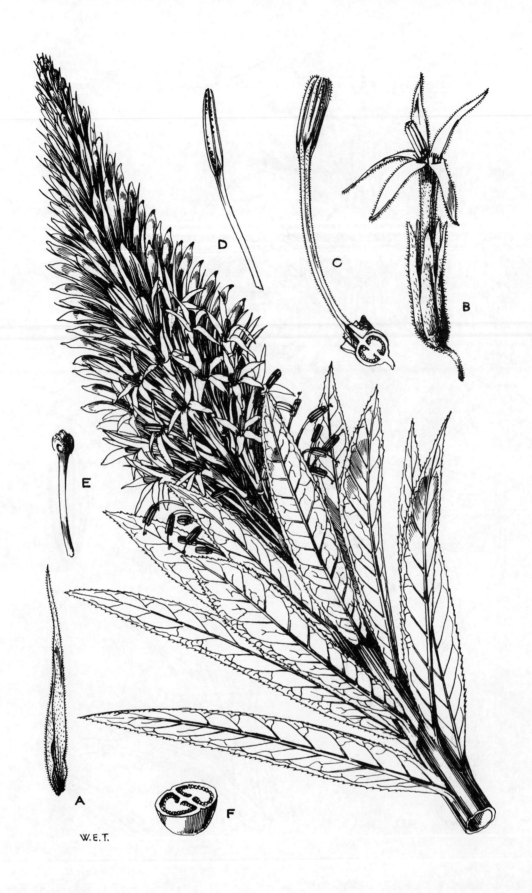

W.E.T.

Fig. 11. *Lobelia columnaris* (Campanulaceae)
Drawn by W.E. Trevithick

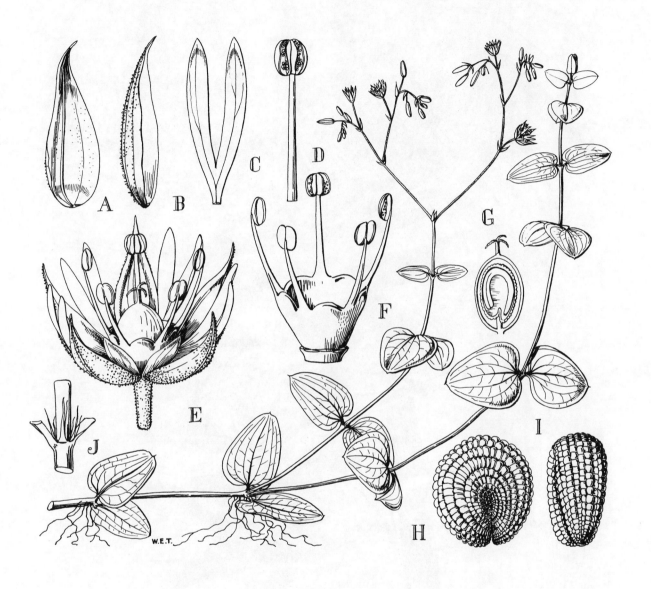

Fig. 12. *Drymaria cordata* (Caryophyllaceae)
Drawn by W.E. Trevithick

Fig. 13. *Helichrysum cameroonense* (Compositae)
Drawn by W. Fitch

Fig. 14. *Solanecio mannii* (Compositae)
Drawn by W.E. Trevithick

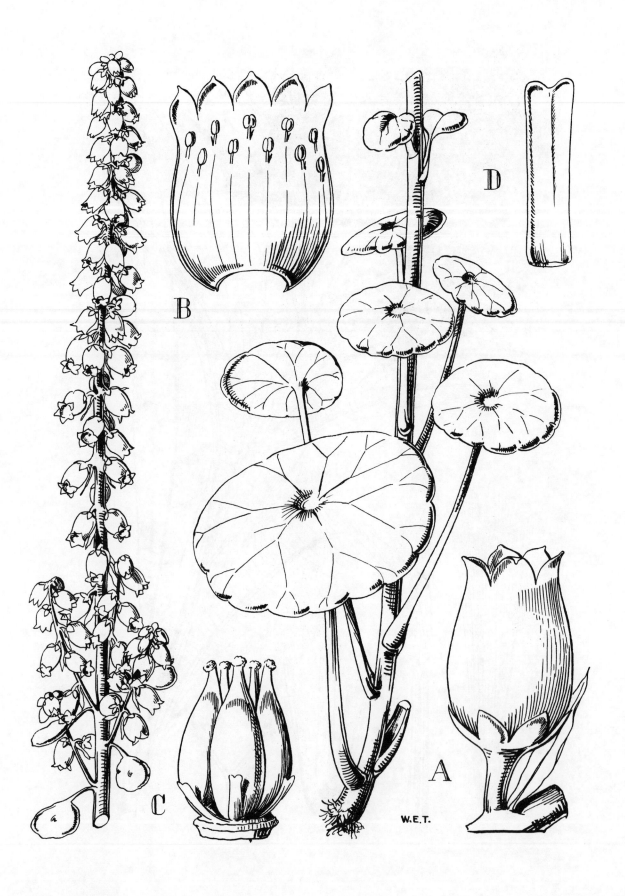

Fig. 15. *Umbilicus botryoides* (Crassulaceae)
Drawn by W.E. Trevithick

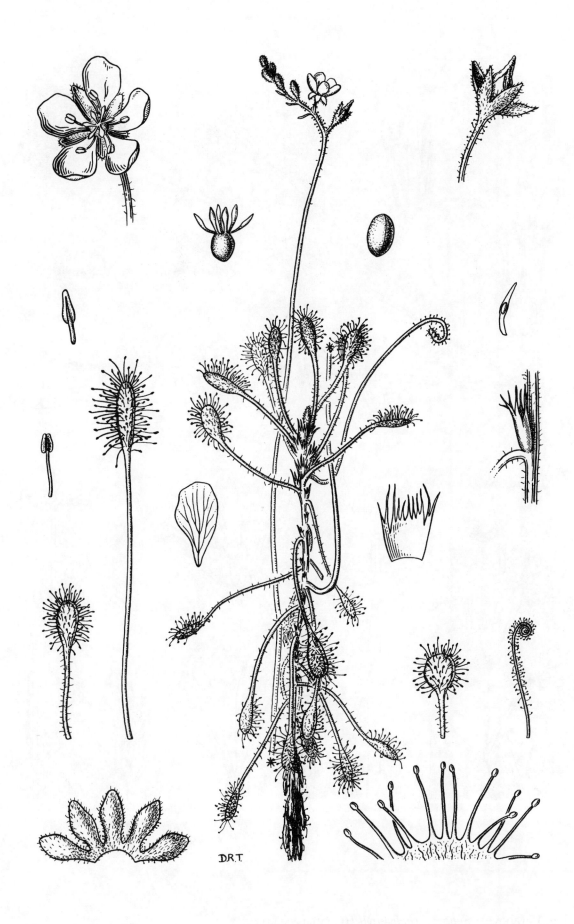

Fig. 16. *Drosera madagascariensis* (Droseraceae)
Drawn by W.E. Trevithick

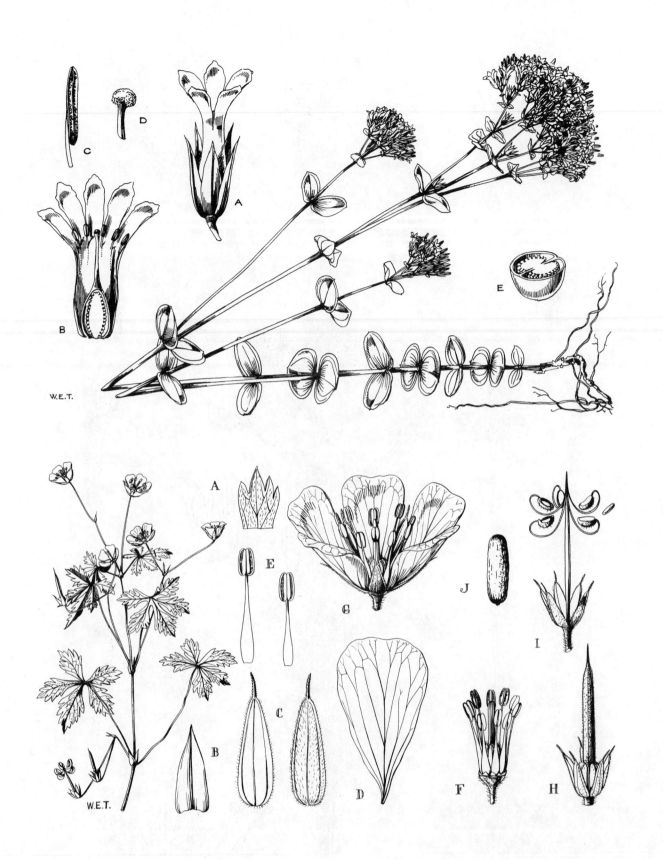

Fig. 17 (above). *Sebaea brachyphylla* (Gentianaceae)
Drawn by W.E. Trevithick
Fig. 18 (below). *Geranium arabicum* (Geraniaceae)
Drawn by W.E. Trevithick

Fig. 19 (above). *Haumaniastrum alboviride* (Labiatae)
Drawn by W.E. Trevithick
Fig. 20 (below). *Maesa lanceolata* (Myrsinaceae)
Drawn by W.E. Trevithick

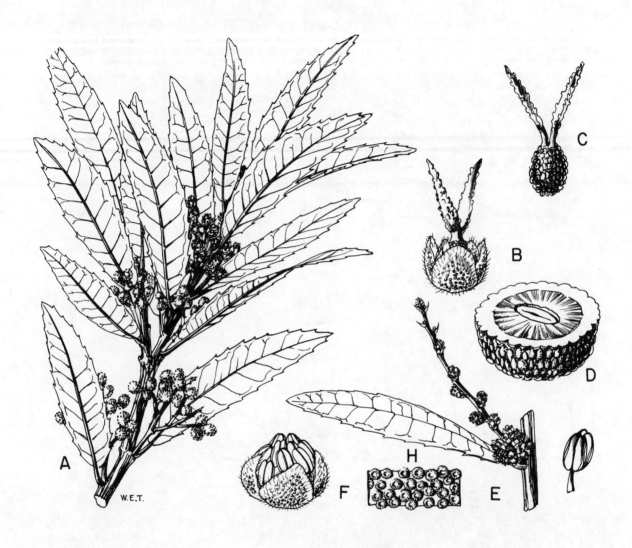

Fig. 21. *Morella arborea* (Myricaceae)
Drawn by W.E. Trevithick

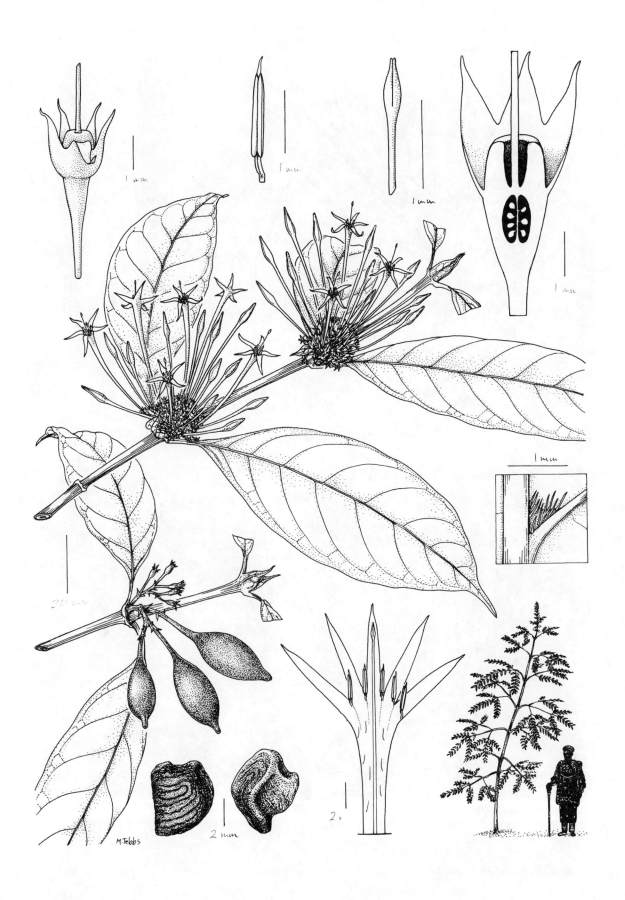

Fig. 22. *Oxyanthus okuensis* (Rubiaceae)
Drawn by Margaret Tebbs

103

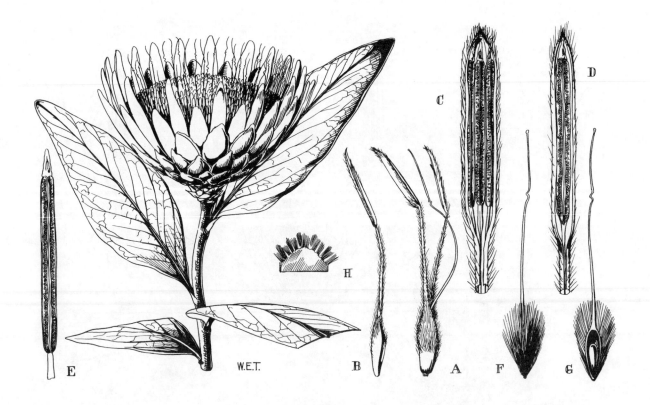

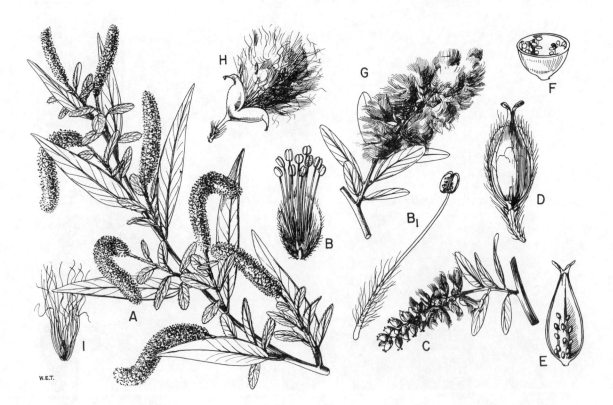

Fig. 23 (above). *Protea madiensis* (Proteaceae)
Drawn by W.E. Trevithick
Fig. 24 (below). *Salix ledermannii* (Salicaceae)
Drawn by W.E. Trevithick

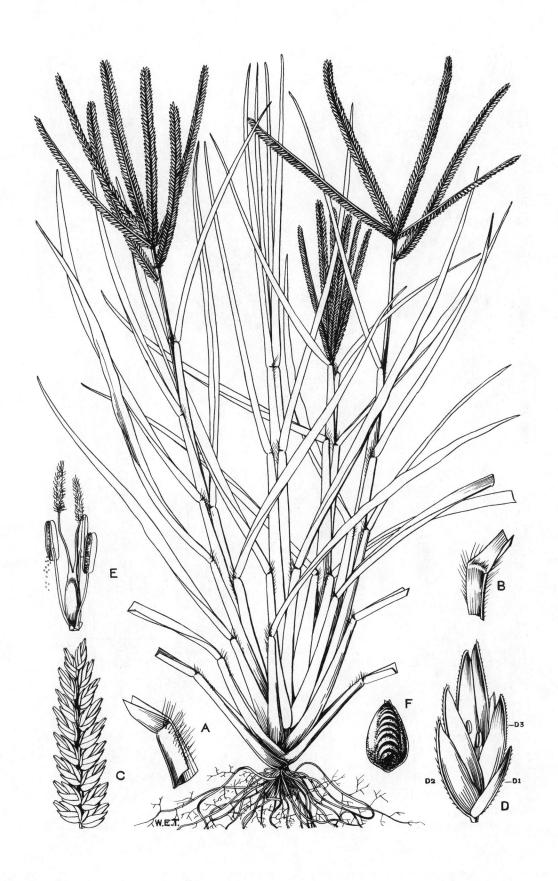

Fig. 25. *Eleusine indica* (*Gramineae*)
Drawn by W.E. Trevithick

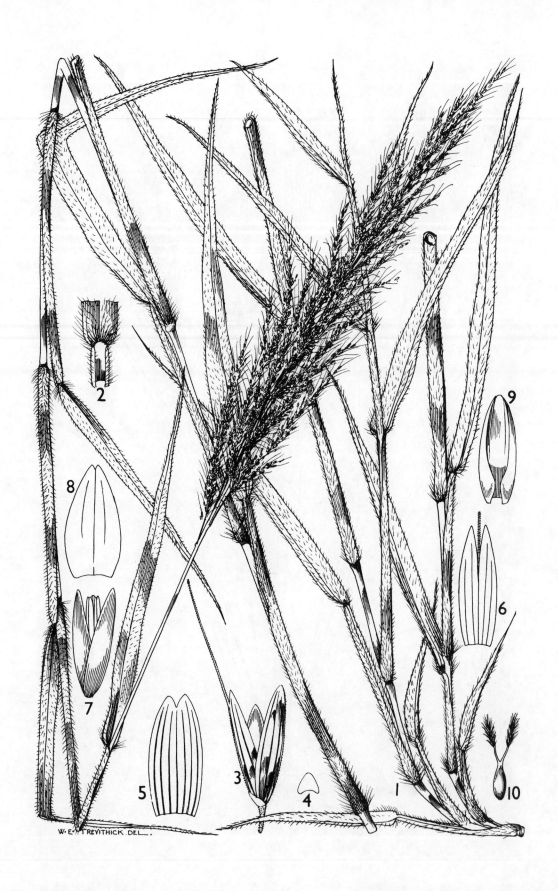

Fig. 26. *Melinis minutiflora* (*Gramineae*)
Drawn by W.E. Trevithick

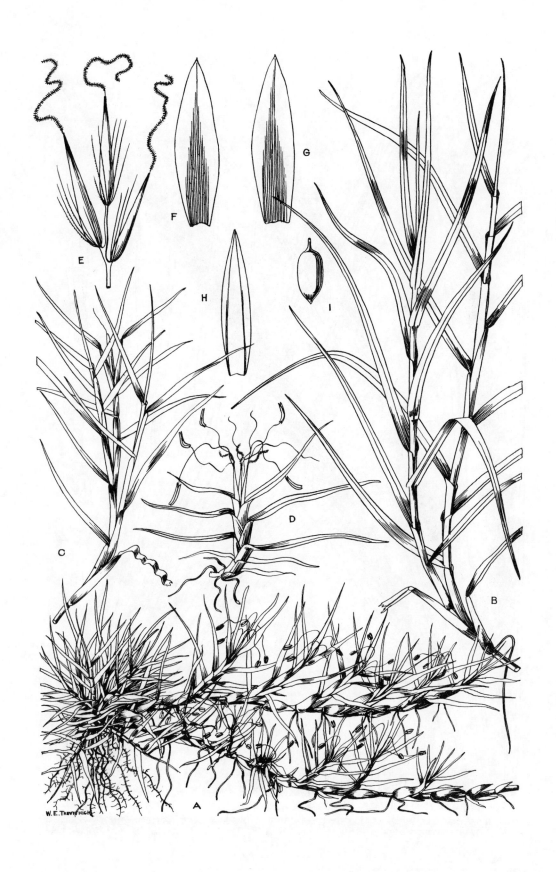

Fig. 27. *Pennisetum clandestinum* (*Gramineae*)
Drawn by W.E. Trevithick

Fig. 28. *Sporobolus africanus* (*Gramineae*)
Drawn by W.E. Trevithick

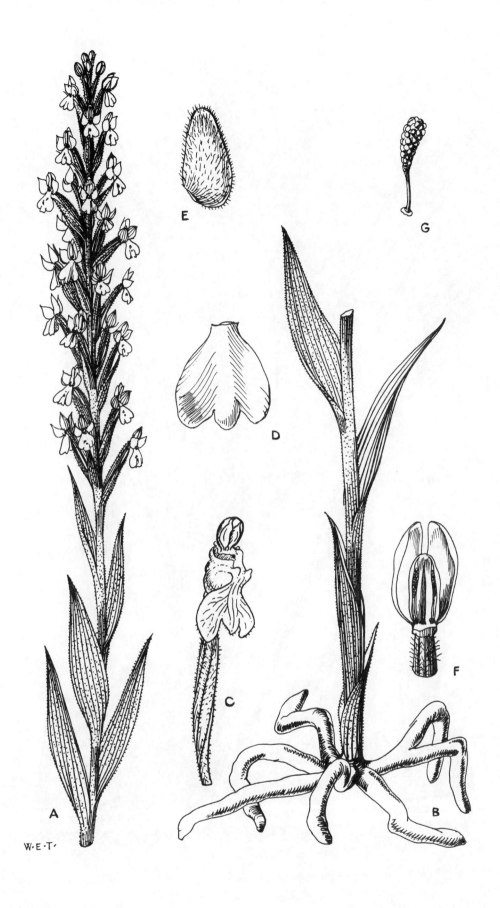

Fig. 29. *Brachycorythis pubescens* (*Orchidaceae*)
Drawn by W.E. Trevithick

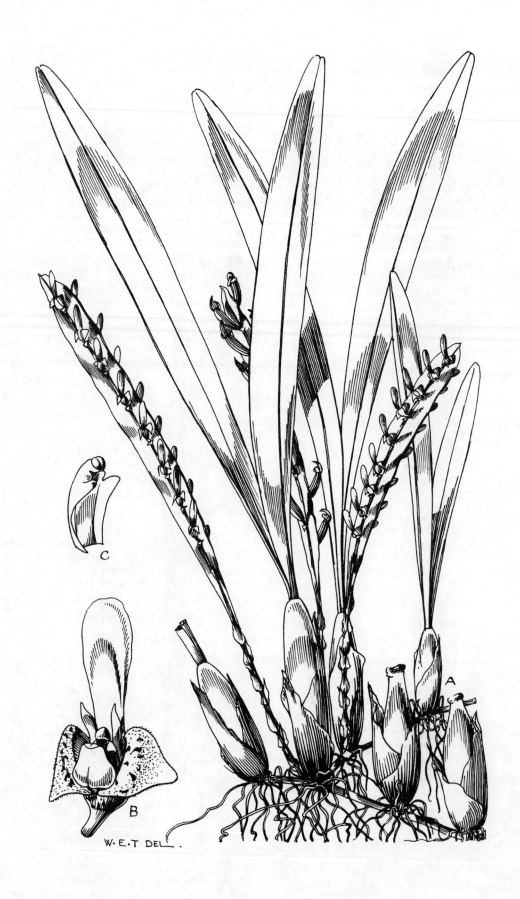

Fig. 30. *Bulbophyllum falcatum* var. *velutinum* (*Orchidaceae*)
Drawn by W.E. Trevithick

ANGIOSPERMAE

DICOTYLEDONAE

ACANTHACEAE
Det. K.Vollesen (K) & J-P.Ghogue (YA)

Acanthopale decempedalis C.B.Clarke
Mass flowering herb to 3 m, nearly glabrous; leaves very unequal in each pair, the larger obovate, long-acuminate, attenuated at the base into winged petiole, to 20 × 8 cm; racemes axillary, to 5 cm long, with about 4 flowers; corolla widely funnel-shaped, white, c. 3 cm long; calyx lobes c. 1 cm long. Forest.
Kilum: Elak, KJ, 2300m, fr., 28 Oct. 1996, Buzgo 606.

Acanthus montanus (Nees) T.Anderson
Local name: Eghaly (Oku) – Etuge 3356.
Herb up to 2 m; stems woody, sparsely branched; leaves dark glossy green above, paler beneath, papery, scabrid; bracts, 2–2.5 cm long, mostly glabrescent or sparsely puberulous, spinously toothed, 10–12 teeth, largest always in the middle; calyx-lobes whitish with green veins. Forest edges.
Kilum: Main road from Ijikijem to Oku, 2350m, 7 Nov. 1996 Etuge 3356.
Ijim: Fundong, near Touristic House, near the water fall, fl., 23 Nov. 1996, Onana 592.

Asystasia gangetica (L.) T.Anderson
Herb, 1–1.3m, erect or half-straggling, slightly pubescent; corolla tubular 1.5 cm long, calyx segments 5 mm long, shortly pubescent with descending hairs. Widespread weed of lower altitudes.
Kilum: Elak-Oku, near the village, fl., 6 Nov. 1996, Onana 504.

Barleria villosa S.Moore
Woody shrub; stems and leaves villous; corolla blue-purple, 3.5 cm long; adaxial sepal broadly ovate-lanceolate, to 17 × 8 mm, abaxial sepal about same size (sometimes larger), at its apex broadly 2-cleft for about 1/3 length. Forest edges, gregarious.
Ijim: Zitum road, 1930m, fl., 21 Nov. 1996, Etuge 3547; Belo to Afua, 1700m, fl., 7 Dec. 1998, Cheek 9826.

Brillantaisia lamium (Nees) Benth.
Coarse herb, erect, branched, 0.6–1.3 m, more or less pilose at first; leaves broadly ovate; inflorescence a glandular-pubescent panicle; flowers 2.5–3 cm long, blue or violet-purple; fruit glabrous, linear, 3–3.5 cm long, many-seeded. In damp shady places; weedy.
Ijim: Fundong, town centre, 1480m, fl., 21 Nov. 1996, Cheek 8697.

Brillantaisia owariensis P.Beauv.
Local name: Kevungua (Oku) – Munyenyembe 841; Anviaghui (Kom) – Cheek 8665; Evengeyu (Kom at Laikom

– Yama Peter) – Pollard 315.
Erect herb to 4 m; stems glandular pubescent at first, later glabrous; petiole winged in the upper half but not to the base; flowers purple, 2.5–3.5 cm long. Forest edges and gaps.
Kilum: Elak, KJ, 2400m, fl., 28 Oct. 1996, Buzgo 623; Above Mboh, path above village to forest through Wambeng's farm, 2200m, fl., 1 Nov. 1996, Cheek 8570; Elak, 2500m, fl., 1 Nov. 1996, Munyenyembe 841; Elak, 2400m, fl., 1 Nov. 1996, Munyenyembe 855.
Ijim: Main road from Atubeaboh (Gikwang Foe) towards Oku, 2260m, fl., 15 Nov. 1996, Etuge 3383; Turbo path, 1 km after its splitting from the main road at the border of the Ijim Mountain Forest Reserve, North from there, 1800–2100m, fl., 18 Nov. 1996, Buzgo 779; Above Aboh village, 1.5 hours on road to Lake Oku on Gikwang road, 2450m, fl., 19 Nov. 1996, Cheek 8665; Boundary of Akwamofu sacred forest, Laikom, 1760m, fl., 5 Dec. 1998, Pollard 315.

Dicliptera laxata C.B.Clarke
Local name: None known (Ernest Kiming) – Cheek 8430; None known – Cheek 8688.
Stout upright herb; stems dark green; leaves elliptic, to 8 × 3.5 cm; inflorescence of axillary cymes, spikelets nearly all pedicelled, bracts to 15 mm, usually dark tinged; flowers white. Forest edge.
Kilum: Elak to the forest at KA, 2200m, fl., 27 Oct. 1996, Cheek 8430; KA path, 2500m, fl., 29 Oct. 1996, Onana 450.
Ijim: Above Aboh village, 1.5 hours on road to Lake Oku on Gikwang road, track towards transect TA, 2400m, fl., 20 Nov. 1996, Cheek 8688; Laikom, 2000m, fl., 3 Dec. 1998, Etuge 4516.

Dyschoriste nagchana (Nees) Bennet
Herb to 50 cm; leaves rhombic-spathulate, c. 3 × 0.8 cm, glabrous; flowers pink, 5 mm long. Forest edge.
Kilum: 6.15N, 10.29E, 1800m, fl., 6 Nov. 1996, Buzgo 714.
Ijim: Chuhuku River near Anyajua Forest, 1600m, fl., 28 Nov. 1996, Kamundi 732.

Eremomastax speciosa (Hochst.) Cufod.
Local name: Nyikteh (Oku) – Cheek 8602.
Herb, 1.5 m; leaf-blade ovate, about 13 × 10 cm, purple below, petiole c. 6 cm long; flowers pale blue, c. 4 cm long, lower lip 5-lobed, c. 3 cm wide. Fallow.
Syn. ***Eremomastax polysperma*** (Benth.) Dandy, F.W.T.A. 2: 397 (1963).
Kilum: Elak, forest edge, 2200m, fl., 1 Nov. 1996, Munyenyembe 846; Lake Oku to Jikijem, 2200m, fl., 5 Nov 1996, Cheek 8602.
Ijim: Aboh village, along the main road, 1700–1800m, 24 Nov. 1996, Onana 604.

Hypoestes aristata (Vahl) Roem. & Schult.
Erect herb 1 m; leaves ovate-elliptic, gradually acuminate, broadly cuneate at base, to 15 × 5 cm; inflorescences of whorls or clusters; bracts with long slender tails at apex, slightly pubescent, to 1.5 cm long; corolla-tube 1.5 cm long; pale mauve with darker markings; stamens long-exserted. In forest.
Kilum: Oku-Elak to the forest at KA, 2200m, fl., 27 Oct.

1996, Cheek 8437.
Ijim: Main road from Atubeaboh (Gikwang Foe) towards
Oku, 2260m, fl., 15 Nov. 1996, Etuge 3381; Turbo path, 1
km after its splitting from the main road at the border of the
Ijim Mountain Forest Reserve, north from there, 2000m, fl.,
18 Nov. 1996, Buzgo 768; Aboh, Anyajua, 2400m, fl., 19
Nov. 1996, Satabie 1069.

Hypoestes forskaolii (Vahl) R.Br.
Local name: Fekin (Kom) – Cheek 8658.
Herb, 1–2 m; leaves narrowly elliptic, c. 10 × 2.5 cm; petiole
2 cm; inflorescences in axillary spikes c. 5 m long; flowers
1.5 cm long, white mottled purple. Fallow.
Kilum: KA, 2600m, fl., 27 Oct. 1996, Buzgo 580; Oku-Elak
to the forest at KA, 2200m, fl., 27 Oct. 1996, Cheek 8452.
Ijim: Gikwang, towards Nyasosso bush, 2260m, fl., 18 Nov.
1996, Etuge 3453; Above Aboh village, 1.5 hours walk along
Gikwang road to Lake Oku, 2450m, fl., 19 Nov. 1996,
Cheek 8658.

Hypoestes triflora (Vahl) Roem. & Schult.
Straggling herb, 0.6–1 m, laxly branched, sometimes with
viscid hairs; floral leaves rounded at apex, c. 10 × 4 mm,
flowers in whorls, almost whitish, upper lip sometimes
spotted with purple. Montane forest.
Ijim: Gikwang road towards Oku, 2260m, fl., 15 Nov. 1996,
Etuge 3390; Above Aboh village, 1.5 hours walk along
Gikwang road to Lake Oku, track towards transect TA,
2400m, fl., 20 Nov. 1996, Cheek 8685; Oku-Elak, above
Fon's Palace, about 30 mins. walk along path to Fulani
settlement, 2050m, fl., 21 Nov. 1996, Cheek 8720; Forest
along the top of the ridge, second gate from Aboh Village,
2380m, fl., 21 Nov 1996, Kamundi 672; Abandoned farms,
Laikom near village, 1950m, fl., 8 Dec. 1998, Etuge 4573.

Isoglossa glandulifera Lindau
Erect, slender, branched herb, 1.2–3.2 m; leaves very thin,
ovate, to 7 × 3.5 cm, small rod-like cystoliths very
conspicuous on the nerves; petiole slender, to 3 cm long, the
upper leaves becoming sessile; inflorescence a lax slender
panicle; flowers few, calyx-lobes 2.5 mm long, fringed with
gland-tipped hairs; corolla 2 cm long, light red. Forest edge.
Kilum: KJ, 2400m, fr., 28 Oct. 1996, Buzgo 621.

Isoglossa sp. aff. glandulifera Lindau
= Morton K860 etc. (fide Vollesen)
Kilum: Summit, 2950m, fl., 29 Oct. 1996, Munyenyembe
794.

Justicia flava (Vahl) Vahl
Local name: Fokai (Kom at Laikom) – Pollard 313.
Erect or straggling herb, 0.2–1.3 m; stems pubescent,
sulcate; leaves elliptic-oblong, hairy, to 12 × 5 cm;
inflorescence spicate, pubescent, 5–14 cm long; bracts
oblanceolate, narrowed to base, c. 1 cm long; corolla c. 2 cm
long, yellow with dark streaks. Forest edge.
Ijim: Boundary of Akwamofu sacred forest, beside path,
Laikom, 1760m, fl., 5 Dec. 1998, Pollard 313; Grass ridges
above Akwamofu sacred forest, on the way to waterfall
below western edge of Ijim ridge grassland plateau, Laikom,
1800m, fr., 10 Dec. 1998, Pollard 362.

Justicia insularis T.Anderson
Straggling, extremely variable herb 0.25–1 m; leaves
variable (from linear to ovate, subsessile to long-petiolate,
acute or acuminate), up to 12 × 4 cm, setulose to nearly
glabrous; corolla to 2.5 cm long, bright pink, red-purple or
yellowish, with purple markings, pubescent on outside.
Widespread, and complex species. Forest edge.
Ijim: Mutef, 6.15N, 10.26E, fl., 22 Nov. 1996, Onana 582.

Justicia striata (Klotzsch) Bullock
Herb, c. 30 cm, leaves unequal in each pair, lanceolate, 8 × 2
cm, petiole 1–3 cm long; flowers white with purple
markings, c. 1 cm long. Forest edge.
Ijim: Gikwang road towards Oku, 2260m, fl., 15 Nov. 1996,
Etuge 3397; Below Fon of Kom's Palace, Laikom, 1820m,
fl., 11 Dec. 1998, Cheek 9909.

Justicia sp. nov. ?
Justicia Sect. **Harnieria**, near *Justicia heterocarpa* subsp.
praetermissa
RED DATA CANDIDATE
Scandent herb c. 60 cm, stems angular; leaves sessile,
lanceolate, c. 2 × 0.7 cm; flowers purple, c. 0.8 cm long.
Swamp.
Mbi Crater: Ndawara area, Mbi swamp, 1900m, 6.05N,
10.21E, fl. fr., 9 Dec. 1998, Etuge 4580.

Mimulopsis solmsii Schweinf.
Mass flowering coarse herb or undershrub to 4 m; stems 4-
angled, pithy, glabrous; leaves very long-petiolate, ovate
triangular, cordate at base, 8–14 × 5.5–8.5 cm, doubly
dentate, sparingly pubescent; calyx densely viscid-pillose,
lobes linear; corolla 2–2.5 cm long, subregular, glabrous,
widely funnel-shaped, lilac or pale violet. Forest.
Kilum: Path above water-tower, KJ, about 1 km into forest,
2250m, fl., 9 Jun. 1996, Cable 3004; KJ, 2300m, fl., 28 Oct.
1996, Buzgo 604; Above Mboh, path above village to forest
through Wambeng's farm, 2200m, fr., 1 Nov. 1996, Cheek
8568; Lake Oku, 2200m, fl., 5 Nov. 1996, Pollard 45.
Ijim: Gikwang road towards Oku, 2260m, fl. fr., 15 Nov.
1996, Etuge 3386.

Oreacanthus mannii Benth.
Local name: Bum (Oku) – Munyenyembe 729.
Mass flowering shrub 2.5–4 m; stems shortly pubescent,
constricted, brittle above the nodes; leaves subequal in each
pair, elliptic, narrowed at both ends, long-acuminate, 15–20
× 4–7 cm; inflorescence a panicle, large and pyramidal;
calyx lobes viscid, linear, 4 mm long; corolla 2-lipped, 1.3
cm long, white or light purple. Forest.
Kilum: KJ, 2300m, fl., 28 Oct. 1996, Buzgo 597; Elak,
2400m, fl., 28 Oct. 1996, Munyenyembe 729; KJ, 2400m,
fl., 29 Oct. 1996, Pollard 3.

Phaulopsis angolana S.Moore
Herb, c. 60 cm; leaves unequal, in a pair, lanceolate-elliptic
8 × 2.5 cm, obscurely serrate, petiole 1.5 cm long; flowers
white, subtended and almost concealed by ovate bracts, 0.7
cm long. Forest edge.

Ijim: Fundong, near touristic house, near the waterfall, 22 Nov. 1996, <u>Onana</u> 585.

Thunbergia cf. fasciculata Lindau

Climber to at least 4 m; leaves opposite, ovate, c. 8 cm long; flowers purple, 5 cm diam. Forest.
Ijim: Tum, 1800m, 4 Nov. 1999, <u>Cheek</u> 9950.

ALANGIACEAE

Alangium chinense (Lour.) Harms

Deciduous tree to 23 m, fast-growing, soft-wooded, branchlets glabrous; leaves ovate-elliptic, very oblique at base, acuminate, 15–17 × 9–10 cm, 5-nerved from base, glabrous; petiole 2.5–3 cm long, puberulous; cymes axillary; calyx a mere toothed rim; flowers c. 10 or more, pedicels jointed at top, c. 4 mm long, petals 1 cm long, linear from broader base, creamy-white to pale yellow, fragrant. Forest margins.
Ijim: Fundong near touristic house, near waterfall, Lat 6.15N Long 10.26E, 23 Nov. 1996, <u>Onana</u> 593.

AMARANTHACEAE
Det. C.C.Townsend (K) & J-P.Ghogue (YA)

Achyranthes aspera L. var. *aspera*

Much–branched herb, 0.25–2 m, variable, sometimes subscandent, sometimes woody; leaves broadly elliptic or broadly obovate to almost round, not exceeding 10 × 5 cm, long silky pilose to almost glabrous; inflorescence an elongating spike; flowers well separated towards base, 4.5 mm long, green or pinkish. Weed of waste places.
Kilum: Path above water-tower, KJ, about 1 km into forest, 2250m, 9 Jun. 1996, <u>Cable</u> 2964 & 2975.

Achyranthes aspera L. var. *pubescens* (Moq.) C.C.Towns.

Local name: Ngiy (Oku) – <u>Cheek</u> 8491.
As var. *aspera*, but leaves oblong to elliptic, not silvery canescent below; flowers 5–7mm long. Forest edge.
Kilum: Elak, 2100m, fl., 8 Jun. 1996, <u>Zapfack</u> 770; KD, fl., 9 Jun. 1996, <u>Etuge</u> 2243; Elak, 2500m, fl. fr., 9 Jun. 1996, <u>Zapfack</u> 800; Oku-Elak, lower parts of KA, 2200m, 6.1349N, 10.3112E, fr., 28 Oct. 1996, <u>Cheek</u> 8491; Elak-Oku, KA path, 6.15N 10.26E, 2500m, fl., 29 Oct. 1996, <u>Onana</u> 460; Path from BirdLife HQ to KA, 2800m, 6.1349N, 10.3112E, fr., 30 Oct. 1996, <u>Cheek</u> 8519; KA, 2600m, fl. fr., 31 Oct. 1996, <u>Buzgo</u> 689.

Achyranthes aspera L. var. *sicula* L.

Local name: None known – <u>Cheek</u> 8679.
As var. *aspera*, but leaves oblong to elliptic, silvery canescent below, flowers 3–4.5 mm long. Forest edge.
Kilum: KA, 2600m, fl., 27 Oct. 1996, <u>Buzgo</u> 589; Elak, KJ, 2300m, fl., 28 Oct. 1998, <u>Buzgo</u> 595.
Ijim: Gikwang road, track towards TA, 2400m, 6.1112N, 10.2529E, fl. fr., 20 Nov. 1996, <u>Cheek</u> 8679; Aboh, Gikwang road towards Nyasosso forest (grassland), 2280m, 6.1116N, 10.2543E, fl., 20 Nov. 1996, <u>Etuge</u> 3505.

Achyranthes aspera L. var. *sicula* L. vel var. *pubescens* (Moq.) C.C.Towns

Local name: Fikein (Bekom) – <u>Kamundi</u> 620.
Intermediate between the two varieties.
Ijim: Aboh, 2160m, 6.11N, 10.25E, fl., 19 Nov. 1996, <u>Kamundi</u> 620.

Cyathula cylindrica Moq. var. *cylindrica*

Local name: None known – <u>Cheek</u> 8690.
Scandent herb; stems to 2.5 m long, densely and softly hairy; leaves elliptic to ovate-elliptic, acute, 5–7 × 2.5–3.5 cm; inflorescence a dense cylindrical spike to 18 cm long; calyx green; corolla white. Forest edge.
Kilum: KJ, 2300m, 28 Oct. 1996, <u>Buzgo</u> 599; Lower parts of KA, 2200m, 6.1349N, 10.3112E, fl., 28 Oct. 1996, <u>Cheek</u> 8470; Elak, 2420m, 28 Oct. 1996, <u>Munyenyembe</u> 725.
Ijim: Above Aboh village, 1.5 hours to Lake Oku on Gikwang road, track towards TA, 2400m, 6.1112N, 10.2529E, fl., 20 Nov. 1996, <u>Cheek</u> 8690; Afua swamp, grassland, 1950m, 6.09N, 10.24E, fl., 7 Dec. 1998, <u>Etuge</u> 4567.

Cyathula prostrata (L.) Blume var. *prostrata*

Much-branched, +/- erect herb, 0.25–1 m; stem relatively stout, to 3mm diam., densely pilose; leaves green sometimes tinged red, also pilose, 2–5 × 1.5–4 cm; inflorescence a slender elongated raceme to 6mm broad, green, often galled. Weed of waste places and forest margins.
Kilum: Lowland forest patches at lower altitude than the BirdLife HQ, 1800m, 6 Nov. 1996, <u>Buzgo</u> 713.

Cyathula uncinulata (Schrad.) Schinz

Climbing herb; stems to 5.3 m, pilose, densely so when young with weak spreading hairs; leaves ovate to elliptic, acute, 5–10 × 2.5–7 cm, densely appressed pilose; flowers in globose heads, 2–3 cm diam., greenish-yellow.
Ijim: Afua swamp, 1950m, 6.09N, 10.24E, fl., 7 Dec. 1998, <u>Etuge</u> 4565; Laikom, grassland near swamp, half a mile W of Ardo's Fulani compound, Ijim ridge plateau, 2100m, 6.1615N, 10.2051E, fl., 7 Dec. 1998, <u>Pollard</u> 344.

Sericostachys scandens Gilg & Lopr.

Climber to 5 m or more; leaves elliptic-rhombic, 10–15 cm long; inflorescence axis densely pubescent, twice branched, spikes bright white, c. 2–3 cm long. Forest edge.
Kilum: Elak, KC, 2600m, fr., 30 Oct. 1996, <u>Zapfack</u> 1110; Lake Oku, 2400m, fl., 26 Nov. 1996, <u>Satabie</u> 1096.

ANACARDIACEAE

Lannea kerstingii Engl. & K.Krause

Deciduous tyree 4–15 m; leaves alternate; leaflets 7–9, bolong-elliptic, 6–8 × 4 cm, apex shortly acuminate, base rounded to shallowly cordate, thickly felty below; inflorescences of pendulous racemces, arising when leafless, c. 10 cm long, densely-flowered; flowers white, c. 3 mm wide; fruit ellipsoid, 1 × 0.5 cm. Savanna.
Babungo-Mbi Crater: 1600m, 8 Nov. 1999, <u>Cheek</u> 9996.

Sorindeia grandifolia Engl.
Shrub or tree to 15 m; leaflets 3–7, alternate or subopposite, ovate-oblong to lanceolate, cuneate, obtuse or more rarely rounded at base, acuminate, 12–30 × 3.7–13 cm; inflorescence paniculate, 16–30 cm long; flower petals imbricate, whitish, tinged pink; fruits ellipsoid, 2 cm long. Forest.
Ijim: Waterfall near Ijim Project HQ, Anyajua, 1300m, 6.11N, 10.22E, fl., 12 Dec. 1998, Cheek 9921.

ANNONACEAE
Det. B.J.Pollard (K)

Annona chrysophylla Bojer
Shrub; leaves alternate, broadly ovate, 10 × 6 cm, apex retuse, base truncate, felty below. Savanna.
Babungo: 1200m, 21 Mar. 1962, Brunt 234.

Annona senegalensis Pers. subsp. *oulotricha* Le Thomas
Shrub c. 4 m; leaves broadly elliptic, c. 8 × 4 cm, apex and base rounded; petiole 1 cm. Savanna.
Kumbo: 35 km NNW at Lassin, 7 Jul. 1973, Mbenkum 404.

Monanthotaxis littoralis (Bagsh. & Baker f.) Verdc.
Erect or climbing shrub; leaves oblong-elliptic, rounded to subcordate at base, obtuse at apex, 3.5–10 × 1.5–4.5 cm; flowers solitary, extra-axillary; pedicel 2.5–4 cm long; fruit yellow. Forest.
Syn. *Popowia littoralis* Bagsh. & Baker.f., F.W.T.A. 1: 44 (1954).
Bambui: 1500m, fl., Jun. 1931, Maitland 1618.

Xylopia acutiflora (Dunal) A.Rich.
Shrub or small tree; leaves ovate-elliptic to lanceolate, obtusely acuminate, obtuse or rounded at base, 4.5–10 × 2–3.7 cm, thinly pilose beneath, glabrescent, midrib above hirsute; flowers solitary, sessile, outer petals c. 4.5 cm long; fruits scarlet. Forest.
Ijim: Anyajua, waterfall near Ijim Project HQ, 1300m, 6.11N, 10.22E, fl. fr., 12 Dec. 1998, Cheek 9923.

APOCYNACEAE
Det. A.J.M.Leeuwenberg (WAG)

Carissa spinosa L.
Shrub 2 m, spiny; leaves ovate, 5–6 cm long, petiole 1–2mm long; flowers white, c. 1 cm wide. Savanna.
Jakiri: Tan, Al Hadji Gey's land, grasslands, 1600m, st., 6 Nov. 1998, Maisels 184.

Landolphia dewevrei Stapf
Liana to 7 m; stems with white exudate, tendrillate; leaves opposite, elliptic-oblong, 5–9 × 2.5 cm, petiole 2–4 mm long. Forest.
Ijim: From Aboh, path towards Tum, 2100m, 6.1116N, 10.2543E, 22 Nov. 1996. Etuge 3585.

Landolphia sp. 1
Kilum: Mwal village, near Ajung village (north of Oku), 1200m, fr., 28 Nov. 1998, Maisels 193.

Rauvolfia vomitoria Afzel.
Local name: Eptong (Oku) – Munyenyembe 831.
Small tree or shrub to 16 m; leaves whorled, elliptic-acuminate to broadly lanceolate, 8–24 × 3–10 cm, 8–16 pairs of lateral nerves with narrowly triangular acute acumen; inflorescence branches vertillicate, ample, distinctly puberulous; corolla 8–10mm long, white to greenish. Forest.
Kilum: Elak, 2300m, fl. fr., 8 Jun. 1998, Zapfack 746, Elak, 2600m, 31 Oct. 1996, Munyenyenbe 831; Ethiale, near the stream, 2000m, fr., 4 Nov. 1996, Zapfack 1174.
Ijim: Laikom, forest slope above Fulani village, 2000m, 6.17N, 10.20E, fr., 3 Dec. 1998, Etuge 4521.
Plot voucher: Forest above Oku-Elak, Jun. 1996, Oku 26.

Strophanthus sp.
More material needed. Forest.
Kilum: Lake Oku, 2300m, fr., 5 Nov. 1996, Pollard 41.

Tabernaemontana cf. ventricosa Hochst. ex A.DC.
Tree c. 8 m; bole with striking, large white lenticels; fruits bilobed, c. 10 × 8 cm, deeply ridged and furrowed, marbled white and green. Forest.
Ijim: Laikom, head of valley, 1900m, Nov. 1999, Tadjouteu s.n.

AQUIFOLIACEAE

Ilex mitis (L.) Radlk.
Tree to 13 m; leaves dark green, oblong, apiculate, obtuse to subacuate at base, to 10 × 4 cm, entire or remotely denticulate, glabrous; inflorescence subfasciculate; flowers on short axillary peduncles, pedicels to 1 cm long; sepals minutely ciliolate; petals shortly connate, cream or white; fruit subglobose, red when ripe, 5 mm diameter. Montane forests by streams.
Ijim: Laikom, 2100m, 6.15N, 10.26E, fr., 21 Nov. 1996, Onana 560; Tum, Mbalabo-Hill, 2000m, fr., 26 Nov. 1996, Etuge 3635.

ARALIACEAE

Cussonia djalonensis A.Chev.
Tree to 10m; leaves digitate, leaflets 7–10, 9–25 cm long, glabrescent; inflorescence spicate to 50 cm long, pubescent. Savanna.
Syn. *Cussonia barteri* Seem., F.W.T.A. 1: 750 (1958).
Babungo-Mbi Crater: 1600m, 8 Nov. 1999, Cheek 9987.

Polyscias fulva (Hiern) Harms
Tree 6–16m, rarely 30m, evergreen; leaves bunched at the top of rather few branches, imparipinnate, to 80 cm or more long, leaflets 4–7 (–10) pairs, opposite, lateral leaflets ovate, ovate-elliptic or ovate-oblong, rounded to cordate at base, 7–20 × 3.5–10 cm, white-tomentose beneath, dark green above;

flowers shortly pedicellate, subsessile or sessile, cream; fruits ellipsoid, ribbed, c. 5 mm long.   Montane forest.
Kilum: KC, 2500m, fr., 30 Oct 1996, Zapfack 1105.

Schefflera abyssinica (Hochst. ex A.Rich.) Harms
Tree to 18 m, deciduous; leaflets serrate; flowers yellowish. Montane forest.
Kilum: Oku evergreen forest, on road to Oku from the 'Ring Road', 2000m, fl. fr., 21 Jun. 1962, Brunt 606; Foot of Lake Oku, fl. fr., 27 Jun. 1973, Mbenkum 359; Forest around Lake Oku, NE Side, 2230m, 6.13N, 10.28E, fl. fr., 17 Feb. 1985, Thomas 4373; Elak, 2300m, fl. fr., 8 Jun. 1996, Zapfack 754; Elak, Path above water-tower, KJ, about 1 km into forest, 2250m, fr., 9 Jun. 1996, Cable 3001; Elak, KD, fr., 9 Jun. 1996, Etuge 2238; Lake Oku, 2300m, fr., 11 Jun. 1996, Zapfack 848.

Schefflera barteri (Seem.) Harms
Tree to 10m, evergreen; leaflets entire; flowers green, tinged pink. Forest.
Syn. *Schefflera hiernana* Harms, F.W.T.A. 1: 750 (1958).
Ijim: Laikom, forest, 2000m, fr., Apr. 1931, Maitland 1623; Path from Belo to Oku, 2500m, fl. fr., 7 Jan. 1951, Keay & Lightbody 28519; Aboh village, in Anyajua, 1800m, fl. fr., 24 Nov. 1996, Satabie 1091.

Schefflera mannii (Hook.f.) Harms
Local name: Uwos (Oku) – Munyenyembe 743.
Tree, 15-20m, evergreen; leaflets entire; inflorescence of erect panicles, crowded at ends of branches; flowers green; anthers yellow. Forest.
Kilum: Path above water-tower, KJ, about 1 km into forest, 2200m, fr., 9 Jun. 1996, Cable 2947; Elak, KA, 2200m, fl., 11 Jun. 1996, Etuge 2316; Elak, 2420m, fl., 28 Oct. 1996, Munyenyembe 743.
Ijim: Ijim Mountain Forest, along top of ridge, second gate from Aboh Village, 2250m, fl., 21 Nov. 1996, Kamundi 676; Laikom, plot between Fon's Palace and Ardo's compound, 2000m, 6.16N, 10.21E, fl., 8 Dec. 1998, Cheek 9866.
Plot voucher: Forest above Oku-Elak, Jun. 1996, Oku 61.

ASCLEPIADACEAE
Det. D.J.Goyder (K) & U.Meve (MSTR)

Ceropegia nigra N.E.Br.
A pubescent twiner; flowers 1 to 1.5 cm long; corolla-tube curved above the obliquely inflated base, the lobes quite free from each other, dark blue-green, turning black. Forest edge.
Kilum: Elak, above Mboh, path above village to forest through Peter Wambeng's farm, 2200m, 6.1146N, 10.2856E, fl., 1 Nov. 1996, Cheek 8558.
Ijim: Main path from Ijim to Tum, 6.1116N 10.2543E, 2000m, fl., 25 Nov. 1996, Etuge 3626.

Cynanchum praecox Schltr. ex S.Moore
Precocious herb; inflorescence 5 cm long; flowers brownish, greenish inside; corona white. Grassland.
Kilum: Dry grassland, 3 miles along Kumbo to Oku road, 1850m, fl., 15 Feb 1958, Hepper 2011.

Dregea schimperi (Decne.) Bullock
Climber; flowers whitish. Forest edge.
Kilum: Around Lake Oku, 2200m, fr., 5 Nov. 1996, Etuge 3339; Near to stream around savanna zone, 2200m, fr., 5 Nov. 1996, Zapfack 1201.
Ijim: Gikwang,towards Nyasosso bush, 2280m, fr., 18 Nov. 1996, Etuge 3458; Aboh Village, 6.15N, 10.26E, fr., 18 Nov. 1996, Onana 528; Afua swamp, 1950m, 6.09N, 10.24E, fr., 7 Dec. 1998, Etuge 4566.

Pentarrhinum abyssinicum Decne. subsp. *angolense* (N.E.Br.) Liede & Nicholas
Twining climber; leaves ovate-oblong up to 10 cm long, inflorescence umbelliform. Forest.
Kilum: In forest above Oku village, 2200–3000m, fl., Apr. 1986, Thomas & McLeod 5984.

Pentarrhinum abyssinicum Decne. subsp. *ijimense* Goyder
Local name: Ntang (Kom) – Cheek 9943.
RED DATA LISTED
Resembling subsp. *angolense* but flowers smaller; corolla not ciliate; corona shorter with short teeth; gynostegium more clearly stipitate. Forest.
Ijim: Buzgo 798; Zitum road, 1900m, 6.1116N, 10.2543E, fl., 21 Nov. 1996, Etuge 3565; Laikom, base of upper waterfall, medicinal forest, below basaltic grassland plateau, 1960m, 6.1630N, 10.1951E, fl. fr., 11 Dec. 1998, Pollard 368; Tum, 1810m, 6.1087N, 10.2467E, fl. fr., 4 Nov. 1999, Cheek 9943; Mbingo "Back Valley", 1750m, 10 Nov. 1999, Cheek 10063.

Periploca nigrescens Afzel.
Twiner, often herbaceous but becoming woody; leaves leathery, glossy; flowers greenish outside. Forest edge.
Ijim: From Aboh, path towards Tum, 6.1116N 10.2543E, 2100m, fr., 22 Nov. 1996, Etuge 3589.

Sphaerocodon caffrum (Meisn.) Schltr.
Suberect perennial 30–90 cm, branched from base; leaves 5–10 cm long; inflorescence of lateral umbelliform cymes; flowers very dark purple. Grassland.
Ijim: Grassy plateau, Laikom, 2000m, fl., May 1931, Maitland 1365.

Tylophora cf. *oblonga* N.E.Br.
Local name: Fu isus fako (Kom at Laikom – Yama Peter) – Gosline 207.
Twining climber; leaves obtusely cuneate, truncate or cordate, 10–15 cm long; flowers greenish yellow, 7 mm diam.; corona of 5 laterally elongated, contiguous fleshy tubercles. Forest.
Kilum: Around Lake Oku, 2200m, 5 Nov.1996, Etuge 3337.
Ijim: Aboh, Gikwang road towards Nyasosso forest, 2300m, fl., 19 Nov. 1996, Etuge 3502; Laikom, in forest below Ardo's compound, 2100m, 6.16N, 10.21E, fl. fr., 10 Dec. 1998, Gosline 207; Anyajua, John DeMarco's bat cave below Ijim HQ, 1400m, 6.11N, 10.23E, fl., 12 Dec. 1998, Cheek 9918.
Plot voucher: Forest above Oku-Elak, Jun. 1996, Oku 66.

BALSAMINACEAE

Impatiens burtonii Hook.f.
Erect or ascending herb, 60 cm; leaves alternate; flowers pink, white-hairy, with green spur veined with purple, throat orange. Forest edge.
Kilum: Elak, 2000m, fl., 8 Jun. 1996, Zapfack 794; Path above water-tower, KJ, about 1 km into forest, 2250m, fl., 9 Jun. 1996, Cable 2976; KD, 9 Jun. 1996, Etuge 2226; Lake Oku, 2200m, 11 Jun. 1996, Zapfack 865; KA, 2600m, fl., 27 Oct. 1996, Buzgo 59; Elak, 2600m, fl., 27 Oct. 1996, Munyenyembe 708.

Impatiens filicornu Hook.f.
Slender, glabrous herb, to 60 cm; leaves alternate; flowers light purple. Damp cliffs.
Ijim: Zitum road, 1850m, 6.1116N, 10.2543E, fl.fr., 21 Nov. 1996, Etuge 3526.

Impatiens hians Hook.f. var. *hians*
Local name: Mboleluk (Oku) – Munyenyembe 889.
Erect herb, to 60 cm; leaves alternate; flowers dark red and greenish. Forest.
Kilum: Mbokei Ba Rock, 2200m, fl., 6 Nov. 1996, Munyenyembe 889.

Impatiens kamerunensis Warb. subsp. *kamerunensis*
Local name: Kimbas (Oku) – Munyenyembe 849; Itwah uchvu = 'itua ichufu' (Kom at Laikom) – Etuge 4548.
Erect herb, to 90 cm; leaves opposite; flowers pink or deep mauve. Damp cliffs.
Kilum: KA, 2760m, fl., 10 Jun. 1996, Asonganyi 1312; KJ, 2500m, fl., 29 Oct. 1996, Pollard 4; KA, 2400m, fl., 30 Oct. 1996, Pollard 15; KA, 2500m, fl., 31 Oct. 1996, Buzgo 685; Cliff-face on Kidzem waterfall, 2200m, fl., 1 Nov. 1996, Munyenyembe 849.
Ijim: Laikom, Akwamofu medicinal forest, 1760m, 6.17N, 10.20E, fl., 5 Dec. 1998, Etuge 4548.

Impatiens mackeyana Hook.f.
Terrestrial herb to 1 m; leaves alternate, lanceolate to elliptic, generally glabrous; flowers up to 5 cm long, pink to purple. Forest edge.
Kilum: Route to lake, 2200m, fl., 11 Jun. 1996, Zapfack 843; Lake Oku, 2200m, fl., 12 Jun. 1996, Etuge 2343.

Impatiens mannii Hook.f.
Straggling herb, 60–90 cm; leaves alternate; petals white with pink tinge, spur yellow-green at tip, purplish above with transverse bars on the lip. Damp cliffs.
Kilum: Above Mboh, path above village to forest through Peter Wambeng's farm, 2200m, 6.1256N, 10.2911E, 1 Nov. 1996, Cheek 8547.

Impatiens sakeriana Hook.f.
Local name: Kibise (Oku) – Cheek 8434; Kimbas kikie – Munyenyembe 788; Ambusakou (Kom) – Cheek 8642.
Stout erect herb, to 3m; stems straggling at the ends; leaves whorled; flowers deep red, with yellow standard petal

marked with two deep crimson blotches. Forest.
Kilum: Elak, 2300m, fl., 8 Jun. 1996, Zapfack 752; Path above water-tower, KJ, about 1 km into forest, 2250m, 9 Jun. 1996, Cable 2978; KD, fl., 9 Jun. 1996, Etuge 2244; Shore of Lake Oku, 2250m, fl., 12 Jun. 1996, Cable 3115; Oku-Elak to the forest at KA, 2200m, 6.1349N, 10.3112E, fl., 27 Oct. 1996, Cheek 8434; 2950m, fl., 29 Oct. 1996, Munyenyembe 788; KJ, 2300m, fl., 29 Oct. 1996, Zapfack 1045.
Ijim: Main road from Atubeaboh (Gikwang Foe) towards Oku, 2260m, fl. fr., 15 Nov. 1996, Etuge 3382; Above Aboh village, 1.5 hours walk along Gikwang road, 2350m, 6.1112N, 10.2529E, fl., 18 Nov. 1996, Cheek 8642; Ijim Mountain Forest, 2160m, fl., 19 Nov. 1996, Kamundi 622; Aboh-Anyajua, 2500m, fl., 20 Nov. 1996, Satabie 1078; Laikom, stream 100m S of pathtop, grassland plateau above Fon's Palace, 2000m, 6.1642N, 10.1950E, fl., 2 Dec. 1998, Pollard 261.
Plot voucher: Forest above Oku-Elak, Jun. 1996, Oku 49; 71.

Impatiens sp.
Kilum: Path above water-tower, KJ, about 1 km into forest, fl., 9 Jun. 1996, Cable 3005.
Plot voucher: Forest above Oku-Elak, Jun. 1996, Oku 8; 40.

BASELLACEAE

Basella alba L.
Climber, glabrous, with slender fleshy branches; leaves broadly ovate, petiolate, to 10 cm long; inflorescence an axillary spike; bracts white, broadly ovate, half the length of the flower; flowers white, 5–10 mm long. Villages.
Kilum: KA, 2200m, fl., 11 Jun. 1996, Etuge 2320; Oku-Elak, between Fon's Palace and Manchok, bank along road in village, 6.1447N 10.3058E, 2000m, fl., 31 Oct. 1996, Cheek 8539.

BEGONIACEAE
Det. V.Plana (E) & M.Cheek (K)

Begonia oxyanthera Warb.
Local name: Mbol Feykak (Oku) – Cheek 8441.
RED DATA LISTED
Epiphytic climbing shrub; stems reaching 7–10m long; leaves ovate to oblong-ovate, 7–13 cm long; flowers red and white. Forest.
Kilum: Elak, 2200m, fr., 8 Jun. 1996, Zapfack 778; Path above water-tower, KJ, about 1 km into forest, 2200m, fl. fr., 9 Jun. 1996, Cable 2948 & 2997; KD, fl. fr., 9 Jun. 1996, Etuge 2240; Lake Oku, 2300m, fl., 11 Jun. 1996, Zapfack 863; Shore of Lake Oku, 2250m, fl. fr., 12 Jun. 1996, Cable 3090 & 3094; Oku-Elak to the forest at KA, 2200m, fl., 27 Oct. 1996, Cheek 8441; Lake Oku, 2800m, fl., 5 Nov. 1996, Pollard 39.
Plot voucher: Forest above Oku-Elak, Jun. 1996, Oku 60.

Begonia poculifera Hook.f.
Epiphytic herb; leaves brittle-succulent, glossy, strongly asymmetric, ovate-dimidiate, 7–9 × 4–5 cm, veins purple red; flowers white; perianth lobes about 1.5 cm long. Forest.
Ijim: Tum, 1800m, 4 Nov. 1999, Cheek 9948.

Begonia schaeferi Engl.
RED DATA LISTED
Terrestrial herb 10–20 cm; flowers bright yellow. Damp cliffs.
Ijim: From Aboh, path towards Tum, 2100m, fl., 22 Nov. 1996, Etuge 3582.

BIGNONIACEAE

Kigelia africana (Lam.) Benth.
Local name: Athem (Kom at Aboh) – Pollard 75.
Tree, 9–25 m; leaves imparipinnate to 50 cm long, leaflets in 3–6 pairs, up to 20 cm long; inflorescence a panicle; flowers 6–13 cm long, colour variable, purplish-red, yellow, orange to greenish-yellow; fruits sausage-like to 45 cm long. Forest and savanna.
Ijim: Ijim Mountain Forest, 1800m, fr., 21 Nov. 1996, Pollard 75; Aboh village, 25 Nov. 1996, Onana 632.

Stereospermum acuminatissimum K.Schum.
Tree to 35 m; leaflets 5–6 pairs, oblong, 8–14 cm long; inflorescence an ample corymbose-pyramidal panicle; flowers pink or pale purple, 3–5 cm long. Forest edge.
Ijim: Fundong, near touristic house, near the waterfall, fr., 23 Nov. 1996, Onana 598.

BORAGINACEAE

Cordia cf. africana Lam.
Shrub or small tree to 8 m; leaves broad, strongly-nerved beneath; inflorescence a paniculate cyme; flowers conspicuous, funnel-shaped, white; fruits succulent, yellow, 1–2 cm diameter. Forest.
Sterile voucher for checklist record.
Kilum: Lowland forest close to Oku village, 2010m, 2 Nov. 1996, Etuge 3334.

Cynoglossum amplifolium A.DC. var. *subalpinum* (T.C.E.Fr.) Verdc.
Local name: Imbanene – Munyenyembe 780; Aghum (Bekom at Nguakon) – Buzgo 753.
Erect herb to 1 m, hispid; leaves ovate, over 5 cm broad, acuminate, narrowed at base with short winged petiole; flowers conspicuous, bluish or white, 3–4 mm across. Grassland.
Syn. *Cynoglossum amplifolium* Hochst. ex A.Rich. fa. *macrocarpum* Brand, F.W.T.A. 2: 324 (1963).
Kilum: KD, 2900m, fl., 10 Jun. 1996, Cable 3056; KD, 2800m, fr., 10 Jun. 1996, Etuge 2278; Summit, 3000m, fl., 29 Oct. 1996, Munyenyembe 780; Path from Project HQ to KA, 2800m, fr., 30 Oct. 1996, Cheek 8514; KA, 2700m, fl., 31 Oct. 1996, Buzgo 659; KA, 2700m, fl, fr., 31 Oct. 1996, Buzgo 662;
Ijim: Turbo path, 500m after its splitting from the main road

at the border of the Ijim Mountain Forest Reserve, 2000m, fr., 18 Nov. 1996, Buzgo 753, 760 & 764; Gikwang road towards Nyasosso forest, Ijim Mountain Forest Reserve, 2090m, fl., 19 Nov. 1996, Etuge 3467; Laikom, rocky 'pavement' in grassland plateau above Fon's Palace, 2100m, 6.1642N, 10.1950E, fl. fr., 3 Dec. 1998, Pollard 301.

Cynoglossum coeruleum A.DC. subsp. *johnstonii* (Baker) Verdc. var. *mannii* (Baker & C.H.Wright) Verdc.
Local name: Imbane (Oku) – Munyenyembe 720.
As *Cynoglossum amplifolium*, but leaves lanceolate, c. 3 cm broad. Grassland.
Syn. *Cynoglossum lanceolatum* Forssk. subsp. *geometricum* (Baker & Wright) Brand, F.W.T.A. 2: 324 (1963).
Kilum: Elak, KD, fl. fr., 9 Jun. 1996, Etuge 2229; Junction of KJ and KN above Elak, 2200m, fl., 10 Jun. 1996, Cable 3028; Summit, 2700m, fl., 10 Jun. 1996, Cable 3068; Summit, 2980m, fl., 10 Jun. 1996, Cable 3074; Lake Oku, 2300m, fl. fr., 11 Jun. 1996, Zapfack 874; Shore of Lake Oku, 2250m, fl., 12 Jun. 1996, Cable 3121; Elak, 2400m, fl. fr., 28 Oct. 1996, Munyenyembe 720; Elak, KC, 2600m, fl., 30 Oct. 1996, Zapfack 1109.
Ijim: Ijim Mountain Forest, 1800m, fl., 21 Nov. 1996, Pollard 73.

BUDDLEJACEAE
Det. A.J.M.Leeuwenberg (WAG)

Buddleja davidii Franch.
Shrub to 2 m; leaves oblong-lanceolate; inflorescence a raceme, about 15 cm long; flowers purple. Forest edge.
Kilum: Elak, 2000m, fl., 8 Jun. 1996, Zapfack 767; Elak, KD, fl., 9 Jun. 1996, Etuge 2248.

Nuxia congesta R.Br. ex Fresen.
Tree to 27 m; bark fluted, sometimes whitish-grey; leaves opposite or in whorls of 3, elliptic, 5–13 cm long; flowers creamy-white, scented. Forest.
Kilum: Elak, KJ, *Podocarpus* forest, 2500m, fl., 29 Oct. 1996, Buzgo 632.
Ijim: Aboh, Gikwang towards Nyasosso bush, 2260m, fl., 18 Nov. 1996, Etuge 3460.

BURSERACEAE

Canarium schweinfurthii Engl.
Tree c. 35m with aromatic resinous exudate; leaves imparipinnate to 60 cm long; fruits fleshy, dark purple, c. 2 cm long. Forest.
Ijim: Akeh, 1570m, 11 Nov. 1999, Cheek 10071.

CALLITRICHACEAE

Callitriche stagnalis Scop.
Terrestrial herb, sometimes aquatic, forming carpets c. 2 cm; leaves opposite, elliptic, c. 1 × 0.5 cm; flowers inconspicuous. Open, damp ground.

Ijim: Above Aboh village, 1.5 hours walk along Gikwang road towards Lake Oku, track towards TA, 6.1112N 10.2529E, 2400m, fl. fr., 20 Nov. 1996, Cheek 8694; Laikom, swamp E of Ardo's compound, 2000m, 6.16N, 10.22E, fl., 5 Dec. 1998, Cheek 9796; Laikom, swamp half a mile W of Ardo's Fulani compound, Ijim ridge grassland plateau, 2100m, 6.1609N, 10.2051E, fl., 7 Dec. 1998, Pollard 339.

CAMPANULACEAE
M.Thulin (UPS), M.Cheek (K) & J-M.Onana (YA)

Lobelia adnexa E.Wimm.
Annual herb c. 5 cm; leaves mostly basal; peduncle angular; flowers white, c. 5 mm long. Cliffs, grassland.
Kilum: Cliff, one hours walk above village at a place called Asu, 2200m, fl. fr., 22 Nov. 1996, Cheek 8729.

Lobelia columnaris Hook.f.
Local name: Ndyaa (Oku) – Munyenyembe 713; Tongidzi (Bekom) – Kamundi 632.
RED DATA LISTED
Erect simple herb, 2–3 m; leaves sessile, lanceolate, acute, to 20 × 3 cm; flowers blue or lilac. Forest edge and fallow.
Kilum: Elak, forest edge, 2200m, fl., 28 Oct. 1996, Munyenyembe 713.
Ijim: Gikwang, towards Nyasosso bush, 2280m, fl., 18 Nov. 1996, Etuge 3455; Ijim Mountain Forest, 2160m, fl., 19 Nov. 1996, Kamundi 632.

Lobelia hartlaubii Buchenau
Decumbent or ascending herb, 30–60 cm; stems branched, sparsely pilose, often rooting below; flowers white, blue or mauve,margins white. Fallow.
Ijim: From Aboh, path towards Tum, 2100m, fl., 22 Nov. 1996, Etuge 3575.

Lobelia minutula Engl.
Small herb; stems creeping; leaves ovate-rounded, 4–5 mm; flowers deep blue or white-spotted. Grassland; swamp.
Kilum: Ntogemtuo swamp, 2950m, fl., 14 May 1997, Maisels 47.
Ijim: Laikom, 2000m, fl., Apr. 1931, Maitland 1773; Sappel, below Ardo's uncle's compound, 2300m, 6.16N, 10.24E, fl., 5 Dec. 1998, Cheek 9805.

Lobelia neumannii T.C.E.Fr.
Local name: Unknown – Cheek 8637.
Annual herb, 5–8 cm; leaves cauline, elliptic up to 1 cm long; peduncle terete, sparsely white, hairy; flowers dull blue, c. 5 mm long. Between tussocks in grassland.
Kilum: Elak, 3000m, fl., 29 Oct. 1996, Munyenyembe 755; Path from Project HQ to KA, 2200m, fl., 30 Oct. 1996, Cheek 8497; Above Mboh, path above village to forest through Wambeng's farm, 2200m, fl., 1 Nov. 1996, Cheek 8556.
Ijim: Above Aboh village, 1.5 hours walk along Gikwang road, 2380m, fl., 18 Nov. 1996, Cheek 8637; Gikwang road

towards Nyasosso forest, 2090m, fl., 19 Nov. 1996, Etuge 3472; Laikom, 2100m, fl., 21 Nov. 1996, Onana 570; Along the ridge towards a point where Lake Oku can be viewed, 2600m, fl. fr., 22 Nov. 1996, Kamundi 688; Laikom, N facing, seasonally damp rockface amongst dense vegetation, by 'gallery' forest along path from Fon's Palace to grassland plateau above Laikom, 2020m, fl., 2 Dec. 1998, Pollard 282; Ijim plateau above Fon's Palace, 1900m, 6.16N, 10.19E, fl., 3 Dec. 1998, Cheek 9727.

Lobelia rubescens De Wild.
Erect, subglabrous herb to 90 cm; stem often decumbent or trailing and rooting at base; flowers blue, 1 cm. Fallow.
Kilum: Elak, path from BirdLife HQ to KA, 2200m, 6.1349N, 10.3112E, fl. fr., 30 Oct. 1996, Cheek 8498.
Ijim: Mbizenaku, 1660m, fl., Jun. 1931, Maitland 1393; Main path from Ijim to Tum, 2100m, 6.1116N, 10.2543E, fl., 25 Nov. 1996, Etuge 3623; Abandoned farms near Laikom, 1950m, 6.16N, 10.20E, fl. fr., 8 Dec. 1998, Etuge 4574.

Lobelia sapinii De Wild.
Glabrous erect branched herb, to 20 cm; leaves ovate to ovate-elliptic, 1.5–5 cm long; flowers blue or white with deep blue stripes, or bright blue with two rows of dark dots across lower lip. Fallow.
Babessi: 2 miles E of village, fl. fr., 22 Oct. 1947, Tamajong FHI 22101.

Monopsis stellarioides (C.Presl) Urb. subsp. *schimperiana* (Urb.) Thulin
Small straggling herb; stems weak, 30–60 cm long; flowers yellow with dull red throat, often reddish outside. Grassland.
Jakiri: Above farm, 1780m, fl. fr., 13 Feb. 1958, Hepper 1957.

Wahlenbergia krebsii Cham. subsp. *arguta* (Hook.f) Thulin
Slender herb to 30 cm; stems straggling; flowers whitish or pale mauve, veins blue. Grassland.
Syn. *Wahlenbergia arguta* Hook.f., F.W.T.A. 2: 309 (1963).
Kilum: Ridge above Lake Oku, grassland, 2560m, fl., 6 Jan. 1951, Keay & Lightbody FHI 28469; Summit, 2900m, fl., 29 Oct. 1996, Muyenyembe 800; Oku-Elak, between Fon's Palace and Manchok, bank along road in village, 2000m, fl., 31 Oct. 1996, Cheek 8537; Road from Rest House E of lake to N of mountain range, 2300m, fl., 5 Nov. 1996, Cheek 8592; Manchock, lowland areas, path leading towards Kumbo road, 2100m, fl., 9 Nov. 1996, Etuge 3362A.
Ijim: Aboh village, 2450m, fl., 19 Nov. 1996, Onana 537.

Wahlenbergia perrottetii (A.DC.) Thulin
Herb, very variable, 10–60 cm; leaves oblong-lanceolate, 1–5 cm; flowers very small, pale mauve. Forest edge.
Syn. *Cephalostigma perrottetii* A.DC., F.W.T.A. 2: 311 (1963).
Ijim: Fundong, near touristic house, near the waterfall, 1400m, fl., 23 Nov. 1996, Onana 596.

Wahlenbergia ramosissima (Hemsl.) Thulin
subsp. *ramosissima*
RED DATA LISTED
Erect branched annual, 7–25 cm; stems slender, ribbed, hispid; leaves 1 cm long; flowers blue. Grassland.
Syn. *Cephalostigma ramosissimum* Hemsl., F.W.T.A. 2: 311 (1963).
Kilum: Above Fon's Palace, about 30 mins. walk on path to Fulani settlement, 2050m, fl., 21 Nov. 1996, Cheek 8717 & fl., 22 Nov. 1996, Cheek 8730.
Ijim: Laikom, Ijim plateau above Fon's Palace, 1900m, 6.16N, 10.19E, fl., 3 Dec. 1998, Cheek 9729.

Wahlenbergia silenoides Hochst. ex A.Rich.
Herb with wiry stems from horizontal perennial base; leaves lanceolate, 1 cm long; flowers pale bluish or white, veins blue. Grassland.
Syn. *Wahlenbergia mannii* Vatke, F.W.T.A. 2: 309 (1963).
Kilum: Ridge above Lake Oku, grassland, fl., 6 Jan. 1951, Keay FHI 28482; Kumbo to Oku, about 3 miles on Oku road, 1850m, fl., 15 Feb. 1958, Hepper 2010; Path above Mboh village to forest through Wambeng's farm, 2200m, fl., 1 Nov. 1996, Cheek 8553; Path from Mboh village, about 6 km from Elak-Oku, 2200–2500m, fl., 1 Nov. 1996, Onana 490; Manchock, lowland areas, path leading towards Kumbo road, 2100m, fl., 9 Nov. 1996, Etuge 3362.
Ijim: Gikwang, towards Nyasosso bush, 2250m, fl., 18 Nov. 1996, Etuge 3461; Aboh village, fl., 18 Nov. 1996, Onana 532; Laikom, Ijim plateau above Fon's Palace, 1900m, 6.16N, 10.19E, fl., 3 Dec. 1998, Cheek 9726.

CAPPARACEAE

Cleome iberidella Welw. ex Oliv.
Local name: Ngei (Oku) – Munyenyembe 895.
Prostrate herb; stems c. 30 cm long, glandular at apex; leaves 3–5-foliolate; flowers zygomorphic, c. 1 cm long, purple. Fallow.
Kilum: Lake route, 2000m, fl. fr., 11 Jun. 1996, Zapfack 839; Lake Oku, 2200m, fl. 12 Jun. 1996, Etuge 2336; Mbokengfish, 2000m, fr., 7 Nov. 1996, Munyenyembe 895.
Ijim: Above Fon's Palace, about 30 mins. walk along path to Fulani settlement, 2050m, fl. fr., 22 Nov. 1996, Cheek 8736; Anyajua Forest, 1850m, fl. fr., 27 Nov. 1996, Kamundi 711; Anyajua Forest, 1800m, fl., 27 Nov. 1996, Kamundi 722; Laikom to Fundong, 1700m, fl., 10 Dec. 1998, Etuge 4590.

CARYOPHYLLACEAE

Cerastium octandrum Hochst. ex A.Rich.
Sticky herb, diffuse or ascending; leaves to 1.5 cm long; inflorescence a leafy cyme; flowers numerous, closely packed white. Grassland.
Ijim: Above Aboh village, 1.5 hours along Gikwang road towards Lake Oku, 2450m, fr., 19 Nov. 1996, Cheek 8666; Gikwang road, Aboh to Akeh, about 1 km after first track, 2400m, fl. fr., 25 Nov. 1996, Cheek 8750; Forest patch towards foot of waterfall above Akwamofu sacred forest, Laikom, 1860m, fr., 11 Dec. 1998, Etuge 4604.

Drymaria cordata (L.) Willd.
Straggling, slightly viscid herb; stems to 90 cm; leaves broadly ovate, 1–2.5 cm long; inflorescence a terminal cyme; flowers few, white. Shady banks.
Kilum: Path above water-tower, KJ, about 1 km into forest, 2250m, fl., 9 Jun. 1996, Cable 2993; Lake Oku, 2200m, fl., 11 Jun. 1996, Zapfack 870; Elak, KJ, 2400m, fl., 29 Oct. 1996, Zapfack 1071; Oku-Elak, between Fon's Palace and Manchok, bank along road in village, 2000m, fl., 31 Oct. 1996, Cheek 8540.
Ijim: Gikwang, towards Nyasosso bush, 2260m, fr., 18 Nov. 1996, Etuge 3449; Ijim Mountain Forest, 2160m, fl., 19 Nov. 1996, Kamundi 612; Aboh village, 2450m, fl. fr., 19 Nov. 1996, Onana 540.

Drymaria villosa Cham. & Schltdl. subsp. *villosa*
Straggling, slightly viscid herb; stems to 90 cm long, sparsely long hairy; leaves broadly ovate, smaller than in *Drymaria cordata*; inflorescence a terminal cyme; flowers few, white. Shady banks.
Kilum: Oku-Elak, between Fon's Palace and Manchok, bank along road in village, 2000m, fl. fr., 31 Oct. 1996, Cheek 8541; Elak, 2100m, fr., 1 Nov. 1996, Munyenyembe 854; Ewook Etele Mbae, 10 km from Oku, 2350m, fr., 6 Nov. 1996, Zapfack 1232.
Ijim: Chuhuku river, near Anyajua Forest, 1580m, fr., 28 Nov. 1996, Kamundi 728.

Sagina abyssinica Hochst. ex A.Rich.
Small, often tufted herb, forming little cushions; leaves linear, acute, to 2.5 cm; sepals green, petals white. Grassland.
Kilum: Path from BirdLife HQ to top of Mt Oku, 2600m, fr., 12 Oct. 1996, Cheek 8418; Path from Project HQ to KA, 2800m, fr., 30 Oct. 1996, Cheek 8513; Summit, 2800m, fr., 31 Oct. 1996, Zapfack 1127.

Stellaria mannii Hook.f.
Diffuse straggling herb, 30-60 cm; leaves long-petiolate; inflorescence a terminal cyme; flowers few, white. Forest.
Kilum: Oku-Elak to the forest at KA, 2200m, fl., 27 Oct. 1996, Cheek 8453; KJ, 2300m, fl., 28 Oct. 1996, Buzgo 594;
Ijim: Gikwang towards Nyasosso, 2300m, fl., 18 Nov. 1996, Etuge 3414; Above Aboh village, 1.5 hours along Gikwang road towards Lake Oku, track towards transect TA, 2400m, fl., 20 Nov. 1996, Cheek 8680; Ijim Mountain Forest, along the top of the ridge, second gate from Aboh village, 2370m, fl., 21 Nov. 1996, Kamundi 665.
Plot voucher: Forest above Oku-Elak, Jun. 1996, Oku 55.

Stellaria media (L.) Vill.
Diffuse herb; branches leafy; inflorescence a lax terminal cyme; flowers white, solitary, in forks of branches. Forest edge.
Kilum: Summit, 2950m, fr., 10 Jun. 1996, Cable 3083.
Ijim: Gikwang road towards Nyasosso forest, 2300m, fr., 19 Nov. 1996, Etuge 3482.

Uebelinia abyssinica Hochst.
Local name: 'Eghang (Oku) – Cheek 8478.
Prostrate herb; stems straggling, hispid, to 45 cm long;

leaves elliptic to orbicular, 7–15 mm long; flowers axillary, white. Grassland.
Syn. *Uebelinia hispida* Pax, F.W.T.A. 1: 130 (1954).
Kilum: Oku-Elak, lower parts of KA, 2200m, fr., 28 Oct. 1996, Cheek 8478; Summit, 3000m, fr., 29 Oct. 1996, Munyenyembe 754; Summit, 2800m, fr., 31 Oct. 1996, Buzgo 672; Path from Mbo village, 2200–2500m, fr., 1 Nov. 1996, Onana 489; Ewook Etele, Oku-Mbae, 2350m, fr., 6 Nov. 1996, Etuge 3348.
Ijim: Above Aboh village, 1.5 hours along Gikwang road to Lake Oku, 2350m, fr., 18 Nov. 1996, Cheek 8647; Ijim plateau above Fon's Palace, Laikom, 1900m, fr., 3 Dec. 1998, Cheek 9737.

CELASTRACEAE
Det. Sebsebe Demissew (ETH)

Cassine aethiopica Thunb.
Shrub or small tree, 7 m or more; leaves variable, 0.7–13 cm long; petals whitish; ripe fruits reddish. Forest.
Syn. *Mystroxylon aethiopicum* (Thunb.) Loes.
Kilum: Lake Oku, 2200m, fl. fr., 4 Nov. 1996, Satabie 1067.

Maytenus buchananii (Loes.) Wilczek
Local name: Alamse (Kom at Laikom) – Etuge 4509.
Shrub or climber with spines 2–3 cm long; leaves elliptic, 3.5–6 × 1.3–2.5 cm, crenate-dentate, secondary nerves 7 pairs; flowers white; fruits obovoid, c. 1.5 cm long. Forest.
Syn. *Maytenus ovatus* (Wall. ex Wight & Arn.) L. var. *ovatus* fa. *pubescens* (Schweinf.) Blakelock, sensu F.W.T.A. 1: 625 (1958) pro parte.
Ijim: Near Aboh, Anyajua, 2500m, fl., 20 Nov. 1996, Satabie 1082B; Zitum road, 1900m, 6.1116N, 10.2543E, fl., 21 Nov. 1996, Etuge 3527; Laikom, beside a stream, 2000m, fr., 2 Dec. 1998, Etuge 4509; Afua swamp, by forest, 1950m, 6.09N, 10.24E, fl., 7 Dec. 1998, Etuge 4569; Plot between Fon's Palace and Ardo's compound, Laikom, 2000m, 6.16N, 10.21E, fl. fr., 8 Dec. 1998, Cheek 9867.

Maytenus undata (Thunb.) Blakelock
Local name: Lyeyese (Oku) – Munyenyembe 785.
Shrub or tree, to 8m; flowers whitish; fruits mostly reddish when ripe. Forest.
Kilum: KD, 2800m, fr., 10 Jun. 1996, Etuge 2271; KJ to KD, 3000m, fr., 13 Jun. 1996, Etuge 2353; Summit, 2800m, 29 Oct. 1996, Munyenyembe 785.
Ijim: Near Aboh, Anyajua, 2500m, fl., 20 Nov. 1996, Satabie 1082A; Laikom, Ijim forest, 2000m, 6.17N, 10.20E, fl. fr., 3 Dec. 1998, Etuge 4531.

Salacia sp.
Sterile voucher for generic record.
Kilum: Above Mboh, path above village to forest through Peter Wambeng's farm, 2200m, 6.1256N, 10.2911E, st., 1 Nov. 1996, Cheek 8549.

CERATOPHYLLACEAE

Ceratophyllum demersum L. var. demersum fa. demersum

Glabrous perennial aquatic herb; branches elongated, floating; leaves in whorls of 5–12, divided into linear segments; flowers small, axillary, sessile; fruit broadly ellipsoid, with 3 horns. Fresh water lakes.
Kilum: Lake Oku, near Baptist Rest House, 2300m, 4 Nov. 1996, Cheek 8576; Lake Oku, 2200m, fl., 4 Nov. 1996, Satabie 1068.

CHENOPODIACEAE

Chenopodium ambrosioides L.
Herb, to 1.2 m, leaves lanceolate; inflorescence an ample panicle; flowers small, sessile, green, in abbreviated cymes. Weed.
Kilum: Oku, inside of Fon's Palace, area in waste ground, 2000m, fl., 17 Feb. 1958, Hepper 2022; Path from Mbo village, about 6 km from Elak-Oku, 2200–2500m, fl., 3 Nov. 1996, Onana 495.

Chenopodium congolanum (Hauman) Brenan
Herb 30–60 cm, glabrous apart from vesicular hairs; leaves narrowly elliptic, sinuate-lobate, 1.5–3 cm long; flowers in small axillary clusters. Weed.
Jakiri: Inside Fulani cattle-pound, above and to W of village on path to farm, fl. fr., 18 Feb. 1958, Hepper 2079.

COMBRETACEAE
Det. C.C.H.Jongkind (WAG) & B.J.Pollard (K)

Combretum fuscum Planch. ex Benth.
Local name: Ilik (Oku) – Munyenyembe 836.
Scandent shrub or woody forest liane; leaves coriaceous, 10–20 × 5–10 cm; flowers white.
Kilum: Elak, 2100m, fl., 1 Nov. 1996, Munyenyembe 836.
Ijim: Laikom, 6.17N, 10.20E, fl., 3 Dec. 1998, Etuge 4517.

Combretum racemosum P.Beauv.
Scandent shrub or forest liane; bracteate leaves white with green veins or rarely pink; petals and stamens vivid dark red.
Ijim: Anyajua, waterfall near BirdLife HQ, 1300m, 6.11N, 10.22E, fl., 12 Dec. 1998, Cheek 9922.

Terminalia avicennioides Guill. & Perr.
Tree 3–5 m; leaves alternate, lanceolate, narrowly elliptc or obovate, 8–15 × 3–6 cm, apex rounded or notched, base cuneate or obtuse, white or brown, felty below; petiole 1.5–3 cm; fruits dry, dull red, 2-winged, narrowly elliptic, c. 5 × 2 cm. Savanna.
Babungo-Mbi Crater: 1600m, 8 Nov. 1999, Cheek 9947.

COMPOSITAE
H.J.Beentje (K), J-M.Onana (YA), B.J.Pollard (K) & D.J.N.Hind (K)

Acmella caulirhiza Delile
Local name: Fesusngokuyi (Oku) – Munyenyembe 869; Fesus fe ngang (Kom at Laikom) – Cheek 8676; Feyeye fafe – Cheek 8677; Fsusf ngag (Kom at Laikom) – Etuge 4545.
Creeping herb; capitula conical with tiny yellow to orange

flowers; ray flowers present, sometimes tiny. Fallow.
Syn. *Spilanthes africana* DC., F.W.T.A. 2: 236 (1963).
Kilum: KC, 2700m, 6.15N, 10.26E, fl., 31 Oct. 1996, Onana 477; Path above Mboh village to forest through Peter Wambeng's farm, 2200m, 6.1146N, 10.2856E, fl., 1 Nov. 1996, Cheek 8567; Upkim Traditional forest, 2060m, fl., 4 Nov. 1996, Munyenyembe 869.
Ijim: Laikom, Maitland 1655; Above Aboh village, 1.5 hours along Gikwang road to Lake Oku, 2000m, 6.1112N, 10.2529E, fl., 19 Nov. 1996, Cheek 8676 & 8677; Zitum road, 1860m, 6.1116N, 10.2543E, fl., 21 Nov. 1996, Etuge 3537; Akwamofu medicinal forest, 1760m, 6.17N, 10.20E, fl., 5 Dec. 1998, Etuge 4545.

Adenostemma caffrum DC. var. *asperum* Brenan
Erect or decumbent herb to 60 cm; capitula with tiny white flowers. Swamp.
Ijim: Afua swamp, 1950m, 6.08N, 10.24E, fl. fr., 7 Dec. 1998, Cheek 9830.

Adenostemma mauritianum DC.
Annual or short-lived perennial herb to 40 cm, with 3-nerved leaves; capitula small with tiny white flowers. Forest.
Kilum: Hepper 2047; KJ, 2300m, fl., 28 Oct. 1996, Buzgo 600; Elak, 2420m, fl., 28 Oct. 1996, Munyenyembe 731; Path from Project HQ to KA, 2600m, 6.1349N, 10.3112E, fl., 30 Oct. 1996, Cheek 8522.
Ijim: Gikwang road towards Nyasosso, 2290m, fl. fr., 18 Nov. 1996, Etuge 3412.

Adenostemma viscosum J.R.Forst. & G.Forst.
Robust perennial herb with larger leaves than *Adenostemma mauritianum* DC. Roadside.
Syn. nov. *Adenostemma perottetii* DC. , F.W.T.A. 2: 286 (1963).
Ijim: Aboh village, along the main road, 1700–1800m, 6.15N, 10.26E, fl. fr., 24 Nov. 1996, Onana 610.

Ageratum conyzoides L. subsp. *conyzoides*
Annual herb to 1.5 m, smelly, hispid; leaves opposite; capitula small with tiny white or blue flowers. Weed.
Ijim: Gikwang road towards Nyasosso forest, Ijim Mountain Forest Reserve, 2090m, fl. fr., 19 Nov. 1996, Etuge 3468.

Anisopappus chinensis Hook. & Arn. subsp. *buchwaldii* (O.Hoffm.) S.Ortíz, Paiva & Rodr.Oubiña var. *buchwaldii*
Perennial herb to 1.5 m; capitula with yellow flowers; ray flowers present. Fallow; grassland.
Syn. *Anisopappus africanus* (Hook.f.) Oliv. & Hiern, F.W.T.A. 2: 258 (1963).
Syn. *Anisopappus suborbicularis* Hutch. & B.L.Burtt, F.W.T.A. 2: 258 (1963).
Kilum: Kumbo to Oku, Hepper 2006; Oku-Elak to KA, 2200m, fl., 31 Oct. 1996, Buzgo 682; Elak, 2200m, fl., 31 Oct. 1996, Munyenyembe 822.
Ijim: Gikwang main road through grassland to Nyasosso forest, 2250m, 6.1116N, 10.2543E, fl., 20 Nov. 1996, Etuge 3512; Border of Lake Oku , fl., 26 Nov. 1996, Onana 637; Grassland above Laikom, 2000m, 6.17N, 10.20E, fl., 2 Dec.

1998, Etuge 4513; Grassland plateau above Fon's Palace, Laikom, 2040m, 6.1642N, 10.1950E, fl., 2 Dec. 1998, Pollard 277.

Aspilia africana (Pers.) C.D.Adams
Small herb; capitula with yellow flowers; ray flowers present. Weed.
Kilum: Elak, 2000m, 8 Jun. 1996, Zapfack 772.

Bafutia tenuicaulis C.D.Adams var. *tenuicaulis*
RED DATA LISTED
Erect herb to 30 cm; capitula with small pinkish-purple florets. Rock outcrops.
Kilum: Shambai, 2500m, fl. fr., 5 Nov. 1996, Munyenyembe 878; Summit, 2480m, fl. fr., 5 Nov. 1996, Zapfack 1199; Ewook Etele Mbae, 10 km from Oku, 2350m, fr., 6 Nov. 1996, Zapfack 1220 & 1224; Ngvuinkei I, 1700m, fl. fr., 6 Nov. 1998, Maisels 167.
Ijim: Sappel, below Ardo's uncle's compound, 6.16N, 10.24E, fl., 5 Dec. 1998, Cheek 9821; Watering hole by Fulani settlement, 11 km E of Laikom, 2400m, 6.1615N, 10.2435E, fl., 9 Dec. 1998, Pollard 358.

Bafutia tenuicaulis C.D.Adams var. *zapfackiana* Beentje & B.J.Pollard
RED DATA LISTED
Differs from the type variety in that the stem, leaves and especially pedicel and involucre are pilose to pubescent, not glabrous; the involucre base is much larger (2.5 to 6.5 mm), there are 8 to 14 involucre lobes (as opposed to 8), with paler zones of tearing below sinus (no zones) and the pappus setae are 0.5 to 0.8mm long, as opposed to 0.1 to 0.5 mm. Fallow and rock outcrops.
Kilum: Elak, fallow, 2200m, fl. fr., 31 Oct. 1996, Munyenyembe 825; Summit, 2800m, fl. fr., 31 Oct. 1996, Zapfack 1130.

Berkheya spekeana Oliv.
Shrubby thistle-like herb to 2.5 m; capitula with yellow ray flowers and orange-yellow disc flowers. Grassland.
Ijim: Aboh, 2250m, fl., 19 Nov. 1996, Pollard 53.

Bidens barteri (Oliv. & Hiern) T.G.J.Rayner
Local name: NJoh (Kom – Yama Peter) – Gosline 199.
Annual herb to 90 cm; capitula with golden-yellow flowers; ray flowers present. Grassland
Syn. *Coreopsis barteri* Oliv. & Hiern, F.W.T.A. 2: 232 (1963).
Kilum: Elak, KA, 2200m, fl., 11 Jun. 1996, Etuge 2309; Ewook Etele Mbae, 10 km from Oku, 2350m, fl., 6 Nov. 1996, Zapfack 1237.
Ijim: Before Ardo's compound above Laikom, near stream crossing, 2100m, 6.16N, 10.21E, fl. fr., 7 Dec. 1998, Gosline 199.

Bidens camporum (Hutch.) Mesfin
Local name: Vola (Oku) – Munyenyembe 766; Njuo (Kom) – Cheek 8673B.
Perennial herb to 1.2 m with several stems; capitula with golden-yellow ray flowers. Forest edge.

Syn. *Coreopsis camporum* Hutch., F.W.T.A. 2: 232 (1963).
Kilum: Summit, 3000m, fl., 29 Oct. 1996, Munyenyembe
766; Fallow, 15 minutes from Manchok to KA, 2200m, fl.,
31 Oct. 1996, Buzgo 684.
Ijim: Above Aboh village, 1.5 hours along Gikwang road to
Lake Oku, 2450m, 6.1112N, 10.2529E, fl., 19 Nov. 1996,
Cheek 8673B; Anyajua, 2400m, fl., 19 Nov. 1996, Satabie
1074; Gikwang main road towards Nyasosso forest, 2270m,
6.1116N, 10.2543E, fl., 20 Nov. 1996, Etuge 3507;
Grassland plateau above Fon's Palace, Laikom, 2000m, fl., 2
Dec. 1998, Pollard 266.

Bidens sp. aff. camporum (Hutch.) Mesfin
Differs from *Bidens camporum* in having smaller ray florets
and less divided leaves. Forest edge.
Ijim: Start of path from Fon's Palace to grassland plateau,
Laikom, 2060m, 6.1642N, 10.1950E, fl., 4 Dec. 1998,
Pollard 309.

Bidens mannii T.G.J.Rayner
Woody herb to 1 m; young stems and leaves reddish; capitula
with pale yellow ray flowers. Grassland.
Syn. *Coreopsis monticola* (Hook.f.) Oliv. including var.
pilosa Hutch. & Dalziel, F.W.T.A. 2: 232 (1963).
Kilum: KA to summit, 2800m, 6.1349N, 10.3112E, fl., 30
Oct. 1996, Cheek 8531; Summit, 2800m, 31 Oct. 1996,
Zapfack 1114.

Bidens pilosa L.
Local name: Kijis (Oku) – Munyenyembe 724.
Annual herb to 1.5 m with large pinnate leaves; capitula
usually without ray flowers; disc flowers yellow. Forest
edge.
Kilum: KD, 2400m, fl. fr., 10 Jun.1996, Cable 3053;
Montane forest, 2400m, fl. fr., 28 Oct. 1996, Munyenyembe
724; Towards KJ, 2100m, fr., 28 Oct. 1996, Pollard 2.
Ijim: Gikwang rd. to Nyasosso f.orest, 2095m, fr.19 Nov.
1996, Etuge 3473.

Blumea axillaris (Lam.) DC.
Usually annual, to 90 cm; whole plant sticky and aromatic;
capitula with minute pink to purple flowers. Grassland.
Syn. *Blumea mollis* (D.Don) Merr., F.W.T.A. 2: 261
(1963).
Syn. *Blumea perottetiana* DC., F.W.T.A. 2: 261 (1963).
Ijim: In grassland, Laikom, 1830m, fl. fr., May 1931,
Maitland 1439.

Carduus nyassanus (S.Moore) R.E.Fr. subsp.
nyassanus
Biennial or perennial herb to 3 m, spiny all over; capitula
with many small white or mauve flowers. Swamp edge.
Kilum: Hepper 2719.
Ijim: Laikom, on edge of bog in grassland plateau, 1850m,
Apr. 1931, Maitland 1649; Sappel, below Ardos' uncle's
compound, 6.16N, 10.24E, fl., 5 Dec. 1998, Cheek 9815.

Conyza attenuata DC.
Annual or biennial herb to 3 m. Capitula small with many
tiny white to yellow flowers. Fallow.

Syn. *Conyza persicifolia* (Benth.) Oliv. & Hiern, F.W.T.A.
2: 254 (1963).
Kilum: Elak, 2100m, fl., 8 Jun. 1996, Zapfack 769.

Conyza bonariensis (L.) Cronquist
Annual herb to 2 m; capitula with many tiny white to yellow
flowers. Fallow and grassland.
Syn. *Erigeron bonariensis* L., F.W.T.A. 2: 253 (1963).
Syn. *Erigeron floribundus* (Kunth.) Sch.Bip., F.W.T.A. 2:
253 (1963).
Kilum: Elak to KD, fl., 9 Jun. 1996, Etuge 2202; Path from
KN to summit, 2200m, fl., 10 Jun. 1996, Cable 3042; Elak
to KJ, 2100m, fl., 30 Oct. 1996, Pollard 17.
Ijim: Gikwang road towards Nyasosso forest, Ijim Mountain
Forest Reserve, 2090m, fl., 19 Nov. 1996, Etuge 3469; Aboh
village, 2450m, fl., 20 Nov. 1996, Onana 556.

Conyza clarenceana (Hook.f.) Oliv.& Hiern
Perennial stoloniferous herb to 1.5 m; stems reddish;
capitula with many tiny yellow flowers. Swamp edge.
Syn. *Conyza theodori* R.E.Fr., F.W.T.A. 2: 254 (1963).
Ijim: Laikom, by stream, 2000m, fl. fr., May 1931, Maitland
1378; In grassland, on plateau, Laikom, 2000m, fl. fr., Apr.
1931, Maitland 1480; Sappel, below Ardo's uncle's
compound, 2300m, 6.16N, 10.24E, fl. fr., 5 Dec. 1998,
Cheek 9812.
Mbi Crater: 1950m, 6.05N, 10.21E, fl. fr., 9 Dec. 1998,
Cheek 9904.
Kumbo: Hepper 2698.

Conyza gouanii (L.) Willd.
Annual herb to 1 m, stems often red; capitula with many tiny
white or cream flowers. Fallow.
Syn. *Conyza hochstetteri* sensu Sch.Bip. ex. A.Rich. non
C.D.Adams
Kilum: Path from Project HQ to KA, 2200m, 6.1349N,
10.3112E, fl., 30 Oct. 1996, Cheek 8496; Lumeto forest near
Oku-Elak, 2200m, 6.1412N, 10.2943E, fl., 7 Nov. 1996,
Cheek 8616.

Conyza pyrrhopappa A.Rich.
Shrub or woody herb to 3 m; leaves aromatic; capitula with
tiny white or yellow flowers. Forest edge.
Syn. *Microglossa pyrrhopappa* (A.Rich.) Agnew & Agnew,
Upl. Kenya Wild Fl., ed. 2 (1994).
Syn. *Microglossa angolensis* Oliv. & Hiern, F.W.T.A. 2:
251 (1963).
Kilum: Junction of KJ and KN, 2200m, fl., 10 Jun. 1996,
Cable 3024; Elak to KJ, 2100m, fl., 30 Oct. 1996, Pollard
18; Mbokengfish, 1900m, fl., 7 Nov. 1996, Munyenyembe
905.
Ijim: Gikwang road, Aboh to Akeh, about 1 km after first
track, 2400m, 6.1412N, 10.3112E, fl., 25 Nov. 1996, Cheek
8748.

Conyza subscaposa O.Hoffm.
Subscapose herb; capitula with tiny pale yellow flowers.
Grassland.
Kilum: KD, towards Elak, fl., 9 Jun. 1996, Etuge 2201; Path
from KN to summit, 2900m, fl., 10 Jun. 1996, Cable 3036;

KD, 2800m, fl., 10 Jun. 1996, Cable 3052; Summit, 2900m, fl., 10 Jun. 1996, Cable 3066; Lake Oku to Jikijem, 2200m, 6.1214N, 10.2733E, fl., 5 Nov. 1996, Cheek 8603.
Ijim: Plateau above Fon's Palace, Laikom, 2000m, 6.1642N, 10.1950E, fl., 2 Dec. 1998, Pollard 258 & 274; Laikom, between Fon's Palace and Ardo's compound, 2000m, 6.16N, 10.21E, fl. fr., 8 Dec. 1998, Cheek 9882.

Crassocephalum bauchiense (Hutch.) Milne-Redh.

Local name: Kitunshu (Oku) – Munyenyembe 848; Abul A Mussa Egom (Kom at Laikom) = 'The dust of the sunbird' – Cheek 9753.
Bushy annual herb to 30 cm; capitula small, with tiny bright blue flowers. Grassland.
Kilum: Path from Project HQ to KA, 2200m, 6.1349N,10.3112E, fl., 30 Oct. 1996, Cheek 8492; Forest edge, 2200m, fl., 1 Nov. 1996, Munyenyembe 848; Lake Oku, 2300m, fl., 4 Nov. 1996, Pollard 38.
Ijim: Laikom, grassland, 2100m, 6.15N, 10.26E, fl., 21 Nov.1996, Onana 576; Ijim plateau above Fon's Palace, 1900m, 6.16N, 10.19E, fr., 3 Dec. 1998, Cheek 9753; Forest patch towards foot of waterfall above Akwamofu sacred forest, Laikom, 1860m, 6.17N, 10.20E, fl., 11 Dec. 1998, Etuge 4599.

Crassocephalum bougheyanum C.D.Adams

Straggling herb to 2.5 m; capitula with small yellow flowers. Forest edge.
Kilum: Elak, 2100m, fl., 8 Jun. 1996, Zapfack 781; Elak to the forest at KA, 2200m, 6.1349N, 10.3112E, fl., 27 Oct. 1996, Cheek 8446; Elak, KJ, 2300m, fl., 29 Oct. 1996, Zapfack 1055; Elak, 2400m, fl., 30 Oct. 1996, Munyenyembe 813.
Ijim: Laikom, grassland, 2100m, fl., 21 Nov. 1996, Satabie 1086; Laikom, forest slope, 2000m, 6.17N, 10.20E, fl., 3 Dec. 1998, Etuge 4522; Near Akwamofu sacred forest, 1900m, 6.1633N, 10.1940E, fl., 5 Dec. 1998, Pollard 325.

Crassocephalum crepidioides (Benth.) S.Moore

Annual herb to 1.5 m; capitula with many tiny yellow to red-brown flowers. Fallow.
Kilum: Path from BirdLife HQ to KA, 2200m, 6.1349N, 10.3112E, fl., 30 Oct. 1996, Cheek 8493.

Crassocephalum gracile (Hook.f.) Milne-Redh.

Herb to 90 cm; capitula with tiny yellow to orange flowers. Fallow.
Kilum: Elak, KD, fl., 9 Jun. 1996, Etuge 2234.
Ijim: Swamp E of Ardo's compound, Laikom, 2000m, 6.16N, 10.22E, fl. fr., 5 Dec. 1998, Cheek 9794.

Crassocephalum montuosum (S.Moore) Milne-Redh.

Local name: Tindansi (Oku) – Munyenyembe – 867.
Annual or perennial herb to 1 m; leaves deeply lobed; capitula with many tiny yellow to orange flowers. Forest edge.
Kilum: Upkim traditional forest, 2060m, fl., 4 Nov. 1996, Munyenyembe 867.

Ijim: Mutef, grassland, 2000m, fl., 22 Nov. 1996, Satabie 1089 & 1090.

Crassocephalum rubens (Jacq.) S.Moore

Annual herb to 1 m; capitula few, with many tiny purple or mauve flowers. Fallow and rocky slopes in grassland.
Belo: Wayside weed in grassland, 1500m, fl., Apr. 1931, Maitland 1466 & 1769.
Bambui: On rock in talus, 2100m, fl., 18 Nov. 1953, Adams, C.D. GC 111239; Upper farm, path leading to Babanki market, 1830m, fl. fr., 31 Oct. 1970, Bauer, P.J. 93.
Kumbo: Weed of cultivation below Fon's Palace, 1680m, fl. fr., 14 Feb. 1958, Hepper 1992.

Crassocephalum sarcobasis (DC.) S.Moore

Annual herb to 1.5 m; capitula with many tiny purple or mauve flowers. Grassland.
Kilum: KD, fl., 9 Jun. 1996, Etuge 2254.
Ijim: Laikom, in grassland on plateau, 2000m, fl. fr., Apr. 1931, Maitland 1482; Aboh, 2250m, fl., 19 Nov. 1996, Pollard 55; Aboh village, Gikwang main road towards Nyasosso forest, 2270m, 6.1116N, 10.2543E, fl., 20 Nov. 1996, Etuge 3508; Ijim plateau above Fon's Palace, Laikom, 1900m, 6.16N, 10.19E, fl., 3 Dec. 1998, Cheek 9763.

Crassocephalum vitellinum (Benth.) S.Moore

Sprawling perennial herb to 2 m; capitula with many tiny yellow or orange flowers. Forest edge.
Kilum: Ecomonitoring Plot 146, Maisels s.n.
Ijim: Gikwang road towards Oku, 2260m, fl., 15 Nov. 1996, Etuge 3396; Laikom to Fundong, 1700m, 6.17N, 10.20E, fl., 10 Dec. 1998, Etuge 4592.

Dichrocephala chrysanthemifolia (Blume) DC. var. *chrysanthemifolia*

Local name: Ifiemen (Oku) – Munyenyembe 784.
Annual or perennial herb to 1.5 m, aromatic; leaves pinnatifid; capitula with tiny white to yellow flowers. Forest edge.
Kilum: Elak, montane forest, 2800m, fl., 29 Oct. 1996, Munyenyembe 784; KA path, 2600–2800m, fl., 30 Oct. 1996, Onana 464; Summit, 2800m, fl., 31 Oct. 1996, Zapfack 1123.
Ijim: Gikwang towards Nyasosso, 2280m, fl. fr., 18 Nov. 1996, Etuge 3431; Laikom, 2000m, 6.17N, 10.20E, fl., 3 Dec. 1998, Etuge 4533.

Dichrocephala integrifolia (L.f.) Kuntze subsp. *integrifolia*

Local name: Kafukobi (Oku) – Munyenyembe 714.
Annual herb to 1 m; leaves pinnatifid; capitula small, with tiny white to yellow flowers. Forest edge.
Kilum: Path above water-tower, KJ, about 1 km into forest, 2250m, fr., 9 Jun.1996, Cable 2971; KD, 2400m, fl., 10 Jun. 1996, Cable 3049; Summit, 2900m, fl., 10 Jun. 1996, Cable 3065; Lake Oku, 2300m, fl., 11 Jun. 1996, Zapfack 873; Shore of Lake Oku, 2250m, fl., 12 Jun. 1996, Cable 3106; Oku-Elak, lower parts of KA, 2200m, 6.1349N, 10.3112E, fl. fr., 28 Oct. 1996, Cheek 8462; Elak, forest edge, 2400m, fr., 28 Oct. 1996, Munyenyembe 714.

Ijim: Aboh village, fl. fr., 18 Nov. 1996, <u>Onana</u> 536;
Gikwang main road towards Nyasosso forest, 2260m,
6.1116N, 10.2543E, fl., 20 Nov. 1996, <u>Etuge</u> 3518; Ijim
mountain forest, near Chufekhe stream, 2010m, 6.1124N,
10.2525E, fl., 20 Nov. 1996, <u>Kamundi</u> 659.
Plot voucher: Forest above Oku-Elak, Jun. 1996, <u>Oku</u> 84.

Echinops gracilis O.Hoffm.

Perennial herb to 90 cm; leaves spiny; capitula with small
white or pale mauve flowers. Grassland.
Ijim: Laikom, grassland, 2100m, 6.15N, 10.26E, 21 Nov.
1996, <u>Onana</u> 562.
Jakiri: Ntan, AlHadji Gey's land, 1600m, fr., 11 Nov. 1998,
<u>Maisels</u> 176.

Echinops mildbraedii Mattf.

Perennial herb to 90 cm; leaves spiny; capitula with small
white or blue flowers. Rock outcrop.
Ndop: Ndop to Kumbo, <u>Boughey</u> GC 4701; Granite kopje by
Ndop village rest house, 1280m, 5.45N, 10.15E, fl. fr., Jul
1962, <u>Brunt</u> 858.

Emilia abyssinica (Sch.Bip. ex A.Rich.) C.Jeffrey var. *abyssinica*

Annual herb to 45 cm; capitula small with yellow ray and
disc flowers. Fallow.
Syn. *Senecio abyssinicus* Sch.Bip. ex A.Rich., F.W.T.A. 2:
249 (1963).
Kumbo: <u>Boughey</u> GC 17485.

Emilia coccinea (Sims) G.Don

Straggling herb to 1 m; leaves often purple beneath; capitula
small, with yellow ray and disc flowers. Fallow.
Kilum: Junction of KJ and KN, 2200m, fl., 10 Jun. 1996,
<u>Cable</u> 3029; KJ, 2300m, fl., 28 Oct. 1996, <u>Buzgo</u> 602.
Ijim: Zitum road, 1860m, 6.1116N, 10.2543E, fl., 21 Nov.
1996, <u>Etuge</u> 3536; Mutef, 6.15N, 10.26E, fl., 22 Nov. 1996,
<u>Onana</u> 581.

Galinsoga quadriradiata Ruiz & Pav.

Annual herb to 30 cm, glandular; capitula small, with white
ray flowers and yellow disc flowers. Fallow and grassland.
Syn. *Galinsoga ciliata* (Raf.) Blake, F.W.T.A. 2: 230
(1963).
Kilum: Path above water-tower, KJ, 1km into forest, 2250m,
fl., 9 Jun. 1996, <u>Cable</u> 2968; Junction of KJ and KN, above
Elak, 2300m, fl., 10 Jun. 1996, <u>Cable</u> 3034; Summit,
3000m, fl., 29 Oct. 1996, <u>Munyenyembe</u> 753.
Ijim: Gikwang road towards Nyasosso forest, 2300m, fl., 19
Nov. 1996, <u>Etuge</u> 3481.

Gerbera piloselloides (L.) Cass.

Scapose herb, the scape to 1 m and inflated below the
capitulum; ray flowers white to maroon. Grassland.
Kilum: Elak, 2000m, fl., 9 Jun. 1996, <u>Zapfack</u> 826.
Ijim: Rocky basalt 'pavement' in grassland plateau above
Fon's Palace, Laikom, 2100m, 6.1642N, 10.1950E, fr., 3
Dec. 1998, <u>Pollard</u> 299.

Guizotia scabra (Vis.) Chiov.

Woody herb to 2 m; leaves opposite; ray and disc flowers
yellow. Grassland.
Kilum: Summit, 3000m, fl., 29 Oct. 1996, <u>Munyenyembe</u>
768; Path from Project HQ to KA, 2800m, 6.1349N,
10.3112E, fl., 30 Oct. 1996, <u>Cheek</u> 8520; Ewook Etele
Mbae, 10 km from Oku, 2350m, fl., 6 Nov. 1996, <u>Zapfack</u>
1235.
Ijim: Gikwang, towards Nyasosso bush, 2260m, fl., 18 Nov.
1996, <u>Etuge</u> 3436.

Gynura pseudochina (L.) DC.

Perennial herb to 1 m; capitula with many orange disc
flowers. Grassland.
Syn. *Gynura miniata* Welw., F.W.T.A. 2: 243 (1963).
Kilum: Lake Oku, 2200m, fl., 12 Jun. 1996, <u>Etuge</u> 2346.

Helichrysum cameroonense Hutch. & Dalziel

Local name: Ifiemen (Oku) – <u>Munyenyembe</u> 762.
RED DATA LISTED
Herb to 1.2 m, aromatic; capitula dense with pale yellow
bracts and small yellow flowers. Forest edge.
Kilum: Lake Oku, Jan. 1951, <u>Keay</u> & <u>Lightbody</u> FHI 28511;
Summit, 3000m, fl., 29 Oct. 1996, <u>Munyenyembe</u> 762.
Ijim: Gikwang, towards Nyasosso bush, 2260m, fl., 18 Nov.
1996, <u>Etuge</u> 3463.

Helichrysum foetidum (L.) Moench var. *foetidum*

Local name: Ifyemen (Oku) – <u>Munyenyembe</u> 811.
Annual or perennial herb to 1.5 m; capitula many, with
yellow bracts and small yellow or orange flowers. Forest
edge.
Kilum: Kumbo to Oku, Feb. 1958, <u>Hepper</u> 1993; Forest
edge, 2200m, fl., 30 Oct. 1996, <u>Munyenyembe</u> 811; Between
Manchok, BirdLife HQ and Lumetoh Traditional Forest
(above the palace of the Fon), 2200m, fl., 1 Nov. 1996,
<u>Buzgo</u> 696; Path from Mbo village (about 6km from Elak-
Oku), 2200–2500m, 6.15N, 10.26E, fl., 1 Nov. 1996, <u>Onana</u>
483,
Ijim: Turbo path, 1 km after its splitting from the main road
at the border of Ijim Mountain Forest Reserve, north from
there, 2000m, fl., 18 Nov. 1996, <u>Buzgo</u> 770; Gikwang,
towards Nyasosso bush, 2260m , fl., 18 Nov. 1996, <u>Etuge</u>
3438; Near Aboh-Anyajua, 2400m, fl., 19 Nov. 1996,
<u>Satabie</u> 1070; Rocky 'pavement' in grassland plateau above
Fon's Palace, Laikom, 2100m, 6.1642N, 10.1950E, fl., 3
Dec. 1998, <u>Pollard</u> 300.

Helichrysum forskahlii (J.F.Gmel.) Hilliard & B.L.Burtt var. *forskahlii*

Local name: Ifiyemen (Oku) – <u>Munyenyembe</u> 773.
Perennial herb or shrub to 1 m; capitula small, with yellow
bracts and small yellow or orange flowers. Forest edge and
swamp.
Syn. *Helichrysum cymosum* sensu F.W.T.A. 2: 264 (1963),
non (L.) D.Don
Kilum: Junction of KJ and KN above Elak, 2200m, fl., 10
Jun.1996, <u>Cable</u> 3023; KD, 2500m, fl., 10 Jun. 1996, <u>Cable</u>
3051; Forest, 2800m, fl., 29 Oct. 1996, <u>Munyenyembe</u> 773;
KA, 2700m, fl., 31 Oct. 1996, <u>Buzgo</u> 666; Above Mboh,

path above village to forest through Peter Wambeng's farm, 2200m, 6.1146N, 10.2856E, fl., 1 Nov. 1996, Cheek 8562. **Ijim**: Gikwang main road towards Nyasosso forest, 2250m, 6.1116N, 10.2543E, fl., 20 Nov. 1996, Etuge 3523 & 3524; Mountain forest, near Chufekhe stream, 2010m, 6.1124N, 10.2525E, fl., 20 Nov. 1996, Kamundi 655; Afua swamp, 1950m, 6.08N, 10.24E, fl., 7 Dec. 1998, Cheek 9855. **Mbi Crater**: 1950m, 6.05N, 10.21E, fl., 9 Dec. 1998, Cheek 9905.

Helichrysum globosum A.Rich. var. *globosum*

Local name: Itoke (Oku) – Muyenyembe 894.
Perennial herb to 1 m; capitula small, with white or pale yellow bracts and small yellow flowers. Forest edge and swamp.
Kilum: KD to Elak, 2800m, fl., 10 Jun. 1996, Etuge 2277; Lumetoh Traditional Forest edge, 2250m, fl., 6 Nov. 1996, Munyenyembe 894; Ntogemtuo swamp, summit, 2900m, fl., 14 May 1997, Maisels 55.
Ijim: Above Aboh village, 1.5 hours along Gikwang road to Lake Oku, 2000m, 6.1112N, 10.2529E, fl., 19 Nov. 1996, Cheek 8675A; Ijim plateau above Fon's Palace, 1900m, 6.16N, 10.19E, fl., 3 Dec. 1998, Cheek 9745.

Helichrysum mechowianum Klatt

Perennial herb to 50 cm; capitula many, with brown or yellow bracts and small yellow flowers. Fallow.
Ndop: Ndop to Kumbo, Boughey GC 11160; Small hill by Ndop village rest house, 1280m, 5.45N, 10.15E, fl., 1 Mar.1962, Brunt 25.

Helichrysum nudifolium (L.) Less. var. *nudifolium*

Herb with annual stems to 1 m; capitula in clusters, with yellow or brown bracts and small yellow flowers. Grassland.
Syn. *Helichrysum nudifolium* (L.) Less. var. *leiopodium* (DC.) Moeser, F.W.T.A. 2: 264 (1963).
Ijim: Laikom, grassland, 2000m, fl., Apr. 1931, Maitland 1483.

Helichrysum nudifolium (L.) Less. var. *oxyphyllum* (DC.) Beentje ined.

Similar to var. *nudifolium*, but the capitula with white bracts. Fallow.
Syn. *Helichrysum rhodolepis* Baker, F.W.T.A. 2: 264 (1963).
Syn. nov. *Helichrysum albiflorum* Moeser, F.T.E.A. Compositae Vol. II (in prep.)
Kilum: Elak, 2000m, fl., 8 Jun. 1996, Zapfack 776.
Ndop: Ndop to Kumbo, Boughey GC 11167.

Helichrysum odoratissimum (L.) Sweet var. *odoratissimum*

Local name: Ifyemen (Oku) – Munyenyembe 810.
Perennial herb to 1 m; capitula in clusters, with yellow bracts and small yellow flowers. Fallow.
Kilum: Fallow by forest edge, 2200m, fl., 30 Oct. 1996, Munyenyembe 810.

Helichrysum stenopterum DC.

Perennial herb to 1 m, capitulum bracts pale brown to pale yellow, flowers yellow. Fallow.
Syn. *Achyrocline stenoptera* (DC.) Hilliard & B.L.Burtt
Syn. *Helichrysum odoratissimum* sensu auct non (L.) Sweet, F.W.T.A. 2: 264 (1963).
Kilum: BirdLife HQ to KA, 6.1349N, 10.3112E, fl., 30 Oct. 1996, Cheek 8494.
Ijim: Gikwang main road towards Nyasosso forest, 2250m, 6.1116N, 10.2543E, fl., 20 Nov. 1996, Etuge 3514.

Lactuca glandulifera Hook.f.

Perennial scrambling herb to 2 m; capitula many, with small yellow or white ligular flowers. Fallow.
Syn. *Lactuca glandulifera* Hook.f. var. *calva* (R.E.Fr.) Robyns, F.W.T.A. 2: 293 (1963).
Local name: Iyen (Oku) – Munyenyembe 792.
Kilum: Kumbo to Oku, c. 3 miles on Oku road, 1850m, fl. fr., 15 Feb. 1958, Hepper 2014; Elak, montane forest, 2800m, fl., 29 Oct. 1996, Munyenyembe 792; KC, 2300–2500m, fl., 30 Oct. 1996, Buzgo 651; KA, 2800m, 6.1349N, 10.3112E, fl., 30 Oct. 1996, Cheek 8534.
Ijim: Gikwang road towards Oku, 2260m, fl., 15 Nov. 1996, Etuge 3407; Aboh village, 6.15N, 10.26E, fl., 18 Nov. 1996, Onana 518.

Lactuca inermis Forssk.

Perennial herb to 2 m; capitula with small blue or mauve ligular flowers. Fallow.
Syn. *Lactuca capensis* Thunb., F.W.T.A. 2: 293 (1963).
Kilum: Lake route, 2500m, fl., 11 Jun. 1996, Zapfack 840; Above Mboh, path above village to forest through Wambeng's farm, 2000m, 6.1256N, 10.2911E, fl., 1 Nov. 1996, Cheek 8545B; Manchok, lowland areas, towards Kumbo road, 2020m, fl., 9 Nov. 1996, Etuge 3360.
Ijim: Gikwang, towards Nyasosso, 2260m, fl., 18 Nov. 1996, Etuge 3427; Mutef, 6.15N, 10.26E, fl., 22 Nov. 1996, Onana 584.

Lactuca schulzeana Büttner

Herb to 1.8 m; capitula small, with small white or yellow ligulate flowers. Grassland.
Ijim: In grassland on plateau, Laikom, 2000m, fl., Apr. 1931, Maitland 1479.

Laggera crispata (Vahl) Hepper & J.R.I.Wood

Annual or perennial herb to 2.4 m; stems winged; aromatic; capitula with many small pink to purple flowers. Forest.
Syn. *Blumea crispata* (Vahl) Merxm. var. *crispata*, Fl. Mascareignes, 109, Composées: 61 (1993).
Syn. *Laggera pterodonta* (DC.) Sch. Bip. ex Oliv., F.W.T.A. 2: 262 (1963).
Kilum: Lake Oku, 2800m, fl., 5 Nov. 1996, Pollard 43.

Launaea nana (Baker) Chiov.

Acaulescent perennial herb; capitula with small yellow to purple ligulate flowers. Fallow.
Syn. *Sonchus elliotianus* Hiern, F.W.T.A. 2: 296 (1963).
Kumbo: Kumbo to Oku, Hepper 2013.

Launaea rarifolia (Oliv. & Hiern) Boulos
Herb with annual stems to 90 cm; capitula with small yellow to purple ligulate flowers. Fallow.
Syn. *Sonchus rarifolius* Oliv. & Hiern, F.W.T.A. 2: 296 (1963).
Bambui: Bambui Experimental Station, 1700m, fl., 4 Mar. 1967, Brunt 64.

Micractis bojeri DC.
Annual herb; capitula small, with few tiny yellow ray flowers and yellow disk flowers. Forest edge.
Syn. *Sigesbeckia abyssinica* (Sch.Bip.) Oliv. & Hiern, F.W.T.A. 2: 242 (1963).
Kilum: Elak, forest edge, 2100m, fl., 1 Nov. 1996, Munyenyembe 847; Lumeto forest, 2200m, 6.1412N, 10.2943E, fl., 7 Nov. 1996, Cheek 8615.
Ijim: Above Aboh, 1.5 hrs walk along Gikwang road to Lake Oku, 2380m, 6.1112N, 10.2529E, fl. fr., 18 Nov. 1996, Cheek 8630; Ijim Mountain Forest, 2160m, fl., 19 Nov. 1996, Kamundi 626; Mutef, Above Fon's Palace, about 30 mins. walk along path to Fulani settlement, 2050m, fl. fr., 22 Nov. 1996, Cheek 8735.

Microglossa densiflora Hook.f.
Scandent shrub to 6 m; capitula many, tiny, with minute white and yellow flowers. Forest edge.
Kilum: KD, 2800m, fl., 10 Jun. 1996, Etuge 2264.

Microglossa pyrifolia (Lam.) Kuntze
(Scandent) shrub to 3(–6) m; capitula many, tiny, with minute cream to pale yellow flowers. Forest edge.
Ijim: Laikom, fl., 1931, Maitland 1583.

Mikania chenopodifolia Willd.
Local name: Kefuvyin (Oku) – Munyenyembe 808.
Sarmentose herb to 10 m long; leaves opposite; capitula small, with few small white flowers. Forest; forest edge.
Kilum: Lake Oku, Keay & Lightbody FHI 28670; Oku-Elak, lower parts of KA, 2200m, 6.1349N, 10.3112E, fl., 28 Oct. 1996, Cheek 8464; Elak, forest edge, 2300m, fl., 30 Oct. 1996, Munyenyembe 808; Secondary forest near Manchok, fr., 6 Nov. 1996, Onana 501.
Ijim: Path from Aboh towards Tum, 2100m, 6.1116N, 10.2543E, fl., 22 Nov. 1996, Etuge 3578.

Senecio burtonii Hook.f.
Perennial herb to 1.2 m; capitula many, small, with large yellow ray flowers and small disc flowers. Grassland.
Ijim: Laikom, in bog on plateau, 2000m, fl., Apr. 1931, Maitland 1395 & fl., Jun. 1931, Maitland 1722.

Senecio hochstetteri Sch.Bip. ex A.Rich.
Perennial herb to 0.6 m; capitula many, small, with large pale yellow ray flowers and small disc flowers. Grassland.
Ijim: Laikom, in grassland on plateau, 2000m, fr. Apr. 1931, Maitland 1481.

Senecio purpureus L.
Perennial herb to 2.4 m; capitula small, with red or purple disk flowers; ray flowers absent. Fallow; grassland.

Ijim: Above Aboh village, about 1.5 hours walk along Gikwang road towards Lake Oku, 2400m, 6.1112N, 10.2529E, fl., 20 Nov. 1996, Cheek 8695; Grassland savanna, Ijim Mountain Forest, 2550m, 6.1116N, 10.2543E, fl., 20 Nov. 1996, Pollard 61.

Senecio ruwenzoriensis S.Moore
Perennial herb to 0.6 m; capitula many, small, with yellow ray flowers and small yellow disc flowers. Grassland.
Ijim: Laikom, 2000m, Maitland 1367.

Sigesbeckia orientalis L.
Local name: Eptokatam (Oku) – Munyenyembe 886.
Annual herb to 90 cm, glandular; capitula small, with few minute yellow ray flowers and disk flowers. Fallow.
Kilum: KA, 2600–2800m, 30 Oct. 1996, Onana 465; KA, 2500m, fl., 31 Oct. 1996, Buzgo 688; Oku-Elak between Fon's Palace and Manchok, bank along road in village, 2000m, 6.1447N, 10.3058E, fl., 31 Oct. 1996, Cheek 8538; Shambai, 2300m, fl. fr., 5 Nov. 1996, Munyenyembe 886; Summit, near savanna zone, 2480m, fr., 5 Nov. 1996, Zapfack 1196.
Ijim: Gikwang road towards Nyasosso forest (grassland), 2260m, 6.1116N, 10.2543E fl., 20 Nov. 1996, Etuge 3517.

Solanecio mannii (Hook.f.) C.Jeffrey
Shrub or small tree to 4(–10) m; capitula many, with small yellow disc flowers. Forest edge.
Syn. *Crassocephalum mannii* (Hook.f.) Milne-Redh., F.W.T.A. 2: 246 (1963).
Kilum: Shore of Lake Oku, 2330m, fl. 6 Jan. 1951, Keay FHI 28486; Elak, KD, fl., 9 Jun. 1996, Etuge 2208.
Plot voucher: Forest above Oku-Elak, Jun. 1996, Oku 18.

Sonchus angustissimus Hook.f.
Perennial herb to 3 m; leaves pinnatilobed; capitula with many yellow ligulate flowers. Fallow.
Kilum: Path from Project HQ to KA, 2200m, 6.1349N, 10.3112E, fl., 30 Oct. 1996, Cheek 8501.

Sonchus schweinfurthii Oliv. & Hiern
Perennial herb to 1.8 m; leaves pinnatilobed; capitula with many yellow ligulate flowers, pink beneath. Fallow.
Kilum: Junction of KJ and KN above Elak, 2200m, fl., 10 Jun. 1996, Cable 3027; Path above Mboh to forest through Peter Wambeng's farm, 2000m, 6.1256N, 10.2911E, st., 1 Nov. 1996, Cheek 8546A.

Tithonia diversifolia (Hemsl.) A.Gray
Perennial herb to 5 m; leaves large, lobed; capitulum with swollen stalk and large yellow ray flowers. Fallow.
Ijim: Ijim Mountain Forest, 2160m, 6.11N, 10.25E, fl., 19 Nov. 1996, Kamundi 627.

Vernonia bamendae C.D.Adams
RED DATA LISTED
Herb to 1.5 m; capitula with small purple flowers. Grassland.
Ijim: Laikom, Maitland 1514.

Vernonia biafrae Oliv. & Hiern
Local name: Keliele (Oku) – Cheek 8445.
Scrambling shrub to 6 m; capitula many, with small mauve
or purple flowers. Fallow.
Syn. *Vernonia tufnelliae* S.Moore, F.W.T.A. 2: 276 (1963).
Kilum: Elak to the forest at KA, 2200m, 6.1349N,
10.3112E, fl., 27 Oct. 1996, Cheek 8445.
Kumbo: Streamside, 1680m, fl., 14 Feb. 1958, Hepper
1968.

Vernonia blumeoides Hook.f.
Small shrub to 1.2 m; capitula many, with small mauve or
purple flowers. Fallow.
Ijim: Mutef, 6.15N, 10.26E, fl., 22 Nov.1996, Onana 583;
Main path from Ijim to Tum, 2100m, 6.1116N, 10.2543E,
fl., 25 Nov. 1996, Etuge 3611; Near Ake, 2400m, fl., 25
Nov. 1996, Satabie 1094.

Vernonia calvoana (Hook.f.) Hook.f. subsp.
calvoana var. *acuta* (C.D.Adams) C.Jeffrey
Local name: Keggi (Oku) – Etuge 2301;
Shrub to 4.5 m; capitula large, with bracts with white
appendages; small mauve or purple flowers. Forest edge.
Syn. *Vernonia leucocalyx* O.Hoffm. var. *acuta* C.D.Adams,
F.W.T.A. 2: 276 (1963).
Kilum: KA, 2200m, fl., 11 Jun. 1996, Etuge 2301; Shore of
Lake Oku, 2300m, fl., 12 Jun. 1996, Cable 3117.
Ijim: Laikom, 1931, Maitland 1485; Near Chufekhe stream,
2010m, 6,1124N, 10.2525E, fl., 20 Nov. 1996, Kamundi
638.

Vernonia calvoana (Hook.f.) Hook.f. subsp.
calvoana var. *calvoana*
As for *Vernonia calvoana* var. *acuta*.
Kilum: Elak, 2100m, fl., 8 Jun. 1996, Zapfack 782; KA,
2500m, fl., 31 Oct. 1996, Pollard 27.

Vernonia guineensis Benth. var. *cameroonica*
C.D.Adams
Herb to 1.5 m; capitula many, with bracts with pink
appendages; small mauve or purple flowers. Savanna.
Ndop: Ndop to Kumbo, Boughey GC 11166; Ndop plain,
near Bamessi, 1300m, 6N, 10.30E, fl., 30 Mar. 1962, Brunt
267.

Vernonia holstii O.Hoffm.
Woody herb or shrub to 4.5 m; capitula many, small, with
small white or pale mauve flowers. Forest edge.
Ijim: Main road from Atubeaboh (Gikwang Foe) towards
Oku, 2265m, fl., 15 Nov. 1996, Etuge 3385; Gikwang,
towards Nyasosso bush, 2250m, fl., 18 Nov 1996, Etuge
3435; Zitum road, 1900m, 6.1116N, 10.2543E, fl., 21 Nov.
1996, Etuge 3558; Shore of lake Oku and along the path to
the Lake, 6.15N, 10.26E, fl., 26 Nov. 1996; Onana 638.

Vernonia hymenolepis A.Rich.
Woody herb or shrub to 4 m; capitula large; bracts with pink
appendages; small mauve or purple flowers. Forest edge.
Syn. *Vernonia calvoana* (Hook.f.) Hook.f. var. *mesocephala*

C.D.Adams, in Journ. W. Afr. Sc. Assoc. 3: 118 (1957).
Kilum: Junction of KJ and KN above Elak, 2550m, fl., 10
Jun. 1996, Cable 3017; KJ, 2300m, fl., 29 Oct. 1996,
Zapfack 1044; Secondary forest near the village, 6.15N,
10.26E, fl., 6 Nov. 1996, Onana 502.

Vernonia infundibularis Oliv. & Hiern
Perennial herb to 1.2 m; capitula with fairly large blue or
rich purple flowers. Grassland.
Syn. *Vernonia saussureoides* Hutch., F.W.T.A. 2: 281
(1963).
Kumbo: Boughey GC 17477.

Vernonia ituriensis Muschl. var. *occidentalis*
(C.D.Adams) C.Jeffrey
Local name: Kefuseh (Oku) – Cheek 8588.
Woody herb to 2 m; capitula many, with small mauve, blue
or white flowers. Forest edge and grassland.
Syn. *Vernonia glabra* (Steetz) Vatke var. *occidentalis*
C.D.Adams, F.W.T.A. 2: 280 (1963).
Kilum: Road from Rest House E of lake to N of mountain
range, 2300m, 6.1214N, 10.2733E, fl., 5 Nov. 1996, Cheek
8588.
Ijim: Isebu mountain forest, 1700m, fl., 26 Nov. 1996,
Kamundi 699; Rocky 'pavement' in grassland above Fon's
Palace, Laikom, 2100m, 6.1642N, 10.1950E, fl., 3 Dec.
1998, Pollard 302.

Vernonia nestor S.Moore
Perennial herb to 1.8 m; capitula many, with small purple or
mauve flowers. Grassland.
Ijim: Zitum road, 1980m, 6.1116N, 10.2543E, fl., 21 Nov.
1996, Etuge 3529.

Vernonia purpurea Sch.Bip. ex Walp.
Perennial herb to 2 m; capitula many, with small purple or
mauve flowers. Grassland; fallow.
Kilum: Path from Project HQ to KA, 2200m, 6.1349N,
10.3112E, fl., 30 Oct. 1996, Cheek 8500Elak, 2200m, fl., 31
Oct. 1996, Buzgo 683; Elak, 2200m, fl.,31 Oct. 1996,
Munyenyembe 823.
Ijim: Open grassland, Ijim Mountain Forest Reserve,
2250m, fl., 19 Nov. 1996, Pollard 57; Gikwang main road
towards Nyasosso forest, 2250m, 6.1116N, 10.2543E, fl., 20
Nov. 1996, Etuge 3511.

Vernonia subaphylla sensu Baker non Muschl.
Herb with annual stems to 60 cm; capitula few, with small
mauve flowers. Savanna.
Ndop: Ndop to Kumbo, Boughey GC 11151.

Vernonia turbinata Oliv.
Local name: Ajba – Akuh (Bekom at Ayeah) – Buzgo 771.
Herb to 1.2 m; capitula small, with small purple flowers.
Grassland.
Syn. *Vernonia rugosifolia* De Wild., F.W.T.A. 2: 282
(1963).
Kilum: Summit, near savanna zone, 2480m, fl., 5 Nov.
1996, Zapfack 1195.
Ijim: Turbo path, 1 km after its splitting from the main road

127

at the border of Ijim Mountain Forest Reserve, north from there, 2000m, fl., 18 Nov. 1996, Buzgo 771; Gikwang, towards Nyasosso bush, 2260m, fl., 18 Nov. 1996, Etuge 3434.

CONNARACEAE

Agelaea pentagyna (Lam.) Baill.
Scrambling shrub; leaflets ovate to broadly elliptic, 8–19 × 5–10 cm; white flowers. Forest.
Syn. **Agelaea dewevrei** De Wild. & T.Durand, F.W.T.A. 1: 746 (1958).
Ijim: Laikom, Maitland 1460.

CONVOLVULACEAE

Cuscuta campestris Yunck.
Parasite; leafless twiner; stems slender, tangled, bright yellow or reddish-orange; flowers small, white. Forest edge.
Kumbo: By river Wi, fl., 14 Feb. 1958, Hepper 1980.

Dichondra micrantha Urb.
Perennial herb; stems trailing; leaves reniform, broad, silky, 1–3 cm long; flowers axillary, solitary, yellow. Weed.
Syn. **Dichondra repens** J.R.Forst. & G.Forst., F.W.T.A. 2: 338 (1963).
Kilum: Oku, inside Fon's Palace, between stones at base of throne platform, 2000m, fl., 17 Feb. 1958, Hepper 2024.

Ipomoea involucrata P.Beauv. var. involucrata
Twiner, often widely climbing, pubescent; involucre boat-shaped; flowers red-purple, 3–5 cm long, opening one at a time. Weed.
Ijim: Zitum road, 1950m, fl., 21 Nov. 1996, Etuge 3534.

Ipomoea tenuirostris Choisy subsp. tenuirostris
Climbing in grass; leaves 4–9 cm long; inflorescence cymose; flowers pale mauve with purple centre. Grassland.
Kilum: Elak, 2200m, fl., 1 Nov. 1996, Munyenyembe 842; Near to the stream around savanna zone, 2200m, fl., 5 Nov. 1996, Zapfack 1200;
Ijim: Gikwang road towards Oku, 2265m, fl., 15 Nov. 1996, Etuge 3408; Laikom, 2000m, 6.17N, 10.20E, fl., 3 Dec. 1998, Etuge 4525.

CRASSULACEAE

Crassula alata (Viv.) A.Berger subsp. pharnaceoides (Fisch. & Mey.) Wickens & M.Bywater
Local name: Unknown – Cheek 8670.
Small herb, 5–10 cm; leaves linear-subulate, acute, 3–4 mm long; flowers pedicellate, dull white. Shady banks.
Syn. **Crassula pharnaceoides** (Hochst.) Fisch. & Mey., F.W.T.A. 1: 116 (1954).
Ijim: Above Aboh village, 1.5 hours walk along Gikwang road, 2450m, fl., 19 Nov. 1996, Cheek 8670; Gikwang road towards Nyasosso forest, 2250m, fl., 20 Nov. 1996, Etuge 3525; Above Fon's Palace, about 30 mins. walk along path

to Fulani settlement, 2050m, fl. fr., 22 Nov. 1996, Cheek 8731; Ijim Mountain Forest, montane rain forest along the ridge towards a point where Lake Oku can be viewed, 2500m, fl., 22 Nov. 1996, Kamundi 687.

Crassula alsinoides (Hook.f.) Engl.
Local name: None known – Cheek 8693.
Fleshy prostrate or ascending herb; leaves broadly ovate, abruptly narrowed at base, 1 cm long; flowers white. Forest edge.
Kilum: Path above water-tower, KJ, about 1 km into forest, 2250m, fl., 9 Jun. 1996, Cable 2970; Summit, 2950m, fl., 10 Jun. 1996, Cable 3075; Elak, 2400m, fl., 1 Nov. 1996, Munyenyembe 852.
Ijim: Aboh village, 2450m, fl., 19 Nov. 1996, Onana 541; Above Aboh village, 1.5 hours walk along Gikwang road, track towards TA, 2400m, fl., 20 Nov. 1996, Cheek 8693.

Crassula schimperi Fisch. & Mey. subsp. schimperi
Local name: Iwo (Oku) – Munyenyembe 758; Mbomfesege (Kom) – Cheek 8667.
Small herb resembling Crassula alata, but leaves lanceolate, c. 7 mm long, broadened towards the base; flowers sessile. Shady banks.
Syn **Crassula pentandra** (Edgew.) Schönl., F.W.T.A. 1: 116 (1954)
Kilum: Elak, 3000m, fl., 29 Oct. 1996, Munyenyembe 758; Summit, 2800m, fl., 31 Oct. 1996, Zapfack 1119.
Ijim: Above Aboh village, 1.5 hours walk along Gikwang road, 2450m, fl., 19 Nov. 1996, Cheek 8667; Gikwang road towards Nyasosso forest, Aboh, 2295m, fl., 19 Nov. 1996, Etuge 3503.

Crassula vaginata Eckl. & Zeyh.
Erect herb to 90 cm; stout stem; linear-lanceolate leaves to 10 cm long; white flowers. Grassland.
Syn. **Crassula alba** sensu F.W.T.A. 1: 116 (1954).
Kilum: Above Mboh, path above village to forest through Wambeng's farm, 2200m, fl., 1 Nov. 1996, Cheek 8564; Near to the stream around savanna zone, 2500m, fr., 5 Nov. 1996, Zapfack 1203 & Zapfack 1208.
Ijim: Gikwang main road towards Nyasosso forest, 2250m, fl., 20 Nov. 1996, Etuge 3520; Inside crater of Lake Oku, 2300m, fl., 25 Nov. 1996, Cheek 8770; Seasonally wet N-facing rockface in 'gallery forest' along path from Fon's Palace, Laikom, to Ijim ridge grassland plateau above, 1960m, 6.1642N, 10.1950E, fl. fr., 2 Dec. 1998, Pollard 272.

Kalanchoe crenata (Andrews) Haw.
Local name: Kitule (Oku) – Munyenyembe 718 & 796; Ayuh ajang (Kom at Laikom) – Etuge 4547.
Fleshy herb, to 2 m; leaves distinctly petiolate; inflorescence densely glandular-pilose; flowers yellow to red. Scrub.
Syn. **Kalanchoe laciniata** sensu Hepper, F.W.T.A. 1: 117 (1954).
Kilum: Elak, 2400m, fl., 28 Oct. 1996, Munyenyembe 718; Summit, 2850m, fl., 29 Oct. 1996, Munyenyembe 796; KC, 2600m, 30 Oct. 1996, Buzgo 656; Cliff-face, 2800m, fl. 31 Oct. 1996, Munyenyembe 826; Summit, 2800m, fl., 31 Oct.

1996, Zapfack 1120 & 1122.
Ijim: Laikom, 2400m, fl., 19 Nov. 1996, Satabie 1072; Main path from Ijim to Tum, 1950m, fl., 25 Nov. 1996, Etuge 3630; Laikom, Akwamofu, 1760m, 6.17N, 10.20E, fl., 5 Dec. 1998, Etuge 4547.

Kalanchoe cf. laciniata (L.) DC.
Herb to 2 m; leaves distinctly petiolate; flowers yellow, orange or magenta. Scrub.
Ijim: Gikwang towards Nyasosso bush, 2260m, fl., 18 Nov. 1996, Etuge 3444; Sappel, below Ardo's uncle's compound, 2300m, 6.16N, 10.24E, fl., 5 Dec. 1998, Cheek 9816.

Kalanchoe sp.
Kilum: Wvem, near Kumbo, 'Mama Saka's' garden, 1800m, st., 6 Nov. 1998, Maisels 182.

Umbilicus botryoides Hochst. ex A.Rich.
Local name: Kentongtong (Oku) – Munyenyembe 756.
Succulent herb, sometimes epiphytic, glabrous, to 30 cm; inflorescence an erect raceme; flowers white, corolla ellipsoid, c. 7 mm long. Banks or in trees.
Kilum: Montane *Podocarpus* forest, 2700m, fl., 29 Oct. 1996, Munyenyembe 756; KA, summit, 2800m, fl., 31 Oct. 1996, Buzgo 673; Summit, 2800m, fl., 31 Oct. 1996, Zapfack 1135.
Ijim: Gikwang main road towards Nyasosso forest (grassland), 2280m, fl., 20 Nov. 1996, Etuge 3506.

CRUCIFERAE

Brassica cf. rapa L.
Herb c. 60 cm; leaves obovate, c. 20 cm long, glaucous blue; flowers yellow. Fallow.
Kilum: KD, 2200m, fl., 10 Jun. 1996, Cable 3046.
Ijim: Aboh-Anyajua, 2400m, fl., 25 Nov. 1996, Satabie 1095.

Cardamine africana L.
Weak herb 30–60 cm; leaves trifoliate, long-petiolate, up to 16 cm long; flowers white; fruits 2.5–4.5 cm long. Forest.
Kilum: Path above water-tower, KJ, about 1 km into forest, 2250m, fr., 9 Jun. 1996, Cable 2977; Elak, 2500m, fr., 9 Jun. 1996, Zapfack 798; KJ, above water tank from Elak, 2 km into forest, 2400m, fl., 10 Jun. 1996, Cable 3008; Summit, 2950m, fl., 10 Jun. 1996, Cable 3073; KD, 3000m, fl., 10 Jun. 1996, Etuge 2288; KD, 3000m, fr., 10 Jun. 1996, Etuge 2290; KJ, 2500m, fr., 29 Oct. 1996, Pollard 5; Elak, 2700m, fl. fr., 31 Oct. 1996, Munyenyembe 833.
Ijim: Aboh village, 2450m, fr., 19 Nov. 1996, Onana 544.

Cardamine hirsuta L.
Small annual, often much-branched or tufted at the base; petals white. Grassland.
Kilum: KA, 2760m, fl., 10 Jun. 1996, Asonganyi 1308; Summit, 2900m, fl., 10 Jun. 1996, Cable 3078.

Cardamine trichocarpa Hochst. ex. A.Rich.
Local name: Sasowaz (Kom – Yama Peter) – Pollard 283.
Weakly erect annual herb; leaves pinnate; petals absent,

sepals 1.2–2.5 mm long; fruits 1.5–3 cm. Grassland.
Kilum: KA, 2760m, fr., 10 Jun. 1996, Asonganyi 1301 & 1302; Junction of KJ and KN above Elak, 2550m, fl., 10 Jun. 1996, Cable 3026; KD, 3000m, fr., 10 Jun. 1996, Etuge 2289.
Ijim: Grassland plateau above Fon's Palace, Laikom, 2060m, fr., 3 Dec. 1998, Pollard 283.

Cardamine sp.
Kilum: Summit, 2950m, 10 Jun. 1996, Cable 3076.

CUCURBITACEAE

Gerrardanthus paniculatus (Mast.) Cogn.
Herbaceous climber to c. 4m; leaves slightly 5-lobed, c. 5 cm long; flowers dully glossy orange, c. 1 cm wide, with 2 long and 3 short pedicels; fruits cigar-shaped, part superior, c. 8 cm long. Forest.
Syn. *Gerrardanthus zenkeri* Harms & Gilg ex Cogn., F.W.T.A. 1: 208 (1954).
Ijim: Montane forest understorey, medicinal forest by grass ridges above Akwamofu Sacred forest, Laikom; 1800m, fl. fr., 10 Dec. 1998, Pollard 367.

Lagenaria breviflora (Benth.) Roberty
Robust, densely white hispid trailing herb; leaves 5-lobed, c. 20 cm long, two spine-like glands at junction of blade and petiole; fruit spherical, 10–15 cm diam., mottled green and yellow. Scrub.
Kilum: Secondary forest near the village, 6 Nov. 1996, Onana 510.

Momordica cissoides Benth.
Tall, graceful climber; leaves 3–5-foliolate; flowers white with a large black-purple spot at the base of each petal; fruit fleshy, c. 3.5 cm long, prickly, orange-yellow, with red seeds. Savanna.
Belo: Belo, by a stream, 1500m, fl., Apr. 1931, Maitland 1501.

Momordica foetida Schum. & Thonn.
Slender climber; leaves broadly ovate-cordate, 8–15 cm long; inflorescence racemose or subcorymbose; flowers numerous, yellow. Forest edge.
Kilum: KD, fl., 9 Jun. 1996, Etuge 2216; Lake route, 2600m, fl., 11 Jun. 1996, Zapfack 841; Shore of Lake Oku, 2250m, fr., 12 Jun. 1996, Cable 3104.
Ijim: Above Aboh village, 1.5 hours walk along Gikwang road, 2000m, fl., 18 Nov. 1996, Cheek 8654.

Oreosyce africana Hook.f.
Slender climber, 3–4 m; leaves ovate, 5–7 cm long; flowers 3–4, yellow, in leaf axils. Forest edge.
Kilum: Elak, 2300m, fl., 8 Jun. 1996, Zapfack 743; Path above water-tower, KJ, about 1 km into forest, 2250m, 9 Jun. 1996, Cable 2994; KJ, above water tank from Elak, 2 km into forest, 2400m, fl., 10 Jun. 1996, Cable 3007; Lake Oku, 2300m, fl., 11 Jun. 1996, Zapfack 860; Shore of Lake Oku, 2250m, fl., 12 Jun. 1996, Cable 3097; Lake Oku, 2200m, fl., 12 Jun. 1996, Etuge 2342; Oku-Elak to the forest

at KA, 2200m, fr., 27 Oct. 1996, Cheek 8448; KJ, 2300m, fl., 29 Oct. 1996, Zapfack 1048; KA to summit, 2800m, 30 Oct. 1996, Cheek 8535; KC path, 2700m, fl., 31 Oct. 1996, Onana 481.

Ijim: Above Aboh village, 1.5 hours walk along Gikwang road, 2350m, fl., 18 Nov. 1996, Cheek 8653; Gikwang, towards Nyasosso bush, 2260m, fl., 18 Nov. 1996, Etuge 3445; Ijim Mountain Forest, near Chufekhe stream, 2010m, fl., 20 Nov. 1996, Kamundi 641.

Raphidiocystis phyllocalyx C.Jeffrey & Keraudren

Herbaceous climber; leaves ovate or obovate-oblong, 6–9 × 5–8 cm, apex rounded, base deeply cordate, margin sinuate, denticulate; flowers simple or fasciculate, subsessile, pendulous, c. 1 m long; sepals 3, leafy, 1 cm long; corolla orange. Forest.
Kilum: Path from Kilum Project HQ to top of Mt Oku, 2200m, fr., 12 Oct. 1996, Cheek 8423; Elak, 2200m, fr., 1 Nov. 1996, Munyenyembe 845.

Zehneria minutiflora (Cogn.) C.Jeffrey

Herbaceous climber to 3 m; leaves ovate, weakly trilobed, c. 6–8 × 5–6 cm, acuminate, cordate, upper face slightly scabrid; inflorescence corymbose; peduncle 0.5–1 cm; flowers 2–8 cm, 2.5 mm diam., yellowish white; fruit ellipsoid 1 × 0.5 cm. Forest edge.
Syn. *Melothria minutiflora* Cogn., F.W.T.A. 1: 209 (1954).
Kilum: KA, 2200m, fr., 11 Jun. 1996, Etuge 2303; Elak, 2400m, fl., 28 Oct. 1996, Munyenyembe 719; Elak, KJ, 2300m, fl. fr., 29 Oct. 1996, Zapfack 1049; KC path, 2700m, fl., 31 Oct. 1996, Onana 479; KDH, 2300m, fl. fr., 1 Nov. 1996, Zapfack 1147.
Ijim: Ijim Mountain Forest, near Chufekhe stream, 2010m, fr., 20 Nov. 1996, Kamundi 644.
Plot voucher: Forest above Oku-Elak, Jun. 1996, Oku 32.

Zehneria scabra (L.f.) Sond.

Herbaceous climber; leaves triangular or ovate-deltoid, 2–6 × 2–4.5 cm; long-acuminate, truncate, margin toothed, upper face punctate-scabrid, lower velvety; inflorescence umbellate, 10–20-flowered; flowers 2.5 mm diam., yellowish-white; fruit ovoid, 6–8 mm diam. Forest edge.
Syn. *Melothria mannii* Cogn., F.W.T.A. 1: 209 (1954).
Syn. *Melothria punctata* (Thunb.) Cogn., F.W.T.A. 1: 209 (1954).
Kilum: Lake Oku, 2200m, fl., 12 Jun. 1996, Etuge 2341A; Elak, 2400m, fl., 28 Oct. 1996, Munyenyembe 732; KJ, 2500m, fl., 29 Oct. 1996, Buzgo 629; Above Mboh, path above village to forest through Wambeng's farm, 2200m, fl., 1 Nov. 1996, Cheek 8569; Mbijame forest, 2000m, fl., 4 Nov. 1996, Munyenyembe 863.
Ijim: Above Aboh village, 1.5 hours walk along Gikwang road, 2350m, fl., 18 Nov. 1996, Cheek 8643; Gikwang towards Nyasosso, 2290m, fl., 18 Nov. 1996, Etuge 3413; Ijim Mountain Forest, near Chufekhe stream, 2010m, fl., 20 Nov. 1996, Kamundi 643; Aboh village, 2450m, fl., 20 Nov. 1996, Onana 558.
Plot voucher: Forest above Oku-Elak, Jun. 1996, Oku 56.

DILLENIACEAE

Tetracera alnifolia Willd.

Large liane; leaves oblong to oblong-ovate; 6–16 cm long; panicles terminal, spreading, flowers white. Forest.
Ijim: Mwal village, near Ajung, 1200m, fr., 28 Nov. 1998, Maisels 190.

DIPSACACEAE

Dipsacus narcisseanus Lawalree
RED DATA LISTED
Robust perennial herb 1–2.5m; leaves basal; inflorescences spherical, 5–8 cm diam. on spiny peduncle; flowers white, 1 cm long. Forest edges, cliffs.
Syn. *Dipsacus pinnatifidus* Steud. ex A.Rich., F.W.T.A. 2: 223 (1963).
Kilum: Ridge above Lake Oku, edge above forest, 2560m, fl. fr., 7 Jan. 1951, Keay & Lightbody FHI 28515; KA to summit, 6.1349N 10.3112E, 2800m, fl., 30 Oct. 1996, Cheek 8532.
Ijim: Gikwang road, between junction for Anyajua and lake, 6.13N, 10.26E, 2500m, fl., 25 Nov. 1996, Cheek 8761.

Succisa trichotocephala Baksay
Local name: Ktunsu (Oku) – Munyenyembe 769; Abua Kou (Kom) – Cheek 8631.
RED DATA LISTED
Erect sparingly branched herb to 1.2 m; leaves sparse, lanceolate, 10–12 cm; flowers white or pale mauve. Forest edge.
Kilum: Ridge above Lake Oku, edge above forest, 2560m, fl. fr., 7 Jan. 1951, Keay & Lightbody FHI 28479; Summit, 3000m, fl., 29 Oct. 1996, Munyenyembe 769; Elak, KD, close to top of mountain, N exposition, 2600m, fl., 31 Oct. 1996, Buzgo 679; Oku summit, Ntogemtuo Swamp, 2900m, fl., 14 Mar. 1997, Maisels 54; Mount Oku above Simonkoh, on a ridge, 2700m, fl. fr., Mar. 1998, Maisels 163.
Ijim: Laikom, in grassland on plateau, 2000m, fl., Apr. 1931, Maitland 1667; Above Aboh village, 1.5 hrs along Gikwang road towards Lake Oku, 6.1112N 10.2529E, 2380m, fl., 18 Nov 1996, Cheek 8631; Sappel, below Ardo's uncle's compound, 2300m, 6.16N, 10.24E, fl. fr., 5 Dec. 1998, Cheek 9814.

DROSERACEAE

Drosera madagascariensis DC.

Herb with sticky glandular hairs; leaves evenly spaced along elongated (to 15 cm) stem; inflorescence 10–30 cm long; flowers pink. Open wet flushes.
Kilum: Maisels 106.
Ijim: From Aboh, path towards Tum., 6.1116N 10.2543E, 2100m, fl., 22 Nov. 1996, Etuge 3605;

ERICACEAE

Agauria salicifolia (Comm. ex Lam.) Hook.f. ex Oliv. var salicifolia

Local name: Bang (Oku) – Munyenyembe 786.
Shrub or small tree, to 13 m; leaves oblong-lanceolate, acute at ends; inflorescence a short axillary raceme; flowers pink. Forest.
Kilum: Lake Oku, Keay & Lightbody FHI 28510; KD, 2800m, fl., 10 Jun. 1996, Etuge 2272; 2800m, fl., 29 Oct. 1996, Munyenyembe 786; Oku shrine, 2140m, fr., 3 Nov. 1996, Munyenyembe 861; Lake Oku, 2800m, fl., 5 Nov. 1996, Pollard 42.
Ijim: Laikom, Maitland 1631; Turbo path, 1km after its splitting from main road at border of Ijim Mountain Forest Reserve, north from there, 2000m, fl. fr., 18 November 1996, Buzgo 776; Gikwang road towards Nyasosso forest, 2092m, fl., 19 Nov. 1996, Etuge 3471; Laikom, 2100m, 6.15N, 10.26E , fl. fr., 21 Nov. 1996, Onana 559; Laikom, forest patch beside path from Fon's Palace to grassland plateau above, 2040m, 6.1642N, 10.1950E, fl., 2 Dec. 1998, Pollard 276.

Erica mannii (Hook.f.) Beentje

Local name: Fengwang (Oku) – Munyenyembe 778.
Heath-like shrub to 4 m; leaves ascending to erect, in whorls of 3; few purplish flowers at end of each branchlet. Forest edge.
Syn. *Philippia mannii* (Hook.f.) Alm & T.C.E.Fr., F.W.T.A. 2: 2 (1963).
Kilum: Oku, Hepper 2059; KA, 2760m, fl., 10 Jun. 1996, Asonganyi 1313; KD, 3000m, fl., 10 Jun. 1996, Etuge 2295; & 2297; Elak, 2700m, fr., 29 Oct. 1996, Munyenyembe 778; Summit, 2800m, fr., 31 Oct. 1996, Zapfack 1142.
Ijim: Above Aboh village, 1.5 hours walk along Gikwang road to Lake Oku, 2450m, 6.1112N, 10.2529E, fl., 19 Nov. 1996, Cheek 8661; Gikwang main road towards Nyasosso forest, 6.1116N 10.2543E, 2250m, fr., 20 Nov. 1996, Etuge 3509.

Erica tenuipilosa (Engl. ex Alm & T.C.E.Fr.) Cheek subsp. tenuipilosa

Local name: Njienen (Oku) – Munyenyembe 779; Unknown – Cheek 8662.
Heath-like undershrub 0.3–0.8 m; leaves in whorls of 3; inflorescence a spicate panicle; flowers purplish. Forest edge.
Syn. *Blaeria spicata* Hochst. ex A.Rich. subsp. *mannii* (Engl.) Wickens.
Syn. *Blaeria manni* (Engl.) Engl., F.W.T.A. 2: 2 (1963).
Kilum: Lake Oku, Jan. 1951, Keay & Lightbody FHI 28497; Summit, 3000m, fl., 29 Oct. 1996, Munyenyembe 77; KA, summit, 2800m, fl., 31 Oct. 1996, Buzgo 667.
Ijim: Laikom, Maitland 1421; Above Aboh village, 1.5 hours along Gikwang road towards Lake Oku, 6.1112N 10.2529E, 2450m, fl., 19 Nov. 1996, Cheek 8662; Gikwang main road towards Nyasosso forest, 6.1116N 10.2543E, fl., 20 Nov. 1996, Etuge 3510; Cattle watering hole, 11 km E of Laikom, 2400m, 6.1615N, 10.2435E, fl., 9 Dec. 1998, Pollard 352.

EUPHORBIACEAE
Det. M.Cheek (K), E.Biye (YA), A.Radcliffe-Smith (K) & S.Carter-Holmes (K)

Acalypha brachystachya Hornem.

Small, slender-branched annual herb, 15–30 cm; leaves 2–5 cm long; inflorescence very short; female bracts digitately 3-partite; female flowers 6–9. Damp cliff.
Ijim: Anyajua, John DeMarco's bat cave, below Ijim BirdLife HQ, 1400m, 6.11N, 10.23E, fr., 12 Dec. 1998, Cheek 9919.

Acalypha manniana Müll.Arg.

Climbing shrub to 2 m; flowers pink. Forest edge.
Kilum: Oku, beside river in high forest below village, fl., 17 Feb. 1958, Hepper 2050.
Ijim: Laikom, on edge of forest, 2000m, fl., Apr. 1931, Maitland 1732.

Acalypha ornata Hochst. ex A.Rich.

Shrub to 3.5 m; leaves serrate, 5–15 ×, 3–10 cm, petiole 1–13 cm long; female bracts about 10-toothed, 8–12 mm broad in fruit; styles laciniate. Forest.
Kilum: Elak, Oku shrine, 2140m, fl., 3 Nov. 1996, Munyenyembe 859.

Acalypha psilostachya Hochst. ex A.Rich. var. psilostachya

Herb 30–40 cm; leaves ovate, c. 3 cm long, apex obtuse or rounded, base cordate, margin finely serrate; petiole c. 2 cm long; male inflorescences densely-flowered, 7–8 cm long. Forest edge.
Ijim: Laikom, 2100m, 6.15N, 10.26E, fl., 21 Nov. 1996, Onana 563; Aboh, edge of Ijim mountain forest, 6.1116N, 10.2543E, fl., 21 Nov. 1996, Pollard 71; Laikom, near Akwamofu sacred forest, farmbush, beans and maize plantation, 1870m, 6.1633N, 10.1940E, fl., 5 Dec. 1998, Pollard 319.

Antidesma venosum Tul.

Shrub or small tree; leaves obovate or obovate-oblong, 3–10.5 × 1.5–5 cm; male inflorescence to 8.5 cm long. Forest edge or savanna.
Ijim: Nchan, on edge of forest, 1660m, fl., Apr. 1931, Maitland 1475.
Bambui: Agirculture Department Farm, 1600m, fl., 26 Apr. 1963, Brunt 1100.
Ndop: Babungo Agriculture Department Farm, 1260m, fl., 17 May 1962, Brunt 436.

Bridelia ferruginea Benth.

Shrub or tree, 3–10 m, not spiny; leaves alternate, elliptic or oblong, 8–12 × 4–5 cm, shortly acuminate, base obtuse, veins 8–10 pairs, densely felty below; petiole 0.5–1 cm; inflorescences of dense axillary panicles; flowers green, c. 4 mm wide; fruits ellipsoid, 1 cm long, green. Savanna.
Ndop: 1100m, Mar. 1962, Brunt 52.

Bridelia speciosa Müll.Arg.

Tree to 10 m; trunk spiny; leaves glabrous, oblong-elliptic; flowers greenish-yellow. Forest edge.

Kilum: Path from Mbo village (about 6 km from Elak-Oku, 2200–2500m, fl., 3 Nov. 1996, Onana 498.

Ijim: Fundong, Touristic Hotel, waterfall on Chumni river, 1600m, fl., 24 Nov. 1996, Cheek 8743; Afua swamp, 1950m, 6.09N, 10.24E, fl., 7 Dec. 1998, Etuge 4571.

Croton macrostachyus Hochst. ex Delile

Local name: Apjam (Oku) – Munyenyembe 807.

Tree, to 7–10 m; leaves ovate, cordate or rounded, undersides covered with silvery stellate hairs, 6.5–18.5 cm long; flowers white, fruit slightly 3-lobed. Forest edge.

Kilum: KD, fl., 9 Jun. 1996, Etuge 2241; Elak, forest edge, 2200m, fr., 30 Oct. 1996, Munyenyembe 807; Emoghwo traditional forest, 1900m, fr., 3 Nov. 1996, Zapfack 1157.

Ijim: Gikwang road towards Nyasosso forest, Ijim Mountain Forest Reserve, 2300m, fr., 19 Nov. 1996, Etuge 3496A.

Erythrococca hispida (Pax) Prain

Local name: Fokule (Kom at Laikom) – Etuge 4542; Fuh koo ala (Kom at Laikom) – Cheek 9907.

Shrub or tree to 5 m; leaves ovate-oblong, pilose beneath, 7–20 cm long; flowers yellowish-green. Forest.

Ijim: Zitum road, fr., 21 Nov. 1996, Etuge 3545; Laikom, Akwo-mofu medicinal forest, 1760m, 6.17N, 10.20E, fr., 5 Dec. 1998, Etuge 4542; Laikom, below Fon of Kom's Palace, remnant of montane forest in compound, 1820m, 6.16N, 10.21E, st., 11 Dec. 1998, Cheek 9907.

Euphorbia depauperata Hochst. ex A. Rich.

Local name: Turru bee hee (Fufube – Manu Ibrahim, Fulani) – Cheek 9822.

Perennial, 10–90 cm; rootstock woody; stems 2 to several, herbaceous; leaves numerous, linear-lanceolate to elliptic or subsessile; bracts ovate, 1 cm; capsule far-exserted, 6 mm. Grassland.

Ijim: Laikom, in grassland on plateau, 2000m, fl., Jun. 1931, Maitland 1502; Main path from Ijim to Tum, 2100m, fl., 25 Nov. 1996, Etuge 3618; Sappel, below Ardo's uncle's compound, grassland, 2300m, 6.16N, 10.24E, fl., 5 Dec. 1998, Cheek 9822.

Euphorbia hirta L.

Erect or decumbent herb, to 45 cm; stems sometimes purple-tinged; involucres in dense terminal glomerules, leaves obliquely ovate to lanceolate. Weed.

Kumbo: Wvem, near Kumbo, cultivated in Mama Saka's medicinal herb garden, fl., 6 Nov. 1998, Maisels 171.

Euphorbia schimperiana Scheele var. schimperiana

Local name: Unknown – Cheek 8674A.

Herb, 0.6–1.2 m; leaves alternate, 4–10 cm long, narrowly oblanceolate, with a whorl of 4–10 below the inflorescence; inflorescence an umbel; glands of involucre crescent-shaped; capsule 3–4 mm. Forest.

Kilum: Elak, 2300m, fr., 8 Jun. 1996, Zapfack 751; KD, 2800m, fl., 10 Jun. 1996, Etuge 2260; Elak, 2600m, fr., 27

Oct. 1996, Munyenyembe 704; KC, 2300m, fl. fr., 30 Oct. 1996, Buzgo 647.

Ijim: Above Aboh village, 1.5 hours walk along Gikwang road, 2450m, 19 Nov. 1996, Cheek 8674A; Ijim Mountain Forest, 2160m, fl. fr., 19 Nov. 1996, Kamundi 623; Ijim Mountain Forest, near Chufekhe stream, 2010m, fl., 20 Nov. 1996, Kamundi 658.

Hymenocardia acida Tul.

Shrub 2–6 m; leaves alternate, leathery, elliptic to oblanceolate, 4–10 × 2–4 cm, apex and base rounded, nerves 4–10 pairs; petiole 1–2 cm; inflorescences pendulous, slender, 4–5 cm long, densely covered in minute white to purple flowers; fruits dry, 2-winged, heart-shaped, c. 3 × 2 cm. Savanna.

Ndop: 1000m, fl., 13 Apr. 1962, Brunt 365.

Macaranga occidentalis (Müll.Arg.) Müll.Arg.

Tree, 10–20 m; trunk and branches often spiny; leaves orbicular-cordate, 20–30 cm long, slightly 5-lobed; flowers green; fruits with a yellow granular covering. Forest edge.

Ijim: Aboh village, fl., 25 Nov. 1996, Onana 626; Afua to Belo, 1950m, 6.09N, 10.22E, st., 7 Dec. 1998, Cheek 9851.

Maesobotrya floribunda Benth. var. *vermeuleni* (De Wild.) J.Léonard

Tree c. 4 m; leaves oblong-elliptic, 7.5–17 cm long, glabrous above; stipules small and cadaverous; flowers yellow; seeds blue. Forest.

Bambui: Near Bambui on Njinikom road, fr., 4 Jan. 1951, Keay FHI 28420.

Neoboutonia mannii Benth.

Tree, to 23 m; leaves orbicular-cordate, 10–20 cm long; flowers yellowish-green, styles narrow with 2 linear lobes. Forest edge.

Kilum: N-E rim of crater of Lake Oku, 2200m, fl., 4 Nov. 1996, Buzgo 699; Lake Oku, 2200m, fl., 4 Nov. 1996, Satabie 1066.

Neoboutonia melleri (Müll.Arg.) Prain

Shrub or small tree 4–6 m; leaves orbicular-cordate, 10–20 cm long, completely scurfy or felted beneath; ovary and capsule densely felty. Forest.

Syn. *Neoboutonia velutina* Prain, F.W.T.A. 1: 404 (1958).

Ndop: 3 miles SW of Bamessi, 1260m, fl., 3 Apr. 1962, Brunt 306.

Phyllanthus mannianus Müll.Arg.

Local name: Unknown (Ernest Kiming) – Cheek 8433.

Erect subshrub to 1.5 m; leaves obovate or obovate-elliptic, 1–2.5 cm; flowers dioecious, males 1-3 in each leaf axil, females solitary; sepals white with broad green central line. Forest gaps.

Kilum: Oku, in open area where road cuts through high forest at mile 11, 2000m, fl., 17 Feb. 1958, Hepper 2043; Path above water-tower, KJ, about 1 km into forest, 2200m, fl., 9 Jun. 1996, Cable 2965; KD, 9 Jun. 1996, Etuge 2252; Elak, 2450m, fr., 9 Jun. 1996, Zapfack 804; Oku-Elak to the forest at KA, 2200m, fr., 27 Oct. 1996, Cheek 8433; KC,

2300m, fl., 30 Oct. 1996, Zapfack 1095; Above Mboh, path above village to forest through Wambeng's farm, 2200m, fl., 1 Nov. 1996, Cheek 8551; Lake Oku, 2200m, fl., 5 Nov. 1996, Pollard 44.
Ijim: Gikwang road towards Oku, 2280m, fr., 15 Nov. 1996, Etuge 3394; Aboh village, turbo path, 500m after its splitting fom the main road at the border of the Ijim Mountain Forest Reserve, 2000m, fl., 18 Nov. 1996, Buzgo 758; Aboh, Anyajua, 2400m, fl. fr., 19 Nov. 1996, Satabie 1071.
Kumbo: Banso, in fuel plantation, fl., 7 Oct. 1947, Tamajong FHI 23471.

Ricinus communis L.
Annual herb to 2 m; leaves orbicular, to 60 cm broad, deeply palmately lobed, green or reddish; inflorescence a large panicle; male flowers below, females above; capsule smooth or prickly, 2.5 cm diameter. Cultivated.
Kilum: Elak, 2000m, fl. fr., 8 Jun. 1996, Zapfack 784.
Ijim: Gikwang road towards Oku, abandoned farmbush, 2265m, fl. fr., 15 Nov. 1996, Etuge 3403.

Sapium ellipticum (Hochst.) Pax
Tree, 8–40 m; branches drooping; leaves elliptic or oblong-elliptic, 4–15 cm long; inflorescence spicate, 5-12 cm long; flowers yellow-green; fruits orange-green. Forest edge.
Kilum: Lumetu Traditional Forest edge, 2300m, fr., 6 Nov. 1996, Munyenyembe 893.
Ijim: Fundong, Touristic Hotel, waterfall on Chumni river, st., 24 November 1996, Cheek 8741.

Tragia benthamii Baker
Local name: Mbasitutush (Oku) – Munyenyembe 870.
Herbaceous climber or trailer with stinging hairs on stem, petioles and leaves; leaves ovate, sharply serrate, 5–14 cm long; flowers minute, green. Forest edge.
Kilum: Upkim Traditional forest, 2060m, fl., 4 Nov. 1996, Munyenyembe 870; Secondary forest near the village, fr., 6 Nov. 1996, Onana 500.
Ijim: Ijim Mountain Forest, 2000m, fl. fr., 21 Nov. 1996, Pollard 78; Aboh village, along the main road, 1700–1800m, fr., 24 Nov. 1996, Onana 606; Afua swamp to Afua junction, 1800m, 6.09N, 10.22E, fr., 7 Dec. 1998, Cheek 9859.

Vernicia montana Lour.
Tree to c. 3 m; leaves alternate, ovate, c. 18 × 18 cm, digitately 5-lobed, petiole 15 cm, gland 4 mm wide, apical; inflorescence a small terminal panicle; flowers white, 3 cm long; Cultivated.
Syn. *Aleurites montana* (Lour.) E.H.Wilson
Ijim: Fundong, near touristic house, near the waterfall, 6.15N, 10.26E, fl., 23 Nov. 1996, Onana 597.

FLACOURTIACEAE

Dovyalis sp. nov.
RED DATA LISTED
Tree 5–7 m, spines to 4 cm long; leaves resembling those of *Oncoba sp. nov.*; flowers 1–2 cm diam., green, hermaphrodite (females longer); fruits soft, fleshy, glabrous, smooth, 3–5 cm diam. Forest.

Kilum: Lake Oku, 2200m, fr., 12 Jun. 1996, Etuge 2329; Lake Oku, near Baptist Rest House, 2300m, fl., 4 Nov. 1996, Cheek 8584; N side of lake, by shore, at water level, 2100m, fr., 5 Nov. 1996, Buzgo 706; Around Lake Oku, 2200m, fr., 5 Nov. 1996, Etuge 3341; Cheek 8772; On road above Lake Oku, 2400m, fl., 20 Apr. 1997, Maisels 60.

Oncoba sp. nov.
RED DATA LISTED
Tree 5–7 m with spines 2–4 cm long; leaves papery, elliptic, c. 6 cm long, serrate; flowers c. 4 cm diam., petals white, staminal mass yellow; fruits globular, with equidistant longitudinal lines, c. 5 cm diam. Forest.
Ijim: Laikom, path from Fon's palace to Mbororo settlement, 1900m, 21 Nov. 1996, Cheek 8698; 1600m, fl., 8 Jul. 1998, Maisels 160; Gateway of Yong's compound, Elemighong-Anyajua, 1550m, fr., 9 Dec. 1998, Cheek 9883; Below Fon of Kom's Palace, Laikom, 1820m, fr., 11 Dec. 1998, Cheek 9906.

GENTIANACEAE
Det. Sileshi Nemomissa (ETH) & M.Cheek (K)

Sebaea brachyphylla Griseb.
Local name: Fuflele (Kom – Yama Peter) – Gosline 202.
Erect herb to 60 cm; leaves orbicular to 1.5 cm; inflorescence a dense terminal cyme; corolla bright yellow, 6 mm long. Grassland and swamps.
Kilum: Grassland, 2800m, fl., 1 Feb. 1970, C. N.A.D. 1829; 2400m, fl., 28 Oct. 1996, Munyenyembe 736; KA, grassland, 2800m, fl., 31 Oct. 1996, Buzgo 670; Elak, secondary forest near the village, 6.15N, 10.26E, fl., 6 Nov. 1996, Onana 503.
Ijim: Turbo path, 500m after its splitting from the main road at the border of Ijim Mountain Forest Reserve, 2000m, fl., 18 Nov. 1996, Buzgo 754; Gikwang towards Nyasosso bush, 2280m, fl., 18 Nov. 1996, Etuge 3439; Laikom, in bog below Ardo's compound, 2100m, fl., 7 Dec. 1998, Gosline 202.

Sebaea grandis (E.Mey.) Steud.
Herb to 20 cm; leaves lanceolate to 3 cm long; flowers pale yellow, corolla 2 cm long or more. Grassland.
Ijim: Laikom, 1700m, fl., Jun. 1931, Maitland 1790.

Sebaea longicaulis Schinz
Erect herb to 60 cm; leaves orbicular to 1.5 cm; inflorescence a dense terminal cyme; flowers bright yellow, corolla 12 mm long. Grassland.
Kilum: Kumbo to Oku, 3 miles from Kumbo, near Oku road, 1850m, fl., 15 Feb. 1958, Hepper 2809.

Sebaea microphylla (Edgew.) Knobl.
Annual herb 5–6 cm, yellowish green; leaves reduced, c. 3 mm long; flowers yellow. Basalt pavement near waterfall.
Kilum: Near to the stream around savanna zone, 2500m, fl., 5 Nov. 1996, Zapfack 1209; Ewook Etele Mbae, 10 km from Oku, 2350m, fl., 6 Nov. 1996, Zapfack 1211.

Swertia abyssinica Hochst.
Erect, often much-branched herb, to 45 cm; leaves ovate to

orbicular, to 2.5 cm long; flowers numerous, white. Grassland, cliffs.

Kilum: Lake Oku, <u>Keay</u> & <u>Lightbody</u> FHI 28495; Elak, KA, fl., 31 Oct. 1996, <u>Buzgo</u> 686; Cliff-face, 2800m, fl., 31 Oct. 1996, <u>Munyenyembe</u> 824; Summit, 2800m, fl., 31 Oct. 1996, <u>Zapfack</u> 1121.

Ijim: Inside crater of Lake Oku, 2300m, fl., 25 Nov. 1996, <u>Cheek</u> 8773; N facing, seasonally damp rockface, amongst dense vegetation, by 'gallery' forest along path from Fon's Palace, Laikom to grassland plateau above, 2000m, 6.1642N, 10.1950E, fl., 2 Dec. 1998, <u>Pollard</u> 269.

Swertia mannii Hook.f.

Erect, slender herb c. 12 cm; leaves linear-lanceolate to narrowly elliptic, 1–3 cm long; flowers numerous, small, white. Grassland.

Kilum: Near to the stream around savanna zone, 2200m, fl., 5
Nov. 1996, <u>Zapfack</u> 1202; Ewook Etele Mbae, 10 km from Oku, 2350m, fl., 6 Nov. 1996, <u>Zapfack</u> 1219; Laikom, Ijim plateau, 1900m, 6.16N, 10.19E, fl., 3 Dec. 1998, <u>Cheek</u> 9751.

Swertia usambarensis Engl.

Annual herb c. 10 cm, resembling *Swertia mannii*, but leaves broadly ovate or broadly elliptic. Grassland.

Kilum: Summit, 3000m, fl., 29 Oct. 1996, <u>Munyenyembe</u> 752.

GERANIACEAE

Geranium arabicum Forssk.

Local name: Agang (Oku) – <u>Munyenyembe</u> 748; Fehkeh (Bekom at Nguakon) – <u>Buzgo</u> 749.
Herb to c. 30 cm, softly hairy; leaves digitately divided; flowers pale pink. Forest edge.
Syn. *Geranium simense* Hochst. ex A. Rich., F.W.T.A. 1: 157 (1954).

Kilum: KJ, above water tank, 2 km into forest, 2400m, fl., 10 Jun. 1996, <u>Cable</u> 3009; Path from KN to summit, 2900m, fl., 10 Jun. 1996, <u>Cable</u> 3040; KD, 3000m, fl., 10 Jun. 1996, <u>Etuge</u> 2286; Forest above Oku-Elak, 2600m, fl., 10 Jun. 1996, <u>Zapfack</u> 838; Elak, 2200m, fl., 28 Oct 1996, <u>Munyenyembe</u> 748; Elak, 2800m, fl., 30 Oct. 1996, <u>Satabie</u> 1061; Ewook Etele, Oku-Mbae, 2350m, fl., 6 Nov. 1996, <u>Etuge</u> 3343.

Ijim: Turbo path, 500m after its splitting from main road at the border of Ijim Mountain Forest Reserve, 2000m, fl., 18 Nov. 1996, <u>Buzgo</u> 749; Above Aboh village, 1.5 hrs along Gikwang road to Lake Oku, 6.1112N 10.2529E, 2350m, fl., 18 Nov. 1996, <u>Cheek</u> 8638; Ijim mountain forest, 2200m, fl., 18 Nov. 1996, <u>Pollard</u> 51; Gikwang main road towards Nyasosso forest, 6.1116N 10.2543E, 2280m, fl., 20 Nov. 1996, <u>Etuge</u> 3505; Ijim plateau above Fon's palace, Laikom, 1900m, 6.16N, 10.19E, fl., 3 Dec. 1998, <u>Cheek</u> 9759.

Geranium mascatense Boiss.

Herb to c. 30 cm, softly hairy; leaves digitately divided; flowers pink with a deep blackish-purple central 'eye'. Forest edge.
Syn. *Geranium ocellatum* Cambess., F.W.T.A. 1: 157

(1954).

Ijim: Grassland below Ijim ridge basaltic plateau, 1960m, 6.1630N, 10.1951E, fl. fr., 11 Dec. 1998, <u>Pollard</u> 370.

GESNERIACEAE

Streptocarpus elongatus Engl.

Local name: Mbdokwah (Oku) – <u>Munyenyembe</u> 850.
Erect herb, 30–45 cm; lower leaves long-petiolate, upper sessile, ovate elliptic, 8–12 × 4–6 cm; inflorescence a lax pedunculate cyme; flowers white. Damp cliffs.

Kilum: Elak, 2400m, fl., 1 Nov. 1996, <u>Munyenyembe</u> 850; Lake Oku to Jikijem, 6.1214N 10.2733E, 2200m, fl., 5 Nov. 1996, <u>Cheek</u> 8599.

Ijim: Fundong to Belo road, cliff, dripping water, 6.17N, 10.20E, 2050m, fl., 21 Nov. 1996, <u>Cheek</u> 8723; Zitum road, 6.1116N 10.2543E, 1860m, fl. fr., 21 Nov. 1996, <u>Etuge</u> 3555.

GUTTIFERAE
Det. M.Cheek (K) & F.Tadjouteu (YA)

Garcinia smeathmannii (Planch. & Triana) Oliv.

Local name: Ekaam (Oku) – <u>Etuge</u> 3357.
Tree 8–10 m, exudate yellow; leaves opposite, dark green, lanceolate-oblong, c. 20 cm long. Forest.
Syn. *Garcinia polyantha* Oliv., F.W.T.A. 1: 294 (1954).

Kilum: Oku, 2010m, 2 Nov. 1996, <u>Etuge</u> 3332; Main road from Ijikijem to Oku, 2350m, fr., 7 Nov. 1996, <u>Etuge</u> 3357; North slope of Oku Mountain, Ngashi area, 2500m, fr., 12 May 1997, <u>Maisels</u> 45.

Ijim: Ijim Mountain Forest, 1950m, fr., 21 Nov. 1996, <u>Pollard</u> 65; Laikom, 2000m, fr., 3 Dec. 1998, <u>Etuge</u> 4519.

Njinikom: <u>Maitland</u> 1489.

Harungana madagascariensis Lam. ex Poir.

Tree or shrub covered with fine stellate hairs, sap orange-coloured; leaves opposite, ovate or ovate-elliptic, 10–20 cm long; inflorescence a terminal corymbose cyme; flowers numerous, small, whitish with black glands, fragrant. Fallow.

Kilum: Mbokengfish, 1900m, fl., 7 Nov. 1996, <u>Munyenyembe</u> 906.

Ijim: Fundong, near touristic house, near the waterfall, fr., 23 Nov. 1996, <u>Onana</u> 595.

Hypericum peplidifolium A.Rich.

Local name: Febanghabus (Kom) – <u>Cheek</u> 8671.
Herb; stems prostrate, to 60–90 cm long; leaves elliptic or obovate, 6–15 mm long; flowers yellow; fruit fleshy and indehiscent. Forest edge.

Kilum: KA, 2760m, fl., 10 Jun. 1996, <u>Asonganyi</u> 1303; KD, 3000m, fl., 10 Jun. 1996, <u>Etuge</u> 2291; KJ to KD, 3000m, fl., 13 Jun. 1996, <u>Etuge</u> 2351.

Ijim: Ijim Mountain Forest, 2200m, fl., 18 Nov. 1996, <u>Pollard</u> 52; Above Aboh village, 1.5 hours walk along Gikwang road, 2450m, fl., 19 Nov. 1996, <u>Cheek</u> 8671.

Belo: <u>Maitland</u> 1468.

Kumbo: Banso, <u>Keay</u> FHI 28694.

Hypericum revolutum Vahl subsp. *revolutum*

Local name: Febangha bus (Kom) – Cheek 8649; Fembangalum (Oku) – Munyenyembe 738.
Shrub or tree to 12m; leaves narrowly elliptic, 1–2(03.5)cm long; petals yellow, 2–3 cm long. Forest edge.
Syn. *Hypericum lanceolatum* Lam., F.W.T.A. 1: 287 (1954).
Kilum: Elak, 2300m, fl., 8 Jun. 1996, Zapfack 777; Path from KN to summit, 2900m, fl., 10 Jun. 1996, Cable 3035; Elak, 2200m, fl., 28 Oct. 1996, Munyenyembe 738.
Ijim: Turbo path, 500m after its splitting from the main road at the border of the Ijim Mountain Forest Reserve, 2000m, fl., 18 Nov. 1996, Buzgo 748; Above Aboh village, 1.5 hours on road to Lake Oku, 2350m, fl., 18 Nov. 1996, Cheek 8649; Laikom, Ijim ridge, about 30 mins. walk above Fon of Kom's palace, 2050m, fl., 2 Dec. 1998, Cheek 9770.

Hypericum roeperianum Schimp. ex A.Rich.

Local name: Weh (Bekom at Ayeah) – Buzgo 766.
Shrub to 3m; leaves elliptic, ovate or lanceolate, 3-8 cm long; petals yellow 2.5–3.5 cm long. Forest edge.
Syn. *Hypericum riparium* A.Chev., F.W.T.A. 1: 287 (1954).
Ijim: Laikom, Maitland 1435; 1488 & 1632; Turbo path, 1 km after its splitting from the main road at the border of the Ijim Mountain Forest Reserve, N from there, 2000m, fl., 18 Nov. 1996, Buzgo 766; Aboh village, fl., 18 Nov. 1996, Onana 521.

Psorospermum aurantiacum Engl.

Tree to 5m; leaves rusty tomentose; petals cream. Forest edge.
Kilum: Elak, 2000m, fr., 8 Jun. 1996, Zapfack 757; Above Mboh, path above village to forest through Wambeng's farm, 2000m, 1 Nov. 1996, Cheek 8574.
Ijim: Aboh village, 1700–1800m, 24 Nov. 1996, Onana 602.

Psorospermum densipunctatum Engl.

Shrub or small tree to 3 m; leaves dark glossy green; flowers white or yellowish; fruits red. Forest edge.
Ijim: Laikom, 2000m, fl., Apr. 1931, Maitland 1662.

Psorospermum febrifugum Spach

Local name: Fnkass – Maisels 183.
Tree or shrub to 4 m; bark smooth, grey; leaves broadly ovate to orbicular; flowers cream; fruits red. Forest edge.
Jakiri: Tan or Ntan, Jakiri area, AlHadji Gey's land, 1600m, 6 Nov. 1998, Maisels 183.

Symphonia globulifera L.f.

Tree to 24 m; exudate yellow; leaves elliptic-oblanceolate, 8–10 × 2–3 cm; flowers bright red, c. 1 cm long; fruits ellipsoid, 2.5 cm diameter. Forest.
Ijim: Ajung, 1900m, 12 Nov. 1999, Cheek 10098.

HALORAGACEAE

Laurembergia tetrandra (Schott) Kanitz subsp. *brachypoda* (Welw. ex Hiern) A.Raynal var. *brachypoda*

Creeping herb; stems reddish; leaves linear or linear-oblanceolate, 10–13 mm long; flowers minute, up to 11 in each leaf-axil, greenish-yellow or brownish. Swamp.
Syn. *Laurembergia engleri* Schindl., F.W.T.A. 1: 171 (1954).
Ijim: Sappel, below Ardo's uncle's compound, 2300m, 6.16N, 10.24E, fl., 5 Dec. 1998, Cheek 9804.

Myriophyllum cf. spicatum L.

Submerged, bottom rooting, aquatic herb; stems 30–90 cm long; leaves alternate, circular, deeply dissected and comb-like; flowers inconspicuous. Lake.
Kilum: Lake Oku, near Baptist Rest House, 2300m, 4 Nov. 1996, Cheek 8575.

LABIATAE
B.J.Pollard (K) & A.J.Paton (K)

Achyrospermum

Several species of the genus *Achyrospermum* are very closely related and it is difficult, in some cases, to decide on the separating line, and to obtain definite key characters. It would appear that there are two or three distinct species from our checklist area, but the delimitation of these is not clear and merits further investigation. It is likely that the following two species occur on Mount Oku and the Ijim ridge, and that *sp. 3* may represent a separate entity. Our specimens cannot be attributed to any one of these taxa until a detailed morphometric analysis is undertaken.

Achyrospermum africanum Hook.f. ex Baker

An erect herbaceous undershrub to about 3 m; inflorescences spicate; flowers purplish-pink; stamens ascending; in semi-shade. Forest.

Achyrospermum oblongifolium Baker

An erect, herbaceous, little-branched, undershrub 30–70 cm; stems little-branched, tomentose; inflorescence spicate; flowers greenish-white; stamens ascending. Forest.

Achyrospermum sp. 3

Similar to *Achyrospermum africanum*, but flowers white; stamens ascending. Forest.

Haumaniastrum alboviride (Hutch.) P.A.Duvign & Plancke

Perennial suffrutex or shrub usually with a single stem arising from a thick woody rootstock, 1–1.3 m; stems dark purple; leaves sessile, ovate, 2–4 cm long, acutely acuminate, sharply serrate; with flower-heads arranged in a loose panicle or corymbose; flowers pale violet; stamens declinate. Grassland.
Ndop: Ndop, 1660m, Adams 1513.

Haumaniastrum buettneri (Gürke) J.K.Morton

Probably perennial herb, 0.6–1 m with usually a single stem; stems densely silver-pubescent to sericeous; leaves entire or shallowly serrate, glanular-punctate; veins raised below, straight, soon becoming ± parallel with midvein; corolla

white or white marked pink; stamens declinate. Grassland.
Kilum: Etuge 3644.
Ndop: Ndop, 1800m, fl., Dec. Boughey GC 10477.

Haumaniastrum caeruleum (Oliv.) J.K.Morton

Perennial herb 0.2–1 m; stems one to several, sparsely pubescent; leaves almost glabrous to densely pubescent, leaf-base sometimes clasping stem; petioles 0–2 mm long; upper leaves subtending heads, 5–25 mm long, caudate or apiculate, white or bluish, apex green, often inrolled; corolla white, pink, blue or purple; stamens declinate. Grassland.
Ijim: Nchan, 1660m, fl., May 1931, Maitland 1522; Mwal village, near Ajung village, 1200m, fl. fr., 28 Nov. 1998, Maisels 192.

Hemizygia welwitschii (Rolfe) Ashley

A bushy, aromatic perennial, to 0.6 m; inflorescence with large, pink bracts at apex (or enveloping flower buds); stamens declinate. Savanna.
Jakiri: By the roadside to Kumbo, on motor road, fl., 25 Apr. 1951, Ujor FHI 30085.
Ndop: Ndop, near Bamessi, fl., 30 Mar. 1962, Brunt 264; Ndop plain, Baba village, 1160m, fl., 17 May. 1962, Brunt 426.

Isodon ramosissimus (Hook.f.) Codd

Local name: Ndogntan (Oku) – Munyenyembe 897; Tung Tung (Bekom at Ayeah) – Buzgo 750; Fengi Fkfing (Kom at Laikom – Yama Peter) – Pollard 262.
An erect or straggling herb to 4 m; stems hollow, strongly quadrangular, pilose; inflorescence an axillary panicle of many-flowered dichotomous cymes; calyx tube declinate, ventricose, teeth subequal; corolla 5 mm long, white, speckled purple in throat; upper lip very small, recurved; stamens declinate. Forest edge.
Syn. *Homalocheilos ramosissimus* (Hook.f.) Codd, F.W.T.A., 2: 460 (1963).
Kilum: Mbokengfish, fallow, 1900m, fl., 7 Nov. 1996, Munyenyembe 897; BirdLife Plot No. 7, Maisels 146.
Ijim: Main road from Atubeaboh (Gikwang Foe) towards Oku, 2230m, fl., 15 Nov. 1996, Etuge 3375; 500m along turbo path, from Ijim Mountain Forest Reserve, 2000m, fl., 18 Nov. 1996, Buzgo 750; Laikom, stream 100m S of top of path from Fon's palace to grassland plateau, 2000m, 6.1642N, 10.1950E, fl., 2 Dec. 1998, Pollard 262.

Leonotis nepetifolia L.

Herb 2m; flowers dull orange, 3–4 cm long; infructescence forming a spherical mass c. 5 cm diam. along the robust, erect main axis; stamens ascending. Savanna.
Kilum: DeMarco s.n.
Babungo-Mbi Crater: 1600m, 8 Nov. 1999, Cheek 10007.

Leucas deflexa Hook.f.

Local name: Fengi Fungwor (Kom at Laikom – Yama Peter) – Pollard 340.
A straggling or semi-erect aromatic herb to 2 m; leaves lanceolate, cuneate at base, serrate with an entire, acute tip, petiolate; inflorescence a densely globose axillary whorl; corolla white; stamens ascending; anthers often conspicuously hairy, orange. Forest edge.

Kilum: Forest, 2300m, fl., 8 Jun 1996, Zapfack 750; Lake Oku, 2200m, fl., 11 Jun 1996, Zapfack 869; Shore of Lake Oku, 2250m, fl., 12 Jun 1996, Cable 3133; Lake Oku, 2200m, fl., 12 Jun 1996, Etuge 2332; KJ, 2500m, fl., 29 Oct 1996, Zapfack 1075; KC, 2500m, fl., 30 Oct 1996, Buzgo 652; Path from project HQ to KA, 2800m, fl., 30 Oct 1996, Cheek 8517.
Ijim: Aboh village, 6.15N, 10.26E, fl., 18 Nov. 1996, Onana 526; Top of ridge, second gate at Aboh, 2360m, 6.12N, 10.26E, fl., 21 Nov 1996, Kamundi 669; Gallery forest along path up from Fon's Palace to Ijim ridge grassland plateau, 1980m, 6.1642N, 10.1950E, fl., 7 Dec. 1998, Pollard 340.

Leucas oligocephala Hook.f. subsp. *bowalensis* (A.Chev.) sensu J.K.Morton non Sebald

A stout erect herb to 90 cm; leaves linear-lanceolate to lanceolate, usually more than 2.5 cm long; inflorescence of several distinct dense, globose, axillary whorls; flowers densely hairy and purple tinged in bud, white when open; stamens ascending. Grassland.
Kilum: Grassland, summit, 3000m, fl., 29 Oct 1996, Munyenyembe 772; Summit, near savanna zone, 2480m, fl., 5 Nov 1996, Zapfack 1197.
Ijim: Laikom village by Fundong, rocky mountain grassland, 2100m, 6.15N, 10.26E, fl., 21 Nov 1996, Onana 561; Grass plateau above Fon's Palace, Laikom, 2000m, 6.1642N, 10.1950E, fl., 2 Dec. 1998, Pollard 256.

Leucas oligocephala Hook.f. subsp. *oligocephala*

A perennial herb 0.3–1.0 m; stems slender, branched, pilose; leaves elliptic to elliptic-lanceolate, up to 2.3 × 1 cm; inflorescence of one (terminal) or rarely two (subterminal) dense, globose, axillary whorls; flowers densely hairy and purple tinged in bud, white when open; stamens ascending. Grassland.
Kilum: Grassland, summit, 2950m, fl., 10 Jun 1996, Cable 3059 & 3082; Top of KA, grassland, 2800m, fl., 31 Oct 1996, Buzgo 675.
Ijim: Gikwang road to Nyasosso, 2280m, fl., 18 Nov 1996, Etuge 3428; Gikwang road, Aboh to Akeh, 1km after first track, 2400m, 6.15N, 10.26E, fl., 25 Nov 1996, Cheek 8753.

Ocimum gratissimum L. subsp. *gratissimum* var. *gratissimum*

A branched, erect, pubescent shrub to ± 3 m, emanating an aroma similar to that of cloves (*Syzygium aromaticum*); inflorescence of several dense spikes, wider than 1 cm; calyx dull, densely lanate, horizontal or slightly downward pointing in fruit; corolla small, greenish-white; stamens declinate. Forest edge.
Ijim: Etuge 3702.

Ocimum lamiifolium Hochst. ex Benth.

An erect woody perennial herb or shrub to 3 m; stems much-branched; inflorescence lax, verticils 1 – 2 cm distant; calyx often purplish; corolla white, white marked with pink or dull mauve; stamens declinate. Forest edge.
Kilum: Emoghwo traditional forest, 1900m, fl., 3 Nov 1996, Zapfack 1164.

Platostoma rotundifolium (Briq.) A.J.Paton

A stout woody perennial to ± 2 m; stems grooved, densely ferrugineous-pubescent; inflorescences several, dense, cylindrical, 2.5–10 cm long; bracts conspicuous, broadly ovate, white or mauve-tinged; stamens declinate. Forest edge.
Syn. *Geniosporum rotundifolium* Briq., F.W.T.A. 2: 453 (1963).
Kilum: Path from KMFP HQ to summit, 2300m, fl., 12 Oct. 1996, Cheek 8422.
Ijim: Gikwang road to Nyasosso forest, 2300m, fl., 19 Nov. 1996, Etuge 3489.

Plectranthus alpinus (Vatke) Ryding

Local name: Unknown (Oku) – Cheek 8591; Ifyengezvu (Kom at Laikom – Yama Peter) – Pollard 333.
A slender straggling herb; stems much-branched; leaves thin, subglabrous to pubescent, distinctly petiolate; inflorescence racemose, up to 35 × 2–2.5 cm; mature calyx 5–6 mm long; corolla up to 1 cm long, bluish-purple; stamens declinate. Swamp edge.
Syn. *Plectranthus assurgens* (Baker) J.K.Morton, F.W.T.A. 2: 459 (1963).
Kilum: Lake Oku, Keay & Lightbody FHI 28508; KD, 2800m, fl., 10 Jun. 1996, Etuge 2256; Road from Rest House East of lake to N of mountain range, 2300m, 6.1214N, 10.2733E, fl., 5 Nov. 1996, Cheek 8591.
Ijim: Laikom, swamp, half a mile W of Ardo's compound, Ijim ridge grassland plateau, 2070m, 6.1609N, 10.2051E, fl., 7 Dec. 1998, Pollard 333; Source of Ngengal stream by forest patch 1 km E of Fulani settlement before ridge dropping to Mbesa, 2460m, 6.1608N, 10.2459E, fl., 9 Dec. 1998, Pollard 351.

Plectranthus hadiensis Forssk.

A bushy herb; stems woody, densely pubescent; leaves deeply and coarsely crenate; inflorescence a dense broad panicle with lateral branches about 15 cm long; mature calyx 3 mm long; corolla 6 mm long, violet; stamens declinate. Grassland.
Syn. *Plectranthus cyaneus* Gürke, F.W.T.A. 2: 459 (1963).
Sabga: Sabga hill, Ring Road, 10 km E of Bambili, 2000m, fl., 7 Nov. 1970, Bauer 135.

Plectranthus esculentus N.E.Br.

Perennial herb with edible tuberous roots; stems coarsely pilose with whitish hairs; leaves entire to somewhat crenate, setulose on the veins beneath; inflorescence a much-branched panicle; mature calyx 8 mm long; corolla yellow, 1.5 cm long; stamens declinate. Cultivated.
Kumbo: Abandoned cultivation, 1680m, fl. fr., 14 Feb. 1958, Hepper 1975.

Plectranthus glandulosus Hook.f.

Local name: Finele Jung (Oku – Isaac Fokom) – Buzgo 644; Ifyengezvu (Kom at Laikom – Yama Peter) – Pollard 263.
A coarse scrambling to erect, often robust, glandular and strongly aromatic perennial herb to ± 3.5 m; leaves up to 15 cm long, glandular-punctate; leaf margin with very uneven, rather small, double or treble crenations; inflorescence of copious loose panicles to ± 65 cm long; mature calyx 9 mm long, corolla violet; stamens declinate. Forest edge.
Kilum: KA, 2200m, 6.1349N, 10.3112E, fl., 27 Oct., 1996, Cheek 8435; Elak, KJ, 2500m, fl., 29 Oct. 1996, Zapfack 1073; KC, 2200m, fl., 30 Oct. 1996, Buzgo 644.
Ijim: Fallow by road from Atubeaboh (Gikwang foe) to Oku, 2280m, fl., 15 Nov. 1996, Etuge 3380; Laikom, stream 100m S of pathtop, grassland plateau above Fon's Palace, 6.1642N, 10.1950E, fl., 2 Dec. 1998, Pollard 263; Laikom, grassland near swamp, half a mile W of Ardo's Fulani compound, Ijim ridge grassland plateau, 2100m, 6.1615N, 10.2051E, fl. (white form), 7 Dec. 1998, Pollard 343.

Plectranthus gracillimus T.C.E.Fr. ex Hutch. & Dandy

Erect slender annual herb; stems branching; leaves usually falling before onset of flowering, sessile, dentate, up to 10 × 3 cm; inflorescence a lax raceme; pedicels very long and slender, several times as long as the calyx; corolla small, bluish; stamens declinate. Grassland.
Syn. *Englerastrum gracillimum* T.C.E.Fr., F.W.T.A. 2: 465 (1963).
Kilum: Elak, KC, 2300m, fl., 30 Oct. 1996, Zapfack 1101.
Ijim: From Aboh, path towards Tum, 2100m, 6.1116N, 10.2543E, fl., 22 Nov. 1996, Etuge 3573.

Plectranthus insignis Hook.f.

Local name: Dzur (Oku) – Cheek 8431; Dra (Oku) – Munyenyembe 801; Bum (Bekom) – Kamundi 617; Boumelomne (Kom) – Cheek 8664.
A large, soft-wooded, monocarpic undershrub from 3–5 m at maturity, usually leafless at the time of flowering; inflorescence a very lax ample racemose panicle; mature calyx 2 cm long, teeth very unequal; corolla 2 cm long, yellow suffused with purple; stamens declinate. Forest.
Kilum: KJ, 2200m, fl., 9 Jun. 1996, Cable 2951; KJ, 2250m, fl., 9 Jun. 1996, Cable 3002; Lake Oku, 2500m, fl., 12 Jun. 1996, Etuge 2330; KA, 2200m, 6.1349N, 10.3112E, fl., 27 Oct. 1996, Cheek 8431; Grassland, 2400m, fl., 30 Oct. 1996, Munyenyembe 801.
Ijim: Above Aboh village, 1.5 hours along Gikwang road towards Lake Oku, 2450m, 6.1112N, 10.2529E, fl., 19 Nov. 1996; Cheek 8664; 6.11N, 10.25E, fl., 19 Nov. 1996, Kamundi 617.

Plectranthus punctatus L'Hér. subsp. *lanatus* J.K.Morton

Local name: Ijim (Bekom at Ayeah) – Buzgo 755.
RED DATA LISTED
An erect or occasionally scrambling glandular-pubescent herb to 20 – 50 cm; stems sub-succulent, swollen at the nodes, greenish-white, speckled red; leaves sessile or nearly so, up to 7.5 ×4 cm; inflorescence dense in flower, lax in fruit; calyx 8–10mm long at maturity; corolla 1.2–1.5 cm long, pale blue speckled purple to royal blue; stamens declinate. Swamp edge.
Kilum: Ridge N of Lake Oku, 2560m, fl., 6 Jan. 1951, Keay & Lightbody FHI 28464; KD, 3000m, fl., 10 Jun. 1996, Etuge 2296; Path from Kilum Project HQ to summit, 2760m, fl., 12 Oct. 1996, Cheek 8419; Elak, KA path, 2600–2800m, 6.15N, 10.26E, fl., 30 Oct. 1996, Onana 467.

Ijim: Mbesa to Laikom, on plateau, in grassland, fl., Jun. 1931, Maitland 1724; Turbo path, 500m from border of Ijim Mountain Forest Reserve, 2000m, fl., 18 Nov. 1996, Buzgo 755; Gikwang to Nyasosso road, 2270m, fl., 18 Nov. 1996, Etuge 3425; Sappel, below Ardo's uncle's compound, 2300m, 6.16N, 10.24E, fl., 5 Dec. 1998, Cheek 9802; Afua swamp, 1950m, 6.08N, 10.24E, fl., 7 Dec. 1998, Cheek 9843; Swamp, half a mile W of Ardo's Fulani compound, Ijim ridge grassland plateau, Laikom, 2100m, 6.1615N, 10.2051E, fr., 7 Dec. 1998, Pollard 341.
Bambui: Bambui, 2100m, fl., Dec., Adams GC 11322.

Plectranthus sylvestris Gürke
Strongly aromatic herb, 0.6–1.0 m; stem brittle, sub-succulent; inflorescence a terminal panicle; corolla c. 1 cm long, dark blue; stamens declinate. Forest.
Kilum: Above Lake Oku, bamboo brake, 2500m, fl. fr., 7 Jan. 1951, Keay & Lightbody FHI 28508.

Plectranthus sp. nov. ?
Local name: Ifyenge zvu (Kom – Yama Peter) – Pollard 267.
RED DATA CANDIDATE
Epilithic herb 20–50 cm; roots often tuberous; leaf margin very coarsely crenate, almost lobed; inflorescence to 30 cm long; flowers mauve; stamens declinate. Damp rockface.
Ijim: N facing seasonally damp rockface, amongst dense vegetation, by 'gallery forest' along path from Fon's Palace, Laikom, to Ijim ridge grassland plateau above, 1960m, 6.1642N, 10.1950E, fl. fr., 2 Dec. 1998, Pollard 267 & fl. fr., 12 Dec. 1998, Pollard 385.

Pycnostachys eminii Gürke
A pubescent aromatic shrub, 2–4 m; leaves about 10 × 4 cm, distinctly crenate-serrate; shortly pubescent; inflorescence a cylindrical spike, about 1.5 cm broad; calyx with spines, sharp to the touch, densely tomentose at maturity; corolla pale blue; stamens declinate. Forest edge.
Bambui: Bambui, 2100m, Adams GC 11304.

Pycnostachys meyeri Gürke
Local name: Bies (Oku) – Munyenyembe 781; Fungi Tungo (Bekom at Nguakon) – Buzgo 747; Bum (Kom at Laikom) – Pollard 331.
Similar to *Pycnostachys eminii* except spike stouter, about 2.5 cm or more broad; mature calyx shortly and rather thinly pubescent; corolla pale blue to purplish-blue, occasionally even pinkish; stamens declinate. Forest edge.
Kilum: KD, grassland-forest boundary, 2950m, fl., 10 Jun. 1996, Cable 3058; Top of KD, 3000m, fl., 10 Jun. 1996, Etuge 2287; Forest, 2900m, fl., 29 Oct. 1996, Munyenyembe 781; Elak, KA, 2700m, fl., 31 Oct. 1996, Buzgo 661; Summit, 2800m, fl., 31 Oct. 1996, Zapfack 1111.
Ijim: Turbo path, 500m from Ijim Mountain Forest Reserve boundary, 2000m, fl., 18 Nov. 1996, Buzgo 747; Gikwang towards Nyasosso bush, 2250m, fl., 18 Nov. 1996, Etuge 3440; Swamp, half a mile W of Ardo's Fulani compound, Ijim ridge grassland plateau above Laikom, 2060m, 6.1609N, 10.2051E, fl., 7 Dec. 1998, Pollard 331; Grassland near swamp, half a mile W of Ardo's Fulani compound, Ijim ridge grassland plateau above Laikom, 2050m, 6.1615N,

10.2051E, fl. (pink form), 7 Dec. 1998, Pollard 342.

Pycnostachys pallide-caerulea Perkins
Similar to the above two species but differs in: stems strongly sulcate; leaves linear to linear-lanceolate, 30 mm × 2 mm, pseudo-whorled and numerous, serrulate; stamens declinate. Forest edge.
Kumbo: Banso Mts, 1800–2000m, Ledermann 5738.

Salvia leucantha Cav.
Introduced alien; stamens ascending.
Kilum: Elak, lower parts of KA, *Eucalyptus* plantation, fl., 28 Oct. 1996, Cheek 8476.

Satureja pseudosimensis Brenan
Local name: Kechuokentie (Oku) – Munyenyembe 777; Fungi iwu (Kom – Yama Peter) – Pollard 350.
A slender weak straggling perennial herb to 10–60 cm; stems softly pubescent and pilose, branched from the base; leaves subsessile, rounded at base, up to 2 cm long, crenulate; inflorescence of distant axillary whorls; calyx broadly tubular, 6–8 mm long in fruit; corolla about 1.2 cm long, mauve; stamens ascending. Forest edge.
Kilum: KA, 2760m, fl., 10 Jun. 1996, Asonganyi 1311; Elak, forest-grassland boundary, 2900m, fl., 29 Oct. 1996, Munyenyembe 777; KA path, 2600–2800m, 6.15N, 10.26E, fl., 30 Oct. 1996, Onana 470; Grassland, 2800m, fl., 30 Oct. 1996, Satabie 1059; Elak, KA, 2700m, fl., 31 Oct. 1996, Buzgo 665.
Ijim: Gikwang road towards Nyasosso forest, 2290m, fl., 19 Nov. 1996, Etuge 3487; 6.1121N, 10.2525E, fl., 19 Nov. 1996, Kamundi 602.

Satureja punctata (Benth.) R.Br. ex Briq.
A robust or occasionally erect woody herb to ± 30 cm; stems branched, pubescent; leaves ovate, 4 –10mm long, conspicuously glandular punctate beneath, subsessile, entire; flowers few together, in axillary whorls; calyx tubular, 4 mm long; corolla rich pink; stamens ascending. Forest edge.
Kilum: Elak, KD, fl., 9 Jun. 1996, Etuge 2247; Towards KJ, 2100m, fl., 29 Oct. 1996, Pollard 9; Between Manchok and Lumito forest, 2100m, fl., 1 Nov. 1996, Buzgo 695; Ewook Etele Mbae 10 km from Elak, 2350m, fl., 6 Nov. 1996, Zapfack 1234.
Ijim: Turbo path 500m from Ijim Mountain Forest Reserve boundary, 2000m, fl., 18 Nov. 1996, Buzgo 759; Laikom, grassland plateau above Fon's Palace, 2100m, 6.1642N, 10.1950E, fl., 3 Dec. 1998, Pollard 303.

Satureja robusta (Hook.f.) Brenan
Local name: Fungi fakessi (Kom – Yama Peter) – Pollard 255.
An erect robust, strongly aromatic, perennial, 80–140 cm at maturity; stems branched; leaves ovate-rotund,crenate; inflorescence broad, of many dense terminal sessile spikes; mature calyx 4–5 mm long; corolla white with mauve marks on the lip; stamens ascending. Forest edge.
Kilum: Towards KJ, 2100m, fl., 30 Oct. 1996, Pollard 16; Forest/grassland boundary, 2480m, fl., 5 Nov. 1996, Zapfack 1198.
Ijim: Turbo path 1 km after splitting from main road at

border of Ijim Mountain Forest Reserve, 2000m, fl., 18 Nov. 1996, Buzgo 778; Gikwang towards Nyasosso, 2280m, fl., 18 Nov. 1996, Etuge 3429; Laikom, plateau above Fon's Palace, 6.1642N, 10.1950E, fl., 2 Dec. 1998, Pollard 255.

Scutellaria violascens Gürke
Local name: Ijim (Oku) – Maisels 195.
A shrub to ± 1 m; stems woody, pubescent; leaves about 5 × 2.5 cm, petiolate, glandular-punctate beneath; inflorescence densely tomentose and glandular-pubescent, composed of stiff racemes; corolla purple, pubescent, about 1.3 cm long; stamens ascending. Savanna.
Kilum: Hill above Lang village, S of Lake Oku, 2400m, fl., 28 Nov. 1998, Maisels 195.

Solenostemon decumbens (Hook.f.) Baker
A small aromatic herb 10–30 cm; stem bases decumbent; leaves thick and fleshy, coarsely crenate, with a broad cuneate base; inflorescence rather dense, 5–15 cm long; corolla pale blue; stamens declinate. Grassland.
Kilum: Summit, 2850m, fl., 10 Jun. 1996, Cable 3086; KA, 2700m, fl., 31. Oct. 1996, Buzgo 663.
Ijim: Gikwang road, Aboh to Akeh, about 1 km after first track, 2400m, 6.15N, 10.26E, fl., 25 Nov. 1996, Cheek 8755.

Solenostemon mannii (Hook.f.) Baker
A herbaceous or somewhat woody perennial; stems quadrangular, climbing or erect; leaves crenate, acutely acuminate, long-petiolate; inflorescence a copiously flowered, dense raceme, 3–4 cm broad, up to 25 cm or more long in fruit; calyx 4–5 mm long; corolla rich bluish-purple; stamens declinate. Streamside.
Ijim: Laikom, streamside in grassland plateau, 2020m, 6.1642N, 10.1950E, fl., 3 Dec. 1998, Pollard 294.

Solenostemon repens (Gürke) J.K.Morton
A small creeping herb, 5–15 cm; stems pubescent; leaves long-petiolate, 2–3 cm long and broad; inflorescence a terminal umbellate cyme of about 4–6 short racemes; calyx 6 mm long, median teeth deltoid; corolla greenish-white; stamens declinate. Epilithic or on decaying tree-trunks.
Ijim: Rockface beside path in gallery forest, below grassland plateau, above Fon's Palace, Laikom, 1980m, 6.1642N, 10.1950E, fr., 2 Dec. 1998, Pollard 270.

Stachys aculeolata Hook.f. var. *aculeolata*
A perennial prickly herb; stems straggly, scrambling, rather slender with stiff prickly bristles and short glandular hairs; leaves cordate at base, to 5.5 × 4.5 cm; petiole as long as the lamina; inflorescence of distant, several-flowered verticils; corolla white or pale pink with dark pink markings; stamens ascending. Forest edge.
Kilum: Elak, 2300m, fl., 8 Jun. 1996, Zapfack 742; KJ, 2200m, fl., 9 Jun. 1996, Cable 2953; KD, ibid., Etuge 2213; Elak, 2360m, fl., 9 Jun. 1996, Zapfack 806; Summit, 2800m, fl., 10 Jun. 1996, Cable 3071; KD, 2800m, fl., 10 Jun. 1996, Etuge 2276; Lake Oku, 2300m, fl., 11 Jun. 1996, Zapfack 856; Elak, 2600m, fl., 27 Oct. 1996, Munyenyembe 702; KJ, 2400m, fl., 28 Oct. 1996, Buzgo 614; KC, 2300m, fl., 30 Oct. 1996, Buzgo 646; KA, 6.1349N, 10.3112E, fl., 30 Oct.

1996, Cheek 8505; KA, fl., 30 Oct. 1996, Cheek 8515; KA, 2700m, fl., 31 Oct. 1996, Buzgo 664.
Ijim: Laikom, near Akwamofu sacred forest, 1900m, 6.1633N, 10.1940E, fl., 5 Dec. 1998, Pollard 326.

Stachys pseudohumifusa Sebsebe subsp. *saxeri* Y.B.Harv.
RED DATA LISTED
Perennial erect herb to 1 m; stems eglandular, lacking siff, prickly bristles; inflorescence terminal; vericils (4–)6-flowered, 3–28 mm apart; calyx 4.5–5 mm long; corolla white(–blue) with ± red markings; stamens ascending. Swamp edge.
Ijim: Laikom, 2000m, fl., Apr. 1931, Maitland 1652; Etuge 3695; Sappel, below Ardo's uncle's compound, 2300m, 6.16N, 10.24E, fl., 5 Dec. 1998, Cheek 9811; Afua swamp, 1950m, 6.08N, 10.24E, fl., 7 Dec. 1998, Cheek 9842; Source of Ngengal stream by forest patch, 1 km E of Fulani settlement before ridge dropping to Mbesa, 2460m, 6.1608N, 10.2459E, fl., 9 Dec. 1998, Pollard 349; Laikom ridge, 2000m, 3 Nov. 1999, Cheek 9938.

LEEACEAE

Leea guineensis G.Don
Erect or sub-erect soft-wooded shrub to 7m; leaves bipinnate; leaflets opposite, imparipinnate, oblong-elliptic to 18 cm long; flowers bright yellow, orange or red; fruits brilliant red turning black. Forest.
Ijim: Aboh village, along the main road, 1700–1800m, 24 Nov. 1996, Onana 607; Aboh village, fr., 25 Nov. 1996, Onana 625.

LEGUMINOSAE-CAESALPINOIDEAE
Det. J-M.Onana (YA), B.A.Mackinder (K) & B.A.Satabie (YA)

Caesalpinia decapetala (Roth) Alston
Spiny shrub or scrambler to 5m; leaves bipinnate, c. 40 cm long; inflorescence an erect spike; flowers pale yellow, 1.5 cm diameter. Hedges and fallow.
Kilum: Emoghwo traditional forest, 1900m, fl., 3 Nov. 1996, Zapfack 1162.

Chamaecrista kirkii (Oliv.) Standl. var. kirkii
Local name: Ikum – Munyenyembe 751.
Erect shrub to 1.5 m; leaflets 25–30 pairs, oblong-linear, 9–17 × 1.5–4 mm; petiole with an apical sessile gland; flowers yellow, solitary or 2–3; fruits 6–7.5 cm long. Roadside.
Syn. *Cassia kirkii* Oliv., F.W.T.A. 1: 452 (1958).
Kilum: 2200m, fl., 28 Oct. 1996, Munyenyembe 751.
Ijim: Main road from Atubeaboh (Gikwang Foe) towards Oku, 2260m, fl. fr., 15 Nov. 1996, Etuge 3384.

Chamaecrista mimosoides (L.) Greene
Local name: Fehyunteh (Bekom at Ayeah) – Buzgo 774.
Herb or shrub to 1.5 m; leaflets 30–70 pairs, linear, 2–8 × 0.7–2 mm; flowers yellow, solitary or 2–3; fruits 3–5 cm long. Fallow.

Syn. *Cassia mimosoides* L., F.W.T.A. 1: 452 (1958).
Kilum: Ewook Etele, Oku-Mbae, 2352 m, 6 Nov. 1996,
Etuge 3345.
Ijim: Turbo path, 1 km after splitting from the main road at
the border of Ijim Mountain Forest Reserve, N of there,
2000m, fl., 18 Nov. 1996, Buzgo 774; Aboh Village, 6.15N,
10.26E, fl., 18 Nov. 1996, Onana 523.

Piliostigma thonningii (Schum.) Milne-Redh.
Shrub or small tree to 8m; leaves finely pubescent beneath,
7.5–16 × 10–18 cm; inflorescence narrowly paniculate,
petals white. Savanna.
Jakiri: Tan or Ntan, AlHadji Gey's land, 1600m, fr., 6 Nov.
1998, Maisels 181.

Senna occidentalis (L.) Link
Herb or shrub to 1.5 m, glabrous; leaflets 4–5 pairs, top pair
largest, lanceolate or ovate, 3.5–10 × 2–4 cm; fruits flattish,
to 14 cm long. Fallow or cultivated.
Syn. *Cassia occidentalis* L., F.W.T.A. 1: 455 (1958).
Kilum: Elak, 2000m, fl., 8 Jun. 1996, Zapfack 771.

LEGUMINOSAE-MIMOSOIDEAE
Det. J-M.Onana (YA) & B.A.Mackinder (K)

Albizia gummifera (J.F.Gmel.) C.A.Sm. var. *gummifera*
Local name: Klan (Oku) – Munyenyembe 884.
Tree to 33 m; leaves pinnate; leaflets 9–17 pairs, 7–17mm
long; flowers white, staminal tube white below, red above;
fruits 9–19 cm long, densely pubescent. Forest.
Kilum: Shore of Lake Oku, 2250m, 12 Jun. 1996, Cable
3105; Oku, 2010m, 2 Nov. 1996, Etuge 3333; Shambai,
2300m, 5 Nov. 1996, Munyenyembe 884.

Albizia gummifera (J.F.Gmel.) C.A.Sm. var. *ealaensis* (De Wild.) Brenan
Local name: Woum (Kom at Laikom) – Etuge 4539.
Tree differing from var. *gummifera* by: leaflets not auriculate
on the proximal side. Forest.
Ijim: Laikom, Akwa-mofu, medicinal forest, 1760m, 6.17N,
10.20E, fl., 5 Dec. 1998, Etuge 4539.

Entada abyssinica Steud. ex A.Rich.
Tree to 13m; leaves pinnate; leaflets 22–55 pairs, linear, 4–
12 mm long; flowers cream; fruit 15–39 × 5–8 cm. Savanna.
Ijim: Ntungfe, Laikom, 1460m, 6.1178N, 10.23E, fr., 9 Dec.
1998, Etuge 4576.

Newtonia camerunensis Villiers
Local name: Adjwa (Kom- Letouzey 13451).
Tree, 50 cm diam. at base, fluted; leaves c. 25 cm long;
petiole 1.5 cm; pinnae 8-10 pairs, gland between each pair;
leaflets 25 per pinnae, 15 x 3.5 mm, margin ciliate, midrinb
prominent; fruits 19-30 x 1.8-2.3 cm; seeds winged, 4-8 x
1.5-2.2 cm. Forest.
Ijim: Nchain, Maitland 1671; Njinikom, Maitland s.n.;
Acha, c. 1800 m, Letouzey 13451.

LEGUMINOSAE-PAPILIONOIDEAE
Det. J-M.Onana (YA), B.A.Mackinder (K) & B.A.Satabie (YA)

Adenocarpus mannii (Hook.f.) Hook.f. var. *mannii*
Local name: Aghum (Bekom at Nguakon) – Buzgo 752.
Shrub to 2m; leaves often fasciculate, leaflets oblong-
lanceolate, 5–8 mm long; yellow flowers in dense sessile
terminal heads; bristly fruits 2 cm long. Forest edge.
Kilum: Elak, 2200m, fl., 28 Oct. 1996, Munyenyembe 735;
Summit, 2800m, fl., 31 Oct. 1996, Zapfack 1128.
Ijim: Gikwang road towards Oku, 2260m, fl. fr., 15 Nov.
1996, Etuge 3400; Turbo path, 500m after it's splitting from
the main road at the border of the Ijim Mountain Forest
Reserve, 1800–2100m, fl., 18 Nov. 1996, Buzgo 752; Aboh
Village, 6.15N, 10.26E, 18 Nov. 1996, Onana 530; Laikom
to Fundong, 1700m, 6.17N, 10.20E, fl., 10 Dec. 1998, Etuge
4593.

Aeschynomene gracilipes Taub.
Local name: Fegisiwok (Oku) – Munyenyembe 881.
Subshrub 0.2–0.7 m; leaves with 7–12 pairs of leaflets;
leaflets oblong 3–5 × 0.5–1.5 mm, apex rounded or retuse;
inflorescence axillary, 1–2-flowered; flowers yellow, c. 5
mm long; pod 1–2-jointed, 3–6 mm long. Grassland.
Syn. *Aeschynomene* sp A., F.W.T.A. 1: 578 (1958).
Kilum: Shambai, 2400m, 5 Nov. 1996, Munyenyembe 881.
Ijim: Laikom, 2100m, fl. fr., 21 Nov. 1996, Satabie 1083.

Aeschynomene sp. 1
Herb to 25 cm; stem red; flowers solitary; standard yellow,
wings sepia-pink, keel blackish. By stream.
Ijim: Laikom, stream by grassland plateau above Fon's
Palace, 2020m, 6.1642N, 10.1950E, fl., 3 Dec. 1998, Pollard
297.

Antopetitia abyssinica A.Rich.
Local name: None known (Ernest Kiming) – Cheek 8428.
Slender herb to 75 cm; leaves imparipinnate; leaflets 3–4
pairs, narrowly oblanceolate, 1 cm long; umbels 2–4-
flowered; fruits torulose. Grassland.
Kilum: Oku-Elak to the forest at KA, 2200m, 6.1349N,
10.3112E, fl. fr., 27 Oct. 1996, Cheek 8428; Forest edge,
2200m, fl., 28 Oct. 1996, Munyenyembe 712; Ewook Etele
Mbae, 10 km from Oku, 2350m, fl. fr., 6 Nov. 1996, Zapfack
1214.
Ijim: Aboh Village, 6.15N, 10.26E, fl., 18 Nov. 1996,
Onana 529; Gikwang road towards Nyasosso forest - Ijim
Mountain Forest Reserve, 2081 m, fl. fr., 19 Nov. 1996,
Etuge 3465; Laikom, Ijim plateau above Fon's Palace,
1900m, 6.16N, 10.19E, fl. fr., 3 Dec. 1998, Cheek 9730.

Centrosema pubescens Benth.
Introduced creeper with shortly pubescent stems and leaves;
leaves elliptic, 4–6 cm long; racemes 2–3-flowered; flowers
pink-mauve or white with purple markings; fruits linear with
broad margins. Forest edge.
Kilum: Oku Shrine, 2140m, fl., 3 Nov. 1996, Munyenyembe
862; Lake Oku, 2020m, 5 Nov. 1996, Munyenyembe 888.

Crotalaria cf. atrorubens Hochst. ex Benth.
Kilum: Elak, 2400m, 28 Oct. 1996, Munyenyembe 733.

Crotalaria bamendae Hepper
Local name: Fegismfum (Oku) – Munyenyembe 882.
RED DATA LISTED
Branched herb to 60 cm; stems densely pubescent; leaflets oblong-oblanceolate, 2.5–3.5 cm long, silky-silvery beneath; inflorescence a terminal raceme; flowers 5–6 mm long. Grassland.
Kilum: Shambai, 2500m, fl., 5 Nov. 1996, Munyenyembe 882.

Crotalaria incana L. subsp. *purpurascens* (Lam.) Milne-Redh.
Herb 0.6–1.2 m; stems coarsely hairy; leaves obovate, 3–4 cm long; flowers numerous, yellow with purple veins; fruit coarsely hairy. Forest edge.
Ijim: Isebu Mountain Forest, along the ridge, 1650m, fl., 26 Nov. 1996, Kamundi 708.

Crotalaria lachnophora A.Rich.
Local name: Findeng (Oku) – Munyenyembe 883.
Spreading bush, 1–2 m, clothed with brown hairs; leaves 3 or 5-foliolate; leaflets oblanceolate, 3–4 cm long; flowers yellow, becoming reddish. 1.5 cm long; fruits ferruginous-tomentose, 2.5–3.5 cm long. Grassland.
Kilum: Shambai, 2300m, fl., 5 Nov. 1996, Munyenyembe 883.

Crotalaria ledermannii Baker f.
RED DATA LISTED
Erect well-branched annual or short-lived perennial 20–70 cm; stems appressed puberulous; leaves 3-foliolate; leaflets oblanceolate, 7–20 × 1–5 mm, apex rounded or truncate, apiculate; inflorescence 1.5–3 cm long; flowers veined red, 5.5–6.5 mm long. Forest edge and grassland.
Kilum: Mile 11, Oku, 2000m, fl., 17 Feb. 1958, Hepper 2048.
Ijim: Aboh, border of Lake Oku and along the path to the lake, 6.15N, 10.26E, fl. fr., 26 Nov. 1996, Onana 634.

Crotalaria mentiens Polhill
RED DATA LISTED
Erect annual or short-lived perennial 20–40 cm; stems finely strigulose, leaves 3-foliolate; leaflets oblanceolate, 5–13 × 1–3 mm, apex rounded to truncate, appressed pubescent beneath; petiole 2–3 mm long; inflorescence 1–3 cm long; flowers yellow, becoming reddish purple, 5–6 mm long.
Jakiri: Brouwers 2.

Crotalaria ononoides Benth.
Basally woody herb 0.3–1.2 m; stem suberect, lower branches often procumbent, brownish-pilose; leaves 3 or 5-foliolate, oblong-elliptic to oblanceolate, 1–6 cm long; inflorescence terminal, yellow turning orange. Roadside.
Ijim: Fundong to Belo road, 2050m, fl., 21 Nov. 1996, Cheek 8725.

Crotalaria subcapitata De Wild. subsp. *oreadum* (Baker f.) Polhill
Local name: Tung Tung (Bekom at Ayeah) – Buzgo 751; Cha-Cha (Kom) – Cheek 9740.
Annual or perennial erect or straggling herb 0.5–1.3 m; leaves 3-foliolate; leaflets very variable; inflorescence a raceme; peduncle shorter than rhachis; flowers yellow, darkly veined, 0.5–1 cm long. Grassland.
Syn. *Crotalaria acervata* sensu F.W.T.A. 1: 550 (1958).
Kilum: Elak, 2400m, 28 Oct. 1996, Munyenyembe 734.
Ijim: Turbo path, 500m after its splitting from the main road at the border of the Ijim Mountain Forest Reserve, 2000m, fl., 18 Nov. 1996, Buzgo 751; Aboh Village, 6.15N, 10.26E, fl., 18 Nov. 1996, Onana 522; Laikom, Ijim plateau above Fon's Palace, 1900m, 6.16N, 10.19E, fl. fr., 3 Dec. 1998, Cheek 9740.

Dalbergia florifera De Wild.
Liane; leaflets 3–7, glabrous, oblanceolate, 6–7 × 2.5–4 cm, acuminate, rounded at base; flowers white 3–4 mm long; pods oblong, flat, 5–6 × 1.5 cm. Forest.
Kilum: Elak, 2200m, fl., 8 Jun. 1996, Zapfack 758; KA, 2200m, fl., 11 Jun. 1996, Etuge 2302.

Dalbergia lactea Vatke
Scandent shrub, leaves 3–7-foliolate; leaflets oblong to ovate 3–7 cm long; inflorescence a many-flowered panicle up to 30 cm long; flowers white. Forest edge.
Ijim: Afua swamp to Afua junction, 1800m, fl., 7 Dec. 1998, Cheek 9858.

Dalbergia saxatilis Hook.f. var. *saxatilis*
Glabrous climbing or straggling shrub; leaflets 11–13, oblong-elliptic, 2.5–6 cm long; inflorescence a loose axillary or terminal panicle up to 30 cm long; flowers white or pink. Forest edge.
Kilum: Emoghwo traditional forest, 1900m, fl., 3 Nov. 1996, Zapfack 1165.
Ijim: Zitum road, 1900m, 6.1116N, 10.2543E, fl., 21 Nov. 1996, Etuge 3531.

Desmodium adscendens (Sw.) DC. var *adscendens*
Undershrub with slender thinly pubescent branches, straggling, sometimes prostrate and rooting; leaves 3-foliolate; terminal leaflet obovate to obovate-elliptic, 2–4.5 cm long; flowers pink or whitish. Forest.
Kilum: Mbokengfish, 1900m, fl., 7 Nov. 1996, Munyenyembe 904.
Plot voucher: Forest above Oku-Elak, Jun. 1996, Oku 58.

Desmodium hirtum Guill. & Perr.
Pilose herb or subshrub 0.5–2 m long; stem prostrate to erect; leaves 3-foliolate; leaflets obovate, 1.2–2.7 × 1.2–1.7 cm; stipules ovate-attenuate, 4–7 × 1.5–2 mm; inflorescence a lax axillary raceme, flowers whitish or pale pink to purple, c. 4 mm long; fruit 2–6-articled. Grassland.
Kilum: Mbokengfish, 1900m, fl., 7 Nov. 1996, Munyenyembe 904.

Desmodium repandum (Vahl) DC.

Local name: None known – Cheek 8636.

Weak slender undershrub straggling or climbing to 1–2 m; stem branches with thinly spreading pubescence; leaves rhombic, up to 9 cm long; flowers deep red, 8–10 mm long; fruits deeply indented, usually with more than one article. Forest.

Kilum: Elak, 2200m, fl. fr., 8 Jun. 1996, Zapfack 789; KA, 2600m, fr., 27 Oct. 1996, Buzgo 583; Elak, 2600m, fl., 27 Oct. 1996, Munyenyembe 707; Elak, 2400m, fl., 28 Oct. 1996, Munyenyembe 730; Elak, KJ, 2300m, fl. fr., 29 Oct. 1996, Zapfack 1056.

Ijim: Above Aboh village, 1.5 hours walk along Gikwang road, 2380m, 6.1112N, 10.2529 E, fl., 18 Nov. 1996, Cheek 8636; Laikom, Akwa-mofu medicinal forest, 1760m, fl. fr., 5 Dec. 1998, Etuge 4544; Laikom, medicinal forest below basaltic grassland plateau, 1900m, 6.1630N, 10.1951E, fr., 12 Dec. 1998, Pollard 381.

Plot voucher: Forest above Oku-Elak, Jun. 1996, Oku 24.

Desmodium sp. 1

Herb, erect, 20 cm; stems pubescent; calyx lobes red, corolla standard 'washed out' turquoise-bluish-mauve, keel pinkish-purple. Forest edge.

Possibly an introduced weed (fide Verdcourt).

Ijim: Laikom, path from Fon's Palace downhill towards Akwamofu sacred forest, 1900m, fl., 6.1630N, 10.1951E, fl., 12 Dec. 1998, Pollard 375.

Dolichos sericeus E.Mey.

Local name: Ovievmbien (Oku) – Munyenyembe 896; Fghagh faku (Kom) – Etuge 4553.

Perennial climbing or prostrate herb 0.6–2.1 m long; leaflets 3, ovate or ovate-triangular, 0.9–8.5 × 0.5–7.5 cm, petiole 1.5–7 cm; stipules ovate-oblong, 4–5 × 2–3.5 mm, ribbed; inflorescence a short axillary raceme; flowers 1–10, pink to bluish-purple, 1–1.6 cm long; pods oblong, curved, 4–9.5 cm long. Forest edge and fallow.

Kilum: Mbokengfish, 1900m, fl., 7 Nov. 1996, Munyenyembe 896.

Ijim: Main road from Atubeaboh (Gikwang Foe) towards Oku, 2220m, fl. fr., 15 Nov. 1996, Etuge 3376; Aboh Village, 6.15N, 10.26E, fl., 18 Nov. 1996, Onana 517; Laikom, 2100m, 6.15N, 10.26E, fl., 21 Nov. 1996, Onana 571; Above Fon's palace, Laikom, about 30 mins. on path to Fulani settlement, 2050m, 6.1642N, 10.1952E, fl., 21 Nov. 1996, Cheek 8707; Aboh Village, along the main road, 1700m, 6.15N, 10.26E, fl., 24 Nov. 1996, Onana 608; Isebu Mountain Forest. Montane rain forest along the ridge, 1700m, fl. fr., 26 Nov. 1996, Kamundi 703; Laikom, Akwa-mofu, 1760m, 6.17N, 10.20E, fl. fr., 5 Dec. 1998, Etuge 4553; Ndawara area, 1760m, 6.06N, 10.23E, fl. fr., 9 Dec. 1998, Etuge 4579; Laikom, path from Fon's Palace downhill towards Akwamofu sacred forest, 1900m, 6.1630N, 10.1951E, fr., 12 Dec. 1998, Pollard 379.

Eriosema montanum Baker f. var. montanum

Subwoody shrub to 90 cm; stem much-branched; leaves 3-foliolate; leaflets elliptic or oblanceolate, 4–10 cm long; inflorescence a dense raceme, 613 cm long; flowers yellow. Forest edge.

Kilum: Elak, 2200m, fl., 28 Oct. 1996, Munyenyembe 750.

Ijim: Turbo path, 500m after splitting from the main road at the border of the Ijim Mountain Forest Reserve, fl., 13 Nov. 1996, Buzgo 746; Above Aboh village, 1.5 hours walk along Gikwang road, 2350m, 6.1112N, 10.2529E, fl., 18 Nov. 1996, Cheek 8650; Laikom village, by Fundong, 2100m, 6.15N, 10.26E, fl. fr., 21 Nov. 1996, Onana 567.

Eriosema parviflorum E.Mey.

Shrub, either erect from woody base or half-straggling, softly tawny-hairy; leaves 3-foliolate, leaflets ovate-elliptic; inflorescence a 14–30-flowered raceme; flowers reddish-yellow or yellow, 6–9 cm long. Forest edge.

Kilum: Elak, 2200m, fl., 30 Oct. 1996, Munyenyembe 818.

Ijim: Above Aboh village, 1.5 hours walk along Gikwang road, 2350m, 6.1112N, 10.2529E, 18 Nov. 1996, Cheek 8651.

Eriosema scioanum Avetta subsp. lejeunei (Staner & Ronse Decr.) Verdc. var. lejeunei

Herb or subshrub 20–75 cm; leaflets 3, elliptic to round, 0.8–7 × 0.6–3.8 cm; racemes axillary, 2–10 cm long; flowers with standard yellowish inside, blackish outside, 7–8 mm long; pods oval-oblong, 8–12 × 6–9 mm.

Ijim: Laikom, rocky 'pavement' in grassland plateau above Fon's Palace, 2100m, 6.1642N, 10.1950E, fl., 3 Dec. 1998 Pollard 298.

Erythrina sp. aff. excelsa Baker

Tree; bole 20–27 m; bark pale, armed with strong prickles; leaves glabrous, 12–23 cm long; inflorescence a stiff, many-flowered raceme; flowers red, appearing in twos and threes when tree is leafless. Savanna.

Ijim: Afua swamp, 1950m, 6.09N, 10.24E, st., 7 Dec. 1998, Etuge 4572.

Indigofera arrecta Hochst ex A.Rich.

Local name: Inkwekwei (Oku) – Munyenyembe 741.

Erect, soft-wooded shrub to 3m; stem branched; leaflets usually 11 or more. Fallow.

Kilum: Elak, 2400m, 28 Nov. 1996, Munyenyembe 741.

Ijim: Aboh Village, 6.15N, 10.26E, fl., 18 Nov. 1996, Onana 533; Aboh Village, 2450m, fl., 19 Nov. 1996, Onana 539; Laikom, grass ridges above Akwamofu sacred forest, on the way to waterfall below W edge of Ijim ridge grassland plateau, 1850m, 6.17N, 10.20E, fr., 10 Dec. 1998, Pollard 365; Laikom, below Fon of Kom's Palace, 1820m, 6.16N, 10.21E, fl. fr., 11 Dec. 1998, Cheek 9910.

Indigofera atriceps Hook.f. subsp. atriceps

Straggling half-woody herb to 75 cm; inflorescence with blackish-brown hairs; leaflets 5–13. Grassland.

Ijim: Laikom village, 2100m, 6.15N, 10.26E, fl., 21 Nov. 1996, Onana 568; Border of Lake Oku and along the path to the lake, 6.15N, 10.26E, fl., 26 Nov. 1996, Onana 635; Laikom, Ijim ridge, about 30 mins. walk above Fon of Kom's Palace, 6.16N, 10.19E, fr., 2 Dec. 1998, Cheek 9768.

Indigofera heudelotii Benth. ex Baker var. heudelotii

Erect, soft-wooded shrub to 2m; stem branched; inflorescence axis bearing flowers, one above the other. Fallow.

Kilum: Elak, 2200m, fl., 30 Oct. 1996, Munyenyembe 815.

Ijim: Belo to Afua, roadside in cultivated area, 1350m, 6.08N, 10.22E, fr., 7 Dec. 1998, Cheek 9824.

Indigofera mimosoides Baker

Subwoody herb to 1.2 m; stem sparingly glandular, red, branching; leaflets rarely more than 9; inflorescence over 5 mm long. Forest edge and fallow.

Kilum: Above Mboh, path above village to forest through Wambengs farm, 2200m, 6,1146N, 10.2856N, fl., 1 Nov. 1996, Cheek 8560.

Ijim: Turbo path, 1 km after its splitting from the main road at the border of the Ijim Mountain Forest Reserve, N from there, 2000m, 18 Nov. 1996, Buzgo 777; Above Aboh village, 1.5 hours walk along Gikwang road, 2000m, 6.1112N, 10.2529E, fl., 18 Nov. 1996, Cheek 8656; Gikwang main road towards Nyasosso forest, 2250m, 6.1116N, 10.2543E, fl. fr., 20 Nov. 1996, Etuge 3521; Laikom, grass ridges above Akwamofu sacred forest, on the way to waterfall below W edge of Ijim ridge grassland plateau, 1850m, 6.17N, 10.20E, fl. fr., 10 Dec. 1998, Pollard 363.

Indigofera patula Baker subsp. okuensis Schrire & Onana
RED DATA LISTED

Kilum: Elak, 2200m, fr., 30 Oct. 1996, Munyenyembe 814.

Ijim: Aboh, Gikwang road towards Nyasosso forest, 2250m, 6.1116N, 10.2543E, fl. fr., 20 Nov. 1996, Etuge 3522.

Kotschya speciosa (Hutch.) Hepper

Local name: Igum (Oku) – Munyenyembe 837.

Erect shrub 1.2–1.7 m, stems softly pilose, numerous, branching; leaflets 4–5 mm long; flowers 1.5 cm long, blue. Grassland.

Kilum: Above Mboh, path above village to forest through Wambeng's farm, 2000m, 6.1256N, 10.2911E, fl., 1 Nov. 1996, Cheek 8573; Elak, 2100m, fl., 1 Nov. 1996, Munyenyembe 837.

Ijim: Isebu Mountain Forest, along the ridge, 1700m; fl. fr., 26 Nov. 1996, Kamundi 697.

Lotus discolor E.Mey.

Ascending herb; stem of wiry branches arising from woody rootstock; leaflets narrowly oblanceolate, terminal leaflet 9–11 mm long; inflorescence a terminal 5–7-flowered head; flowers yellow, c. 10 mm long. Fallow.

Kilum: Above Mboh, path above village to forest through Wambeng's farm, 2200m, 6.1146N, 10.2856E, fl., 1 Nov. 1996, Cheek 8555.

Millettia conraui Harms

Local name: Ebfume (Oku – Peter Wambeng) – Cheek 8607; Kebalmkan (Oku) – Munyenyembe 898.

Small tree; leaflets about 6 pairs, oblong-elliptic, 5–10 cm long; flowers pale purple; standard markedly furrowed, yellow-pubescent; calyx yellow-pubescent. Forest edge.

Kilum: Near Kumbo, beginning of Oku road, 1850m, 6.12N, 10.40E, fl., 15 Feb. 1958, Hepper 2715; Lake Oku to Jikijem, 2200m, 6.1214N, 10.2733E, 5 Nov. 1996, Cheek 8607; Mbokengfish, 1900m, 7 Nov. 1996, Munyenyembe 898.

Millettia sp. 1

Ijim: Fundong near Touristic Hotel, near the waterfall, 6.15N, 10.26E, 22 Nov. 1996, Onana 587.

Millettia sp. 2

Liane. Sterile voucher for species record.

Ijim: Laikom, Plot between Fon's Palace and Ardo's compound, 2000m, 6.16N, 10.21E, st., 8 Dec. 1998, Cheek 9880.

Neonotonia wightii (Wight & Arn.) J.A.Lackey

Local name: Ovievnanak (Oku) – Munyenyembe 880; Kemeiankwah (Kom) – Cheek 8628.

Climbing herb, pubescent; leaflets 3, ovate, c. 3.5 × 3 cm, petiole 4 cm; axillary racemes c. 6 cm long, 10–15-flowered' flowers white, 5 mm long. Weed.

Syn. *Glycine javanica* L., F.W.T.A. 1: 564 (1958).

Kilum: Elak, 2200m, fl., 30 Oct. 1996, Munyenyembe 817; Shambai, 2300m, fl., 5 Nov. 1996, Munyenyembe 880.

Ijim: Above Aboh Village, 1.5 hours walk along Gikwang road, 2380m, 6.1112N, 10.2529E, fl. fr., 18 Nov. 1996, Cheek 8628; Gikwang towards Nyasosso bush, 2260m, fr., 18 Nov. 1996, Etuge 3446; Anyajua, 2400m, fl., 19 Nov. 1996, Satabie 1073.

Pseudarthria hookeri Wight & Arn.

Shrub 1–3 m; leaves trifoliolate, grey tomentose beneath; leaflets elliptic, c. 6 × 3 cm, mucronate, broadly crenate; stipels persistent; inflorescence c. 40 × 15 cm; flowers red, 1 cm long. Savanna.

Belo: Belo, 1400m, May 1931, Maitland 1434.

Rhynchosia sp.

Herbaceous climber; leaves trifoliate; bracts silvery-gray; flowers brick red, 2 cm long.

Ijim: Ntum, 1800m, Nov. 1999, Ghogue 387.

Sesbania macrantha Welw. ex E.Phillips & Hutch.

Shrub 1.7–2m; stem and branches rough with small thorns; leaflets 10–24 pairs, oblong, 9–24 mm long; flowers yellow tinged pink 1.2–1.8 cm long. Forest edge.

Kilum: KA, 2200m, fl. fr., 11 Jun. 1996, Etuge 2315; Elak, 2000m, fl., 27 Oct. 1996, Munyenyembe 703.

Ijim: Gikwang road towards Oku, 2270m, fl. fr., 15 Nov. 1996, Etuge 3404; Isebu Mountain Forest, along the ridge, 1700m, fr., 26 Nov. 1996, Kamundi 698.

Smithia elliotii Baker f. var. *elliotii*

Local name: Ngei (Oku) – Munyenyembe 874; Feyentim (Kom) – Cheek 8660.
Ascending herb; stem glabrescent, hispid above, 20–70 cm long; leaflets 6–8 pairs, opposite, linear-oblong, about 1 cm long; inflorescence an umbel, flowers erect, pink or blue. Grassland and fallow.
Kilum: Shambai, 2400m, fl., 5 Nov. 1996, Munyenyembe 874; Ewook Etele, Oku-Mbae, 2350m, fl. fr., 6 Nov. 1996, Etuge 3347; Ewook Etele Mbae, 10 km from Oku, 2350m, fl. fr., 6 Nov. 1996, Zapfack 1213.
Ijim: Above Aboh village, 1.5 hours walk along Gikwang road, 2450m, 6.1112N, 10.2529E, fl., 19 Nov. 1996, Cheek 8660; Gikwang road towards Nyasosso forest - Ijim Mountain Forest Reserve, 2090m, fl., 19 Nov. 1996, Etuge 3474; Laikom, fl., 21 Nov. 1996, Satabie 1085; Laikom, Ijim plateau above Fon's Palace, 1900m, 6.16N, 10.19E, fl. fr., 3 Dec. 1998, Cheek 9721.

Tephrosia preussii Taub.

Local name: None known (Peter Wambeng) – Cheek 8589.
Robust erect undershrub to 1.8 m; stems striate, pubescent; leaflets oblanceolate, 2.5–5 cm long, 6–8 lateral nerves; inflorescence terminal; flowers purplish or orange, 15–20 mm long; fruits blackish-pilose. Roadside.
Kilum: Road from Rest House E of lake to the N of mountain range, 2300m, 6.1214N, 10.2733E, fl., 5 Nov. 1996, Cheek 8589.

Tephrosia vogelii Hook.f.

Erect shrub 2–3 m clothed with dense yellowish or rusty tomentum; leaflets of 5 or more pairs, oblanceolate, 4–7 cm long; inflorescence a dense raceme; flowers conspicuous, red or reddish-purple. Fallow.
Kilum: KD, fr., 9 Jun. 1996, Etuge 2246.
Ijim: Anyajua Forest, 1850m, fl. fr., 27 Nov. 1996, Kamundi 710.

Trifolium baccarini Chiov.

Herb to 20 cm; leaflets less than 1.5 cm long; flowers bluish-purple, occasionally white. Grassland.
Kilum: Elak, 2200m, fl., 30 Oct. 1996, Munyenyembe 816.
Ijim: Aboh Village, 6.15N, 10.26E, fl., 18 Nov. 1996, Onana 519; Near Chufekhe stream, 2010m, 6.1124N, 10.2525E, fr., 20 Nov. 1996, Kamundi 651A.

Trifolium simense Fresen.

Straggling perennial to 45 cm; leaflets narrowly oblanceolate; inflorescence more than 1 cm broad; calyx-nerves 15–20; flowers purple. Grassland.
Kilum: KA, 2700m, fl., 31 Oct. 1996, Buzgo 658.
Ijim: Turbo path, 1 km after its splitting from the main road at the border of Ijim Mountain Forest Reserve, N of there, 2000m, fl., 18 Nov. 1996, Buzgo 772; Above Aboh village, 1.5 hours walk along Gikwang road, 2380m, 6.1112N, 10.2529E, fl., 18 Nov. 1996, Cheek 8632A; Near Chufekhe stream, 2010m, 6.1124N, 10.2525E, fl., 20 Nov. 1996, Kamundi 660; Cattle watering hole by Fulani settlement, 11 km E of Laikom, 2400m, 6.1615N, 10.2435E, fl., 9 Dec. 1998, Pollard 356.

Trifolium usambarense Taub.

Straggling herb to 45 cm; leaflets broadly oblanceolate; inflorescence less than 1 cm broad; calyx-nerves 10–12; flowers purple, occasionally white. Grassland.
Kilum: Elak, KJ, 2620m, fl., 29 Oct. 1996, Zapfack 1093; 2800m, fl., 30 Oct. 1996, Satabie 1060 & 1062; Ewook Etele Mbae, 10 km from Oku, 2350m, fl., 6 Nov. 1996, Zapfack 1236.
Ijim: Turbo path, 500m after its splitting from the main road at the border of the Ijim Mountain Forest Reserve, 2000m, fl., 18 Nov. 1996, Buzgo 756; Gikwang road, 2380m, 6.1112N, 10.2529E, fl., 18 Nov. 1996, Cheek 8632B; Near Chufekhe stream, 2010m, 6.1124N, 10.2525E, fr., 20 Nov. 1996, Kamundi 651B; Laikom, Ijim plateau above Fon's Palace, 1900m, 6.16N, 10.19E, fl., 2 Dec. 1998, Cheek 9736 & Cheek 9772; Laikom, grassland plateau, 2020m, 6.1642N, 10.1950E, fl. (white form), 3 Dec. 1998, Pollard 296.

Vigna fischeri Harms

Local name: Ikunokwak (Oku) – Munyenyembe 838.
Twining or trailing herb; stems 1.8–6m long, densely ferruginous-pubescent; leaflets 3, ovate-lanceolate to lanceolate, 2.8–10 × 0.75–4.3 cm; inflorescence several-flowered; flowers yellow, 1.5–2.2 cm long; pods linear, compressed, 3–5.5 cm long, densely ferruginous-pubescent. Grassland.
Kilum: Towards KJ, 2100m, fl. fr., 28 Oct. 1996, Pollard 1; Elak, 2100m, fl., 1 Nov. 1996, Munyenyembe 838; Lake Oku, 2200m, fl., 4 Nov. 1996, Satabie 1065.
Ijim: Gikwang road towards Oku, 2260m, fl. fr., 15 Nov. 1996, Etuge 3395; Above Aboh village, 1.5 hours along Gikwang road to Lake Oku, 2350m, 6.1112N, 10.2529E, fl., 19 Nov. 1996, Cheek 8652; Laikom, Ijim ridge, about 30 mins. walk above Fon of Kom's Palace, 2050m, 6.16N, 10.19E, fl. fr., 2 Dec. 1998, Cheek 9766.

Vigna gracilis (Guill & Perr.) Hook.f.

Slender, extensively twining herb; leaves very variable; leaflets 3, ovate-triangular, usually 3 × 2 cm; flowers pink or bluish, turning yellow; fruits 2 cm long. Forest edge.
Kilum: Elak, 2200m, fl., 28 Oct. 1996, Munyenyembe 749.
Ijim: Turbo path, 1 km after its splitting from the main road at the border of the Ijim Forest Reserve, N of there, 2000m, fl., 18 Nov. 1996, Buzgo 769.

Vigna sp. aff. gracilis (Guill & Perr.) Hook.f.

Compare with Saxer 346, Cameroon.
Kilum: Elak, Ewook Etele, Oku-Mbae, 2350m, fl. fr., 6 Nov. 1996, Etuge 3346.
Ijim: Afua swamp, 1950m, 6.0875N, 10.24E, fr., 7 Dec. 1998, Etuge 4570.

Vigna multiflora Hook.f.

Slender twiner; leaflets broadly ovate, usually 6–7 × 5 cm; flowers pink or bluish-purple quickly turning yellow; fruits up to 4 cm long. Fallow.
Kilum: KD, 2200m, fl., 10 Jun. 1996, Cable 3045.

LENTIBULARIACEAE

Genlisea hispidula Stapf
Slender erect herb 10–25 cm; roots with subterranean inverted Y-shaped traps for the capture of small organisms; leaves spathulate to 5 cm long; flowers pedicellate, violet, 6–9mm long. Swamps and seepages.
Ijim: Cattle watering hole by Fulani settlement, 11 km E of Laikom, 2400m, 6.1608N, 10.2500E, fl., 9 Dec 1998, Pollard 360.

Utricularia appendiculata E.A.Bruce
Herb; leaves linear to narrowly obovate, 30 × 0.6–2 mm; inflorescence twining to the left (viewed from the side) to 60 cm; flowers 1–8; corolla white or pale yellow, 5–9 mm long. Swamps.
Mbingo: Swamp en route to "Back Valley", 10 Nov. 1999, Cheek 10048.

Utricularia livida E.Mey.
Herb; leaves petiolate scattered on stolons, narrowly obovate, 0.2–7 × 0.1–0.6 cm, peduncle not twining, 2–15(–80)cm; flowers 2–8, pale to dark violet, 5–15 mm long. Basalt pavement.
Ijim: Laikom ridge, Palace to Ardo's compound, 2000m, 3 Dec. 1998, Cheek 9783.

Utricularia mannii Oliv.
Epiphyte on mossy tree-trunks; tuber small; leaves 1–2, obovate-spathulate or broadly linear; scape slender, 2.5–10 cm long; flowers 1–3, yellow, 2–2.5 cm long. Forest.
Kilum: Elak, 2300m, fl., 8 Jun. 1996, Zapfack 756.

Utricularia pubescens Sm.
Terrestrial herb; leaves peltate, orbicular, c. 4mm wide; scape usually pubescent, few-flowered; flowers c. 3mm long, white or mauve. Wet places.
Ijim: Laikom ridge, basalt pavement, 2000m, 2 Dec. 1998, Cheek 9758.

Utricularia scandens Benj.
Local name: Angmwo mwosong – Cheek 9750.
Herb, erect or twining to about 15 cm; corolla yellow, 1 cm long; spur slightly curved, slender, acute. Basalt pavement and swamps.
Kilum: Elak, near to the stream around savanna zone, 2500m, fl., 5 Nov. 1996, Zapfack 1205; Ewook Etele Mbae, 10 km from Oku, 2350m, fl., 6 Nov. 1996, Zapfack 1233.
Ijim: Laikom, Ijim plateau above Fon's Palace, 1900m, 6.16N, 10.19E, fl., 3 Dec. 1998, Cheek 9750.

Utricularia striatula Sm.
Epiphyte or epilith 2–5 cm; leaves numerous, orbicular-spathulate; flowers small, pink or white; calyx-lobes very unequal. Forest and damp cliffs.
Ijim: N facing, seasonally damp rockface along path from Fon's palace Laikom to grassland plateau above, 2000m, 6.1642N, 10.1950E, fl., 2 Dec. 1998, Pollard 271.

LINACEAE

Radiola linoides Roth
Annual herb to 7.5 cm, glabrous; leaves opposite, ovate or elliptic, 4 mm × 2.3 mm; inflorescence a leafy dichotmous cyme; flowers very small, numerous; petals white. Grassland, between tussocks.
Ijim: Above Fon's palace, Laikom, about 30 mins. walk on path to Fulani settlement, 2050m, fl., 21 Nov. 1996, Cheek 8716.

LOGANIACEAE
Det. A.J.M.Leeuwenberg (WAG)

Anthocleista scandens Hook.f.
Epiphytic climbing shrub 7–17 m; branchlets square; leaves oblong-elliptic, 6–20 × 2.5–11 cm; flowers white; corolla tube c. 3 cm long. Forest.
Kilum: Ethiale, near the stream, 2000m, fl., 4 Nov. 1996, Zapfack 1175.
Ijim: Zitum road, 1900m, 6.1116N, 10.2543E, fl. fr., 21 Nov. 996, Etuge 3552.

Anthocleista schweinfurthii Gilg
Tree to 18 m; leaves oblanceolate, 7–45(–95) cm × 3.5–18(–30) cm; flowers white; corolla tube c. 3 cm long. Forest edge.
Ijim: Mbesa-Akeh, 1570m, 11 Nov. 1999, Cheek 10076.

LORANTHACEAE
Det. R.M.Polhill (K)

Englerina gabonensis (Engl.) Balle
Local name: Kilimlim (Oku) – Munyenyembe 787; Ebalabala (Kom) – Cheek 8633.
Parasitic shrub; leaves lanceolate, elliptic or ovate, 4.5–16 cm × 1.8–6 cm; inflorescence an umbels; flowers 5-6, yellow, basally red, 4–5 cm long. Forest.
Kilum: Junction of KJ and KN above Elak, 2550m, fl., 10 Jun. 1996, Cable 3019; Summit, 2850m, fr., 10 Jun. 1996, Cable 3069; Elak, 2620m, fr., 10 Jun. 1996, Zapfack 833; Elak, 2850m, fl., 29 Oct. 1996, Munyenyembe 787; Elak, KJ, 2400m, fr., 29 Oct. 1996, Zapfack 1059; Path from Project HQ to KA, 2600m, fr., 30 Oct. 1996, Cheek 8524; Summit, 2800m, fl., 31 Oct. 1996, Zapfack 1131; Summit, 2800m, fr., 31 Oct. 1996, Zapfack 1132.
Ijim: Above Aboh village, 1.5 hours along Gikwang road to Lake Oku, 2380m, fr., 18 Nov. 1996, Cheek 8633.

Globimetula oreophila (Oliv.) Danser
Local name: Ebalabala (Kom) – Cheek 8635.
Parasitic shrub; leaves narrowly ovate to lanceolate, pinnately nerved, 5–14 cm long; umbels 8–20-flowered; flowers red, tipped green. Forest.
Kilum: Path above water-tower, KJ, about 1 km into forest, 2250m, fl., 9 Jun. 1996, Cable 2982; Iwook Etele Mbae, 2500m, fl., 5 Nov. 1996, Zapfack 1190.
Ijim: Laikom, Maitland 1584; Turbo path, 1 km after its splitting from the main road at the border of the Ijim

Mountain Forest Reserve, N from there, 1800–2100m, fl., 18 Nov. 1996, Buzgo 775; Above Aboh village, 1.5 hours on road to Lake Oku on Gikwang road, 2380m, fl., 18 Nov. 1996, Cheek 8635; Gikwang, towards Nyasosso bush, 2250m, fl. fr., 18 Nov. 1996, Etuge 3437; Akwo-Mufu, Laikom, 1760m, fl., 5 Dec. 1998, Etuge 4558.

Phragmanthera polycrypta (Didr.) Balle
Local name: Ibala-bala (Bekom at Ayeah) – Buzgo 762.
Parasitic shrub; leaves lanceolate or ovate; flowers with an orange tube and violet lobes; berries red. Forest.
Kilum: KA, 2200m, fl., 11 Jun. 1996, Etuge 2304; Above Mboh, path above village to forest through Wambeng's farm, 2200m, fr., 1 Nov. 1996, Cheek 8566; Oku shrine, 2140m, fr., 3 Nov. 1996, Munyenyembe 860; Iwook Etele Mbae, 2500m, fr., 5 Nov. 1996, Zapfack 1189.
Ijim: Laikom, Maitland 1651; Turbo path, 500m after its splitting from the main road at the border of the Ijim Mountain Forest Reserve, 1800–2100m, fr., 18 Nov. 1996, Buzgo 762; Ijim Mountain Forest, Aboh, 2250m, fr., 19 Nov. 1996, Pollard 54.

Tapinanthus globiferus (A.Rich.) Tieghem
Parasitic shrub; leaves petiolate, glaucous, usually elliptic, oblong or obovate, 2–9 cm long; flowers shining red. Forest.
Kilum: KD, fl., 9 Jun. 1996, Etuge 2242; Lake Oku, 2200m, fl., 11 Jun. 1996, Zapfack 867; KJ to KD, 3000m, fl., 13 Jun. 1996, Etuge 2348; Emoghwo traditional forest, 1900m, fl., 3 Nov. 1996, Zapfack 1158; Ethiale, near the stream, 2000m, fl., 4 Nov. 1996, Zapfack 1177.
Ijim: Ijim Mountain Forest, Nyasosso forest, Aboh, 2200m, fr., 27 Nov. 1996, Pollard 82.

Tapinanthus letouzeyi (Balle) Polhill & Wiens
RED DATA LISTED
Local name: Eymbenmben-Emlam (Oku) – Cheek 8586.
Parasitic shrub, often on *Gnidia*, glabrous; leaves opposite, elliptic, 6–12 × 1.5–6 cm; umbels axillary, 4–6-flowered; flower-heads white turning black; corolla tube 2.7–3 cm, red, paler above. Forest edge.
Kilum: Elak, 2300m, fl, 8 Jun. 1996, Zapfack 739; Lake Oku, near Baptist Rest House, 2300m, fr., 4 Nov. 1996, Cheek 8586.

MALVACEAE

Hibiscus macranthus Hochst. ex A.Rich
Shrub or woody herb to 2m or more; leaves ovate to pentagonal, stellate-hirsute beneath; flowers large, 5–6 cm diam., yellow with deep purple centre. Forest edge.
Ijim: From Aboh, path towards Tum, 1950m, fl., 22 Nov. 1996, Etuge 3583; Aboh village, 1700–1800m, fl. fr., 24 Nov. 1996, Onana 603; Chuhuku River near Anyajua Forest, 1600m, fl., 28 Nov. 1996, Kamundi 727; Belo to Afua, 1700m, 6.08N, 10.22E, fl., 7 Dec. 1998, Cheek 9829.

Hibiscus noldeae Baker f.
Shrubby perennial herb sometimes scandent; leaves deeply digitately 3–5-lobed, up to 6.5 cm long; flowers yellow with deep red centre. Forest edge.

Kilum: Emoghwo traditional forest, 1900m, fr., 3 Nov. 1996, Zapfack 1163.
Ijim: Aboh village, 1700–1800m, fl., 24 Nov. 1996, Onana 601.

Kosteletzkya adoensis Hochst. ex A.Rich
Local name: Ofiefanen – Munyenyembe 791.
Prostrate or scrambling herb; leaves ovate to 3-lobed, dentate, hirsute, to 10 cm × 8 cm; inflorescence a lax axillary fascicle; flowers pink or white with reddish or purple centre; fruits with spreading bristles. Forest edge.
Kilum: Lake Oku, 2200m, fl., 12 Jun. 1996, Etuge 2328; KJ, 2300m, fl., 28 Oct. 1996, Buzgo 601; Oku-Elak, lower parts of KA, 2200m, fl., 28 Oct. 1996, Cheek 8468; Elak, 2800m, fl., 29 Oct. 1996, Munyenyembe 791; Forest above Elak-Oku, peg KJ–4, 2200m, fl., 6 Nov. 1998, Maisels 180.
Ijim: Aboh village, 2450m, fl., 19 Nov. 1996, Onana 538; Isebu Mountain Forest, montane rain forest along the ridge, 1700m, fl., 26 Nov. 1996, Kamundi 701.
Plot voucher: Forest above Oku-Elak, Jun. 1996, Oku 54; 85.

Pavonia urens Cav. var. *glabrescens* (Ulbr.) Brenan
Local name: Chim (Oku) – Munyenyembe 839; Unknown – Cheek 8639.
Woody herb; stems and leaves pubescent or glabrescent; leaves subovate, c. 4 × 1.5 cm, incised; flowers 1.5 cm long, white with pink or reddish markings.
Kilum: Shore of Lake Oku, 2250m, fl., 12 Jun. 1996, Cable 3138; KC, 2300m, fr., 30 Oct. 1996, Buzgo 645; Above Mboh, path above village to forest through Wambeng's farm, 2200m, fl., 1 Nov. 1996, Cheek 8550; Elak, 2500m, fl., 1 Nov. 1996, Munyenyembe 839.
Ijim: Above Aboh village, 1.5 hours walk along Gikwang road, 2350m, fl., 18 Nov. 1996, Cheek 8639; Ijim Mountain Forest, 2160m, fl., 19 Nov. 1996, Kamundi 618; Border of Lake Oku and along the path to the lake, fr., 26 Nov. 1996, Onana 636.

Pavonia urens Cav. var. *urens*
Local name: Vehsefigini (Bekom at Ayeah) – Buzgo 757; Asejeu (Kom) – Cheek 8641.
Coarse woody herb to 3 m; stem and leaves densely long-hirsute; leaves orbicular, 5-lobed, c.12 cm diam., cordate; flowers 2 cm long, white with pink or purplish markings.
Kilum: KA, 2500m, fl., 29 Oct. 1996, Onana 453.
Ijim: Main road from Atubeaboh (Gikwang Foe) towards Oku, 2250m, fl. fr., 15 Nov. 1996, Etuge 3377; Gikwang road towards Oku, 2260m, fl., 15 Nov. 1996, Etuge 3393; Turbo path, 500m after its splitting from the main road at the border of the Ijim Mountain Forest Reserve, 2000m, fl., 18 Nov. 1996, Buzgo 757; Above Aboh village, 1.5 hours walk along Gikwang road, 2450m, fl., 18 Nov. 1996, Cheek 8641; Near Aboh - Anyajua, 2400m, fl., 19 Nov. 1996, Satabie 1076.

Sida acuta Burm.f. subsp. *carpinifolia* (L.f.) Borss.Waalk.
Shrubby perennial to 1 m; stems much-branched; leaves

lanceolate to 6.5 cm long; flowers c. 1 cm long, pale yellow. Roadside.
Kilum: Fundong to Belo road, 2050m, fl., 21 Nov. 1996, Cheek 8724;

Sida rhombifolia L.
Local name: Nshim (Oku) – Cheek 8451; Cheng (Bekom) – Kamundi 631.
Subshrub 30 cm; leaves rhombic-elliptic, c. 5 × 1.5 cm, finely stellate-tomentellous beneath; flowers pale yellow. Forest edge.
Kilum: Forest above Oku-Elak, 2500m, fl., 10 Jun. 1996, Zapfack 827 & 828; Lake Oku, 2200m, fl., 12 Jun. 1996, Etuge 2337; Oku-Elak to the forest at KA, 2200m, fl., 27 Oct. 1996, Cheek 8451; KJ, 2500m, fl., 29 Oct. 1996, Buzgo 633;
Ijim: Gikwang towards Nyasosso, 2280m, fl., 18 Nov. 1996, Etuge 3426; Ijim Mountain Forest, 2160m, fl., 19 Nov. 1996, Kamundi 631; Laikom, path from Fon's palace downhill towards Akwamofu sacred forest, 1900m, 6.1630N, 10.1951E, fl. fr., 12 Dec. 1998, Pollard 378.
Plot voucher: Foret above Oku-Elak, Jun. 1996, Oku 21; 53.

Urena lobata L. subsp. lobata
Subshrub 0.3–0.6 cm; leaves ovate, 2–5 cm long; flowers pink, 1.5 cm long; fruits with 5 mericarps bearing grapnel hooks. Weed.
Bambili: Bambili, in a stream between two hills, fl. fr., 19 Aug. 1951, Ujor FHI 29984.

MELASTOMATACEAE

Amphiblemma mildbraedii Gilg ex Engl.
Erect shrubby herb to 2.3 m; leaves ovate, 8–14 cm long, densely puberulous beneath; stipules foliaceous; inflorescence a panicle of cymes; petals deep pink; fruit 8–11 mm long, yellowish-brown. Forest.
Kilum: Between Manchok, BirdLife HQ and Lumatu Traditional Forest (above the palace of the Fon), 2100m, fl., 1 Nov. 1996, Buzgo 690; Lometo Traditional Forest, 2050m, fl., 7 Nov. 1996, Zapfack 1238.
Ijim: Laikom village, by Fundong, 2100m, fl., 21 Nov. 1996, Onana 564.

Antherotoma naudinii Hook.f.
Local name: Igangabua (Oku) – Munyenyembe 840.
Annual herb, 5–10(–30) cm; leaves few, lanceolate, 3–4 cm long; inflorescence a close terminal cluster subtended by leaves; flowers pink or mauve. Grassland.
Kilum: Elak, 2200m, fl., 1 Nov. 1996, Munyenyembe 840; Lowland, 1900m, fl., 6 Nov. 1996, Pollard 50.
Ijim: Ijim plateau, Laikom, 1900m, 6.16N, 10.19E, fl., 3 Dec. 1998, Cheek 9728.

Dichaetanthera corymbosa (Cogn.) Jacq.-Fél.
Tree to 10m; branches densely brown setose; leaves ovate to ovate-elliptic, 6–16 cm long; flowers very numerous, magenta-pink. Forest edge.
Syn. **Sakersia calodendron** Gilg & Ledermann, F.W.T.A. 1: 249 (1954).

Bambui: Near Bambui, on Njinikom road, 1500m, fl. fr., 4 Jan. 1951, Keay FHI 28422.
Ndop: Swamp forest on organic alluvium fringing sedge swamp, Ndop plain, 3 miles S of Baba village, 1260m, fr., 5 Mar. 1962, Brunt 126.

Dissotis brazzae Cogn.
Shrubby herb to 2.8 m; leaves ovate, 7–9-nerved, up to 12 cm long; flowers mauve or purple; petals up to 1 cm long. Grassland.
Ijim: 1 km uphill on the main road from the BirdLife Project HQ towards Aboh village and Ijim Mountain Forest Reserve, 1400–1700m, fl., 20 Nov. 1996, Buzgo 783; Mbalabo-Hill, 2100m, fl., 26 Nov. 1996, Etuge 3633.

Dissotis perkinsiae Gilg
Shrub 1–1.8m; leaves ovate to ovate-lanceolate, 5–8 cm long; flowers 11 cm diam.; petals 3 cm long, reddish-violet or deep mauve-purple. Grassland.
Ijim: Nchan, 2000m, fl., Jun. 1931, Maitland 1429.
Jakiri: Low lava, near 'Journey's end' house, Jakiri, 1660m, fl. fr., 19 Jun. 1962, Brunt 541.
Kumbo: Near Banso, in grass by roadside, 1900m, fl. fr., 19 Jun. 1962, Brunt 580.

Dissotis princeps (Kunth) Triana var. princeps
Stout herb to 1.7 m; leaves ovate to ovate-lanceolate, 4–9 × 1–4 cm, long pilose above and beneath; flowers deep pink or violet, 5 cm across. Grassland.
Syn. **Dissotis bamendae** Brenan & Keay, F.W.T.A. 1: 258 (1954).
Kilum: KA, on the way between the forest and the village, within the fields, 15 minutes uphill from Mankok centre, 2200m, fl., 31 Oct. 1996, Buzgo 681.
Ijim: Ijim ridge, about 30 mins. walk above Fon of Kom's palace, 2050m, 6.16N, 10.19E, fl., 2 Dec. 1998, Cheek 9767.

Dissotis irvingiana Hook.
Annual herb 0.5–1 m; stem 4-angled; leaves subsessile, linear-elliptic, c. 8 × 1.5 cm; flowers pink or mauve. Grassland.
Syn. **Dissotis senegambiensis** (Guill. & Perr.) Triana var. senegambiensis
Ijim: Laikom, 2100m, fl., 21 Nov. 1996, Onana 565.
Bambui: Bambui Agric. Dept. Farm, pastures near office,1730m, fl. fr., 12 Nov. 1962, Brunt 885.

Dissotis thollonii Cogn. ex Büttner var. elliotii (Gilg) Jacq.-Fél.
Stout herb or undershrub to 1.8m; leaves lanceolate, 3–5-nerved, to 20 cm long, glossy; flowers bright pink, paniculate. Grassland.
Syn. **Dissotis elliotii** Gilg var. elliotii, F.W.T.A. 1: 258 (1954).
Ijim: Laikom, in grassland, fl. fr., Apr. 1931, Maitland 1677.
Bambui: In fields above Bambui Experimental Station rest house and up to forest remnants in gullies, fl. fr., 4 Mar. 1962, Brunt 65.

Heterotis angolensis (Cogn.) Jacq.-Fél. var.
bambutorum (Gilg & Ledermann ex Engl.) Jacq.-
Fél.
Erect herb to 1 m; stem with long spreading hairs; leaves
ovate or elliptic to 3.5 cm long; flowers bright pink,
paniculate. Swamps.
Syn. *Dissotis bambutorum* Gilg & Ledermann ex Engl.,
F.W.T.A. 1: 258 (1954).
Kilum: Kumbo to Oku, 2 miles on Oku road, 1850m, fl. fr.,
15 Feb. 1958, Hepper 2000.
Ijim: Laikom, about the edge of forest, 2000m, fr., Apr.
1931, Maitland 1679; Laikom, grassland, 2100m, 6.15N,
10.26E, fl., 21 Nov. 1996, Onana 565; Afua swamp, 1950m,
6.08N, 10.24E, fl., 7 Dec. 1998, Cheek 9832.
Bambui: Escarpment on road to Upper Farm, Bambui
Experimental Station, 1830m, fl. fr., 4 Apr. 1962, Brunt 457.

Heterotis canescens (E.Mey. ex Graham) Jacq.-
Fél.
Erect herb, 1–1.2 m; leaves narrowly ovate to lanceolate, to
7 cm long; flowers pink or purple, paniculate.
Syn. *Dissotis canescens* (E.Mey. ex Graham) Hook.f.,
F.W.T.A. 1: 258 (1954).
Ndop: 3 miles S of Baba village, sedge swamp community,
1300m, fr., 5 Mar. 1962, Brunt 120.

Melastomastrum theifolium (G.Don) A.Fern. &
R.Fern. var. *theifolium*
Subshrub to 1.8 m; stem 4-angled; leaves oblong-lanceolate
to linear, 3–7 × 1–3 cm, petiole 2–4 mm; flowers pink.
Bambui: Bambui, in grassland, 1330m, fl., Apr. 1931,
Maitland 1627.

MELIACEAE

Carapa grandiflora Sprague
Local name: Uvar (Oku) – Munyenyembe 745; Ghuine
(Bekom) – Kamundi 616.
Tree to 10m; stem short; crown wide-spreading; leaves to 1.2
m long; leaflets in 4–7 pairs, opposite or subopposite,
oblong, 10–30 cm long; sepals and petals greenish; staminal
tube white; disk orange; stigma white. Forest.
Lebrun & Stork (1992) list this name as a synonym of
Carapa procera DC., but here we follow F.W.T.A.
Kilum: Elak, 2450m, fl., 9 Jun 1996, Zapfack 796; KD, 9
Jun. 1996, Etuge 2245; Elak, 2420m, fr., 28 Oct. 1996,
Munyenyembe 745; KJ, 2400m, fr., 29 Oct. 1996, Zapfack
1072; KD-4, 2300m, fl., 1 Nov. 1996, Zapfack 1145.
Ijim: Nchan, Maitland 1663; Gikwang, towards Nyasosso
bush, 2260m, fl., 18 Nov. 1996, Etuge 3452; Ijim Mountain
Forest, 2160m, 19 Nov. 1996, Kamundi 616.
Plot voucher: Forest above Oku-Elak, Jun. 1996, Oku 9.

Entandrophragma angolense (Welw.) C.DC.
Forest tree to 55 m; bole long and straight; crown open;
buttresses broad, low; bark smooth, pale grey-brown to
orange-brown, scales flaking high up tree; slash dark red and
pink; leaves pinnate, clustered at ends of branches; leaflets
oblong; flowers yellowish. Forest.

Ijim: Nchan, Johnstone 90/31; Aboh village, 6.15N, 10.26E,
fr., 25 Nov. 1996, Onana 629.

MELIANTHACEAE

Bersama abyssinica Fresen.
Local name: Ibiyense (Oku) – Munyenyembe 770.
Tree to 18m; leaves alternate, 30–65 cm long, pinnate;
leaflets 7–13; rhachis often winged; fruits ovoid, 4 cm,
woody, dehiscent; seeds yellow and orange, arillate. Forest.
Kilum: Lake Oku, Keay FHI 28516 & 28518; Elak, 2300m,
fr., 8 Jun. 1996, Zapfack 744; Lake Oku, 2300m, fr., 11 Jun.
1996, Zapfack 859; KJ, 2300m, fr., 28 Oct. 1996, Buzgo
608; Elak, 2300m, fr., 29 Oct. 1996, Munyenyembe 770.
Ijim: Laikom, Maitland 1377 & 1490; Aboh village, 25 Nov.
1996, Onana 630.
Plot voucher: Forest above Oku-Elak, Jun. 1996, Oku 5.

MENISPERMACEAE

Stephania abyssinica (Quart.-Dill. & A.Rich.)
Walp. var. *abyssinica*
Local name: None known – Cheek 8678.
Slender liane, to 10m; leaves ovate to orbicular-ovate, 5-10
cm long, dark green above, glaucous beneath; inflorescences
4-7 cm diam., pseudo-umbels, flowers green or purple.
Forest.
Kilum: KD, fr., 9 Jun. 1996, Etuge 2225; 2450m, fr., 9 Jun.
1996, Zapfack 801; KA, 2760m, fl., 10 Jun. 1996, Asonganyi
1309; KD, 2800m, fl., 10 Jun. 1996, Etuge 2267; Elak,
2600m, fl., 10 Jun. 1996, Zapfack 832; KJ, 2620m, fr., 29
Oct. 1996, Buzgo 637; KA to summit, 2800m, fl., 30 Oct.
1996, Cheek 8533; KA and grassland boundary, 2650m, fr.,
31 Oct. 1996, Pollard 28.
Ijim: above Aboh village, 1.5 hours walk along Gikwang
road, track towards TA, 2400m, fr., 20 Nov. 1996, Cheek
8678.
Plot voucher: Forest above Oku-Elak, Jun. 1996, Oku 62.

MONIMIACEAE

Xymalos monospora (Harv.) Baill. ex Warb.
Local name: Fegei (Oku) – Cheek 8450.
Tree or shrub 3–5 m; leaves obovate to narrowly elliptic, 7–
20 cm long, lemon-scented; fruit orange. Forest.
Kilum: Elak, 2300m, fr., 8 Jun. 1996, Zapfack 738; Path
above water-tower, KJ, about 1 km into forest, 2250m, fl., 9
Jun. 1996, Cable 2990 & fr., Cable 3003; KD, fl. fr., 9 Jun.
1996, Etuge 2237; Oku-Elak to the forest at KA, 2200m, fr.,
27 Oct. 1996; Cheek 8450; Elak, KJ, 2300m, fr., 29 Oct.
1996, Zapfack 1046.
Ijim: Gikwang road towards Oku, 2260m, fr., 15 Nov. 1996,
Etuge 3391.
Plot voucher: Forest above Oku-Elak, Jun. 1996, Oku 90.

MORACEAE

Det. L.Csiba (K), F.Tadjouteu (YA) & M.Cheek (K)

Ficus chlamydocarpa Mildbr. & Burret
Local name: Kelonifam (Oku) – Munyenyembe 891.
Tree to 33m with large aerial roots; crown very wide; leaves oblong-elliptic, 12–33 cm × 3–15 cm. Forest.
Kilum: Lumetu Traditional Forest edge, 2300m, fr., 6 Nov. 1996, Munyenyembe 891.

Ficus cf. craterostoma Mildbr. & Burret
Local name: Kegrum (Oku) – Cheek 8606.
Strangling epiphytic shrub to medium-sized tree 4–20 m; leaves oblong-elliptic, c. 9 × 4 cm, petiole 2 cm long; figs yellow, 1.2 cm diam., scattered along leaf-bearing branches. Forest.
Kilum: Above Mboh, path above village to forest through Wambeng's farm, 2200m, 1 Nov. 1996, Cheek 8552; Lake Oku to Jikijem, secondary grassland, 2200m, 6.1214N, 10.2733E, fr., 5 Nov. 1996, Cheek 8606.
Ijim: Zitum road, 1850m, fr., 21 Nov. 1996, Etuge 3533; From Aboh, path towards Tum, 2100m, fr., 22 Nov. 1996, Etuge 3584; Fundong, Touristic Hotel, waterfall on Chumni river, 1600m, fr., 24 Nov. 1996, Cheek 8742; Aboh, Isebu Mountain Forest, growing in abandoned garden site, 1650m, fr., 26 Nov. 1996, Kamundi 709; Laikom, grass ridges above Akwamofu sacred forest, on the way to waterfall below western edge of Ijim ridge grassland plateau, forest surrounding cool waterfall basin, 1950m, 6.17N, 10.20E, fr., 10 Dec. 1998, Pollard 361.
Plot voucher: Forest above Oku-Elak, Jun. 1996, Oku 3.

Ficus cf. cyathistipula Warb.
Local name: Kilon (Oku) – Munyenyembe 871.
Epiphytic strangling shrub; leaves oblong-oblanceolate, to 18 cm x 6 cm, prominently veined below; stipules triangular, 1 x 1 cm, glabrescent; figs c. 1.8 cm diam., glabrous. Forest.
Kilum: Upkim traditional forest, 2060m, fr., 4 Nov. 1996, Munyenyembe 871; Ethiale, near the stream, 2000m, fr., 4 Nov. 1996, Zapfack 1176.
Ijim: Aboh, path towards Tum, 2200m, fr., 22 Nov. 1996, Etuge 3590; Laikom, forest patch towards foot of waterfall above Akwamofu sacred forest, 1860m, 6.17N, 10.20E, fr., 11 Dec. 1998, Etuge 4601.

Ficus sp. aff. lutea Vahl
Strangler becoming a shrub or tree, 10–20 m; branchlets stout; leaves oblanceolate to narrlowly elliptic, 6–18 × 3–10 cm, 5–8 main lateral nerves each side of midrib, upper surface shiny; figs c. 2 mm diam., green, yellow-spotted. Forest.
Ijim: Above Fon's palace, Laikom, about 30 mins. on path to Fulani settlement, 2050m, 21 Nov. 1996, Cheek 8708; Fundong, Touristic Hotel, Waterfall on Chumi river, 1600m, fr., 24 Nov. 1996, Cheek 8744; Laikom, Akwo-mofu, 1760m, 6.17N, 10.20E, fr., 5 Dec. 1998, Etuge 4541.

Ficus natalensis Hochst.
A shrub or tree to 14 m, sometimes epiphytic at first; levaes obovate, oblanceolate or sometimes elliptic, rounded or very shortly acuminate, 2.5–8 cm, 5–10 main lateral nerves each side of midrib; figs becoming reddish. Forest.
Kilum: Path from KN to summit, 2200m, st., 10 Jun. 1996, Cable 3041.

Ficus ottoniifolia (Miq.) Miq.
Local name: Fengak (Oku) – Munyenyembe 872.
Shrub or tree to 17 m, usually epiphytic at first; branchlets smooth pale; leaves oblong to oblong-elliptic, 10–20 × 4–8.5 cm; receptacles ellipsoid, 1.3–2.3 cm diam.; fruits long-stalked, in pairs or clusters on older branches below leaves. Forest.
Kilum: Upkim Traditional forest, 2060m, fr., 4 Nov. 1996, Munyenyembe 872.

Ficus sur Forssk.
Local name: Akanneh (Akaynii – Kom) – Etuge 4559.
Cauliflorous tree to 20m; leaves papery, ovate or elliptic, 10–17 cm long, coarsely serrate-incised; figs on peduncles c. 30 cm long. Forest.
Ijim: Laikom, Akwa-mofu medicinal forest, 1760m, 6.17N, 10.20E, fr., 5 Dec. 1998, Etuge 4559.

Ficus cf. sur Forssk.
Ijim: Zitum road, 2000m, fr., 21 Nov. 1996, Etuge 3541.

Ficus vallis-choudae Delile
Tree to 20m; bole short; crown spreading; branchlets nearly glabrous; leaves smooth above, glabrescent beneath, broadly ovate 5–15 cm × 4–10 cm; receptacles densely tomentose, 4 cm diam., fruits large, solitary. Forest.
Ijim: Fundong, Touristic Hotel, waterfall on Chumni river, 1400m, st., 22 Nov. 1996, Cheek 8738.

MYRICACEAE

Morella arborea (Hutch.) Cheek comb. nov.
Basionym: *Myrica arborea* Hutch., Kew Bull. 1917: 234 (1917).
Type: Cameroon, Mt Cameroon, Mann 1203 (Syntype K!)
Monoecious tree, 7–10 m; bark rough; branchlets puberulous, glandular; leaves oblong or oblong-lanceolate, toothed, 5–9 cm; male inflorescence an axillary spike or a raceme on long leafless shoots; fruits ellipsoid, 5 mm.
For rationale of this new combination see Killick *et al.*, Kew Bull. 53(4): 993 (1998).
Ijim: Above Fon's palace, Laikom, about 30 mins. on path to Fulani settlement, 2050m, 6.1642N 10.1952E, 21 Nov. 1996, Cheek 8705.

MYRSINACEAE

Ardisia kivuensis Taton
Local name: Echia (Oku) – Cheek 8440.
Shrub, 0.3–2 m; leaves ovate-elliptic or oblong, 13–18 cm long; petioles pink; flowers purple or reddish, in small clusters along branches; fruits bright red. Forest.
Syn. *Afrardisia cymosa* (Baker) Mez, F.W.T.A. 2: 31 (1963).

Kilum: Elak, 2100m, fr., 8 Jun. 1996, Zapfack 764; Elak, 2200m, fl., 8 Jun. 1996, Zapfack 786; Path above water-tower, KJ, about 1 km into forest, 2200m, fl., 9 Jun. 1996, Cable 2950; KD, fl., 9 Jun., 1996, Etuge 2215; Junction of KJ and KN above Elak, 2550m, fl. fr., 10 Jun. 1996, Cable 3032; Shore of Lake Oku, 2250m, fl., 12 Jun. 1996, Cable 3101;
Oku-Elak to the forest at KA, 2200m, fr., 27 Oct. 1996, Cheek 8440; KJ, 2620m, fl., 29 Oct. 1996, Buzgo 638; KJ, 2500m, fr., 29 Oct. 1996, Zapfack 1076.
Ijim: Laikom, forest edge, fl., Apr. 1931, Maitland 1408; Aboh, Gikwang road towards Nyasosso forest, 2300m, fl., 19 Nov. 1996, Etuge 3497; Near Aboh-Anyajua, 2500m, fr., 20 Nov. 1996, Satabie 1077; Laikom, forest, 6.17N, 10.20E, fr., 3 Dec. 1998, Etuge 4525.
Plot voucher: Forest above Oku-Elak, Jun. 1996, Oku 10.

Embelia mildbraedii Gilg & G.Schellenb.
Local name: Ntoh (Oku) – Munyenyembe 830.
Climbing shrub; leaves elliptic, dentate, glabrous, 5–6 cm; inflorescences arise from leaf-axils of current year's growth, racemes 1–1.5 cm long, subumbellate; flowers white or cream, c. 3 mm diam., clustered towards apex. Forest.
Kilum: Path above water-tower, KJ, about 1 km into forest, 2250m, fl., 9 Jun. 1996, Cable 2992; KA, 2200m, fr., 11 Jun. 1996, Etuge 2312; Lake Oku, 2300m, fl., 11 Jun. 1996, Zapfack 857; Lake Oku, 2200m, fl., 12 Jun. 1996, Etuge 2344; 2600m, fr., 31 Oct. 1996, Munyenyembe 830.
Ijim: From Aboh, path towards Tum, 2100m, fr., 22 Nov. 1996, Etuge 3593.
Plot voucher: Forest above Oku-Elak, Jun. 1996, Oku 11.

Embelia schimperi Vatke
Straggling shrub or climber; leaves elliptic to obovate-elliptic, smooth, leathery, 5–9 cm × 2–4 cm; inflorescence a raceme, 2–5.5 cm long on leafless part of previous year's branchlet; flowers whitish, 5 mm diam.; fruits red. Forest.
Kilum: KD, fr., 9 Jun. 1996, Etuge 2236.
Ijim: Aboh village, 25 Nov. 1996, Onana 627; Gikwang towards Nyasosso, 2290m, fr., 18 Nov. 1996, Etuge 3419; Forest below Ardo's compound above Laikom, 2100m, 6.16N, 10.21E, fr., 7 Dec. 1998, Gosline 206; Cheek 9860.

Maesa lanceolata Forssk.
Local name: Sem (Oku) – Munyenyembe 789.
Shrub or small tree to 7m; leaves elliptic-oblanceolate, serrulate; inflorescence profusely branched,
5–9 cm long; flowers numerous, subsessile, greenish-white; fruits globose. Forest.
Kilum: Ridge above Lake Oku, 2600m, fl., 6 Jan. 1951, Keay & Lightbody FHI 28474; Path above water-tower, KJ, about 1 km into forest, 2250m, fl., 9 Jun. 1996, Cable 2989; KD, fl., 9 Jun. 1996, Etuge 2233; Elak, 2500m, fl., 9 Jun. 1996, Zapfack 797; Elak, 2800m, fr., 29 Oct. 1996, Munyenyembe 789; KA path, 2500m, fr., 29 Oct. 1996, Onana 451; Elak, 2400m, fr., 29 Oct. 1996, Zapfack 1057.
Ijim: Ijim Mountain Forest, 2160m, fr., 19 Nov. 1996, Kamundi 608; Near Akeh, 2400m, fr., 25 Nov. 1996, Satabie 1092.

Rapanea melanophloeos (L.) Mez
Local name: Ntoko (Oku) – Munyenyembe 737.
Tree or small shrub 3–16 m; leaves obovate-oblanceolate, 8–12 cm, glabrous, nerves lateral; petioles pinkish; flowers in axillary clusters or below leaves, 3 mm, white; fruits globose, to 5 mm, bright bluish-purple. Forest.
Kilum: Junction of KJ and KN above Elak, 2550m, fl., 10 Jun. 1996, Cable 3033; KD, 2800m, fl., 10 Jun. 1996, Etuge 2279; Elak, 2400m, fl., 28 Oct. 1996, Munyenyembe 737; KR-3, KJ, 2620m, fr., 29 Oct. 1996, Zapfack 1088; Elak, KR-3, following on from KD-10, 2400m, fl., 1 Nov. 1996, Pollard 33.

MYRTACEAE

Eucalyptus sp. 1
There are probably three species of introduced *Eucalyptus* in the Kilum-Ijim Checklist area (pers. comm. DeMarco).
Ijim: Laikom, path from Fon's Palace downhill towards Akwamofu sacred forest, 1900m, 6.1630N, 10.1951E, fr., 12 Dec. 1998, Pollard 373.

Eugenia gilgii Engl. & Brehmer
Local name: Ibelabela (+ndwam) (Kom) – Etuge 4506.
RED DATA LISTED
Shrub or small tree to 8 m; leaves elliptic, 6–10 cm long; inflorescence a very short raceme, central bracteate axis to 4mm long; flowers pink borne below leafy parts of shoots. Forest.
Ijim: Laikom, Maitland 1486 & 1661; Above Fon's palace, Laikom, about 30 mins. on path to Fulani settlement, 2050m, fl., 21 Nov. 1996, Cheek 8703; Fundong, near Touristic House, near the waterfall, fl., 23 Nov. 1996, Onana 590 & 594; Grassland and gallery forest, Ijim, Laikom, 2000m, 6.17N, 10.20E, fl., 2 Dec. 1998, Etuge 4506.
Belo: Maitland 1705.

Syzygium staudtii (Engl.) Mildbr.
Tree to 16 m commonly with a dense basal tangle of adventitious roots; leaves obovate-elliptic or elliptic, 4–9 cm long; flowers 3–4 mm diameter. Forest.
Kilum: Lake Oku, Johnstone 1/31; KJ, 2500m, fl., 29 Oct. 1996, Buzgo 628; KA path, 2500m, fl., 29 Oct. 1996, Onana 454; Elak, 2300m, fr., 30 Oct. 1996, Munyenyembe 805.
Ijim: About 2 km E of Laikom, near plot 146 of Kilum-Ijim Forest Project, 1900m, 6.17N, 10.21E, fr., 11 Dec. 1998, Cheek 9911.
Plot voucher: Forest above Oku-Elak, Jun. 1996, Oku 27; 50; 70.

OLACACEAE

Strombosia scheffleri Engl.
Local name: Ukwam (Kom at Laikom) – Etuge 4554.
Tree to 33 m; branchlets strongly angled; leaves ovate-elliptic or oblong, 5–8 main lateral nerves each side of midrib, venation distinct; 6–20 × 3–13 cm; flowers greenish-yellow or white, 3–5 mm; fruits obconical c. 2 cm. Forest.

Ijim: From Aboh, path towards Tum, 2100m, 22 Nov. 1996, Etuge 3587; Fundong, near Touristic House, near the waterfall, fl., 22 Nov. 1996, Onana 588; Akwamofu medicinal forest, Laikom, 1760m, 6.17N, 10.20E, fl., 5 Dec. 1998, Etuge 4554; Ntungfe, behind Simon Young's compound, Elemighong or Anyajua, 1570m, 6.12N, 10.23E, fl., 9 Dec. 1998, Etuge 4577.

OLEACEAE
Det. P.S.Green (K)

Jasminum dichotomum Vahl
Local name: Ikii-nkii (Kom at Laikom) – Etuge 4552; Ikuh Anding (Kom at Laikom – Yama Peter) – Gosline 200. Scrambling shrub or woody climber to 8 m; leaves ovate or broadly lanceolate, 5–11 cm; inflorescence densely corymbose; flowers numerous c. 2 cm long; corolla tube red; petals 5–9, oblanceolate, white, fragrant. Forest.
Ijim: Above Aboh village, 1.5 hours along Gikwang road to Lake Oku, track towards TA, 2400m, fl., 20 Nov. 1996, Cheek 8689; Laikom village, 2100m, fr., 21 Nov. 1996, Onana 572; Akwo-Mufu, Laikom, 1760m, fl., 5 Dec. 1998, Etuge 4552; Laikom, before Ardo's compound near stream crossing, 2100m, fl., 7 Dec. 1998, Gosline 200.

Jasminum pauciflorum Benth.
Slender scandent shrub; stem and petioles pubescent; leaves glabrous, ovate or ovate-elliptic, 2.5–6 cm long; flowers 1.5–2 cm long, pure white, fragrant; fruits black. Forest.
Kilum: Lake Oku, 2000m, fl., 12 Jun. 1996, Etuge 2347.
Ijim: Laikom, Maitland 1676.

Olea capensis L. subsp. *macrocarpa* (C.H.Wright) I.Verd.
Tree or shrub from shrubby growth about 60 cm or more to trees 2–12 m, occasionally to 30 m; leaves very variable in texture, size and shape; inflorescence many-flowered, not densely so; calyx cupular, 4-toothed or 4-lobed almost to the middle; corolla up to 3 mm long; fruit large, oblong-elliptic, about 1.5 cm long.
Syn. *Olea hochstetteri* Baker, F.W.T.A. 2: 49 (1963).
Local name: Ebwing (Oku) – Cheek 8595.
Kilum: Road from Rest House E of lake to N of mountain range, 2300m, 5 Nov. 1996, Cheek 8595.

ONAGRACEAE

Epilobium salignum Hausskn.
Herb to 1.8 m, herbaceous or thinly woody; stems often branched above, appressed grey-puberulous; leaves lanceolate to oblong-linear, 2–8 cm long; flowers white or cream later pink-tinged, 6–11.5 mm long. Streamsides and swamps.
Kilum: Between Mankok, BirdLife Project HQ and Lumatu Traditional Forest, 2100m, fl. fr., 1 Nov. 1996, Buzgo 691; Summit, Kinkolong swamp, 2900m, fl. fr., 3 Dec. 1998, Maisels 194.
Ijim: Gikwang road, Aboh to Akeh, about 1 km after first track, 2400m, fl. fr., 25 Nov. 1996, Cheek 8746; Main path from Ijim to Tum, 2100m, fr., 25 Nov. 1996, Etuge 3616;

Swamp E of Ardo's compound, Laikom, 2000m, 6.16N, 10.22E, fl. fr., 5 Dec. 1998, Cheek 9787.

Ludwigia abyssinica A.Rich.
Herb or weak shrub to 3m, stems much-branched, struggling or erect; flowers in axillary clusters, 1.5-3.5 mm long. Streamsides and swamps.
Syn. *Jussiaea abyssinica* (A.Rich.) Dandy & Brenan, F.W.T.A. 1: 170 (1954).
Ijim: Laikom, Maitland 1448.
Mbi Crater: Approach to Mbi crater by short cut from Belo-Afua, 1750m, 6.05N, 10.23E, fl. fr., 9 Dec. 1998, Cheek 9885.

OXALIDACEAE

Biophytum umbraculum Welw.
Local name: Ferongti (Oku) – Munyenyembe 900.
Herb to 30 cm; stems slender; leaves pinnate, 2–6 cm; leaflets in 3–8 pairs, rectangular-ovate, glabrous, in terminal crown, sensitive to weather/touch; flowers yellow, orange or red, sessile or on peduncle to 4 cm. Cliff.
Kilum: Mbokengfish, 1900m, fl., 7 Nov. 1996, Munyenyembe 900.

Oxalis corniculata L. var. *corniculata*
Local name: Mbol (Oku) – Munyenyembe 711.
Variable, diffuse, slender herb; leaflets obcordate, green or reddish-brown, 5–12 mm; flowers yellow; peduncles 1–7 cm; fruit tomentellous, 1.5–2 cm long. Grassland edge.
Kilum: Path above water-tower, KJ, about 1 km into forest, 2250m, fl. fr., 9 Jun. 1996, Cable 2969; Summit, 2950m, fl., 10 Jun. 1996, Cable 3070; Elak, forest edge, 2300m, fl., 28 Oct. 1996, Munyenyembe 711.
Ijim: Gikwang towards Nyasosso bush, 2280m, fl. fr., 18 Nov. 1996, Etuge 3448.

PASSIFLORACEAE
Det. M.Cheek (K) & F.Tadjouteu (YA)

Adenia cissampeloides (Planch. ex Benth.) Harms
Climber; leaves rhomboid, marbled above, pale beneath, 7 cm; flowers greenish-yellow. Forest edge.
Kilum: Above Mboh, path above village to forest through Wambeng's farm, 2200m, 1 Nov. 1996, Cheek 8546B.

Adenia rumicifolia Engl. & Harms var. *miegei* (Aké Assi) W.J.de Wilde
Climber; leaves broadly ovate, to c. 12 × 10 cm, cordate; fruit obovoid to globose, c. 3 cm diam., stipe 2 cm long. Forest.
Ijim: Zitum road, 1900m, fr., 21 Nov. 1996, Etuge 3546.

Passiflora edulis Sims fa. *edulis*
Climber; leaves circular, c. 10 cm long, trisect, serrate; flowers c. 3 cm diam., white. Forest edge.
Kilum: Lake Oku, 2200m, fr., 12 Jun. 1996, Etuge 2334; Lowland region, 1900m, fr., 6 Nov. 1996, Pollard 49.

PHYTOLACCACEAE
Det. B.J.Pollard (K).

Phytolacca dodecandra L'Hér.
Scandent shrub or liane to 17m; leaves elliptic or ovate-elliptic, thinly papery, glabrous, 4–15 cm; racemes slender to 40 cm long; male flowers white or yellow-green; ripe fruits red. Forest.
Ijim: Laikom, Johnstone 98/31; Laikom, grass ridges above Akwamofu sacred forest, on the way to the waterfall below western edge of Ijim ridge grassland plateau, 1850m, 6.17N, 10.20E, fl., 10 Dec, 1998, Pollard 366; Laikom, below Fon of Kom's Palace, 1820m, 6.16N, 10.21E, fl., 11 Dec. 1998, Cheek 9908.

PIPERACEAE
Det. M.Cheek (K) & E.Biye (YA)

Peperomia fernandopoiana C.DC.
Stout epiphytic herb to c. 30 cm, resembling *Peperomia vulcanica*; leaves alternate, ovate or lanceolate c. 5 cm long, long-acuminate. Forest.
Kilum: Ethiale, near the stream, 2060m, 4 Nov. 1996, Zapfack 1181B.
Ijim: Aboh village, 25 Nov. 1996, Onana 621; Laikom, Plot between Fon's Palace and Ardo's compound, 2000m, 6.16N, 10.26E, fl., 8 Dec. 1998, Cheek 9862.

Peperomia molleri C.DC.
Epiphytic or terrestrial herb up to 20 cm; leaves opposite, ovate, 5 × 3.5 cm, margin ciliate. Forest.
Kilum: Ethiale, near the stream, 2060m, 4 Nov. 1996, Zapfack 1181A.

Peperomia retusa (L.f.) A.Dietr.
Epiphytic herb; vegetative stems creeping; flowering stems erect, glabrous, c. 5 cm long; leaves alternate, orbicular, uppermost opposite, ovate, c. 1.8 × 1.2 cm. Forest.
Includes *Peperomia retusa* (L.f.) A.Dietr. var. *mannii* (Hook.f.) Düll
Kilum: Lake Oku, fl., Jan. 1951, Keay FHI 28532; Shore of Lake Oku, 2250m, fl., 12 Jun. 1996, Cable 3096; KJ, 2300m, fl., 29 Oct. 1996, Zapfack 1051; KJ, 2400m, fl., 29 Oct. 1996, Zapfack 1067.
Ijim: Zitum road, 1950m, 21 Nov. 1996, Etuge 3556; Ijim Mountain Forest, top of the ridge, second gate from Aboh village, 2370m, fl., 21 Nov. 1996, Kamundi 663; Main path from Ijim to Tum, 2100m, fl., 25 Nov. 1996, Etuge 3620; Laikom, Ijim ridge, about 30 mins. walk above Fon of Kom's Palace, 2050m, st., 2 Dec. 1998, Cheek 9781; Laikom, 2000m, 6.17N, 10.20E, fl., 3 Dec. 1998, Etuge 4526; Afua swamp to Afua junction, 1800m, 6.09N, 10.22E, fr., 7 Dec. 1998, Cheek 9856; Laikom, plot between Fon's Palace and Ardo's compound, 2000m, 6.16N, 10.21E, fl., 8 Dec. 1998, Cheek 9861.
Plot voucher: Forest above Oku-Elak, Jun. 1996, Oku 64.

Peperomia tetraphylla (G.Forst.) Hook. & Arn.
Epiphytic herb, glabrous; leaves obovate-elliptic, whorled in fours, glabrous, shining, 1.5 cm long. Forest.

Ijim: Laikom, 2000m, 6.17N, 10.20E, fl., 3 Dec. 1998, Etuge 4535; Afua to Belo, 1950m, 6.09N 10.22E, st., 7 Dec. 1998, Cheek 9853.

Peperomia thomeana C.DC.
RED DATA LISTED
Epiphytic herb resembling *Peperomia retusa*; leaves obovate to broadly oblanceolatae, 2–3.5 × 1–1.5 cm, apex notched, margin ciliate. Forest.
Kilum: Path above water-tower, KJ, about 1 km into forest, 2250m, fl., 9 Jun. 1996, Cable 2967 & 2981; KJ, above water tank from Elak, 2 km into forest, 2400m, fl., 10 Jun. 1996, Cable 3016; Lake Oku, 2300m, fl., 11 Jun. 1996, Zapfack 878; Shore of Lake Oku, 2250m, fl., 12 Jun. 1996, Cable 3098 & 3134; KJ, 2300m, fl., 29 Oct. 1996, Zapfack 1052.
Plot voucher: Forest above Oku-Elak, Jun. 1996, Oku 63.

Peperomia vulcanica Baker & C.H.Wright
Local name: Mbol (Oku) – Cheek 8458.
Epiphytic or terrestrial herb, glabrous; flowering stems erect, stout, c. 15 cm long; leaves alternate, rhombic, to 6 × 2.5 cm. Forest.
Kilum: Elak, 2300m, fl., 8 Jun. 1996, Zapfack 741; Path above water-tower, KJ, about 1 km into forest, 2300m, fl., 9 Jun. 1996, Cable 2966; KJ, above water tank from Elak, 2 km into forest, 2400m, 10 Jun. 1996, Cable 3014; Summit, 2950m, fl., 10 Jun. 1996, Cable 3084; Shore of Lake Oku, 2250m, 12 Jun. 1996, Cable 3132; Oku-Elak to the forest at KA, 2200m, 27 Oct. 1996, Cheek 8458.
Ijim: Laikom, Ijim ridge, about 30 mins. walk above Fon of Kom's Palace, 2050m, 6.16N, 10.19E, fl., 2 Dec. 1998, Cheek 9779.

Peperomia cf. laeteviridis Engl.
Epiphytic herb c. 15 cm long; leaves whorled, elliptic to obovate, c. 2 cm long. Forest.
Kilum: KJ, 2500m, fl., 29 Oct. 1996, Buzgo 631.

Piper capense L.f.
Local name: Bobo (Oku) – Cheek 8439.
Shrub c. 1 m; leaves orbicular, c. 10 cm long, cordate; peduncle as long as or longer than flowering axis; fruits spicate. Forest.
Kilum: Elak, 2300m, fl., 8 Jun. 1996, Zapfack 748; KD, fl., 9 Jun. 1996, Etuge 2221; Oku-Elak to the forest at KA, 2200m, fl., 27 Oct. 1996, Cheek 8439.
Ijim: Gikwang towards Nyasosso, 2270m, fl., 18 Nov. 1996, Etuge 3432; Ijim Mountain Forest, 2160m, fl., 19 Nov. 1996, Kamundi 624; Aboh village, 2450m, fl., 20 Nov. 1996, Onana 554; Laikom, 2000m, 6.17N, 10.20E, fr., 3 Dec. 1998, Etuge 4530.
Plot voucher: Forest above Oku-Elak, Jun. 1996, Oku 4.

PITTOSPORACEAE

Pittosporum viridiflorum Sims "mannii"
Local name: Ibyerese (Oku) – Munyenyembe 744.
Shrub or small tree; leaves glabrous; flowers whitish-yellow; seeds sticky, scarlet. Forest edge.

Kilum: Elak, 2300m, fl., 8 Jun. 1996, Zapfack 737; KD, fl., 9 Jun. 1996, Etuge 2224; Path from KN to summit, 2900m, fl., 10 Jun. 1996, Cable 3039; Elak, 2600m, fl., 10 Jun. 1996, Zapfack 835; Lake Oku, 2200m, fl., 12 Jun. 1996, Etuge 2335; Elak, 2420m, fr., 28 Oct. 1996, Munyenyembe 744; KA path, 2500m, fr., 29 Oct. 1996 Onana 452; KJ, 2620m, fr., 29 Oct. 1996, Zapfack 1091; Summit, 2800m, fr., 31 Oct. 1996, Zapfack 1112.
Ijim: Gikwang road towards Oku, 2265m, fr., 15 Nov. 1996, Etuge 3387; Ijim Mountain Forest, 2160m, fr., 19 Nov. 1996, Kamundi 605; Ijim Mountain Forest, near Chufekhe stream, 2010m, fr., 20 Nov. 1996, Kamundi 636; Laikom, 2100m, fr., 21 Nov. 1996, Satabie 1087.

PLUMBAGINACEAE

Plumbago zeylanica L.

Climbing shrub c. 1 m, viscid above; branches ribbed, glabrous; leaves glabrous, ovate or ovate-lanceolate, 6–10 cm long; flowers white c. 2 cm long. In thickets and near villages.
Ijim: Etuge 3657.

PODOSTEMACEAE

Saxicolella marginalis (G.Taylor) C.Cusset ex Cheek *comb. nov.*

Basionym: ***Butumia marginalis*** G.Taylor, Bull. Brit. Mus. (Nat. Hist.) (Bot.) 1: 55 (1953).
TYPE: Nigeria, Butum Falls, Keay, Savory & Russell in FHI 25152 (holotype K!; isotype BM).
RED DATA LISTED
Herb; thallus radiating, branching dichotomous, c. 0.75 cm wide, 5–6 cm long, margin bearing at intervals tufts of linear leaves 2–3 mm long, each surrounding a single sessile flower of the same length. Waterfalls.
Ijim: Fundong, Touristic Hotel, 30 m waterfall on Chumni River, 1400m, fl. fr., 22 Nov. 1996, Cheek 8740.

Ledermanniella keayi (G.Taylor) C.Cusset
RED DATA LISTED
Herb; thallus inconspicuous, aerial stems erect, 2–4 cm long, apically bearing numerous short, slender, leaf-bearing branches; leaves ovate c. 2 mm long. Waterfalls.
Syn. ***Inversodicrea keayi*** G.Taylor, F.W.T.A. 1: 127 (1954).
Kilum: Above Mboh, path above village to forest through Wambeng's farm, 2000m, fl. fr., 1996, Cheek 8546A.

Ledermanniella cf. musciformis (G.Taylor) C.Cusset
RED DATA CANDIDATE
Herb; thallus resembles *Saxicolella marginalis*; aerial stems 1–2 cm long, produced from the centre of the thallus; leaves ovate, c. 2mm long. Waterfalls.
Ijim: Anyajua, waterfall near Project HQ, 1300m, 6.11N, 10.22E, fl. fr., 12 Dec. 1998, Cheek 9920.

POLYGALACEAE

Polygala albida Schinz subsp. *stanleyana* (Chodat) Paiva

Herb to 45 cm; leaves linear to oblong-linear, 2–6 cm long; inflorescence racemose, oblong; sepals mauve-pink, glabrous, c. 5 mm long; petals white. Grassland.
Mbi Crater: Approach to Mbi Crater by short cut from Belo-Afua, 1750m, fl., 9 Dec. 1998, Cheek 9888.

Polygala tenuicaulis Hook.f. subsp. *tayloriana* Paiva
RED DATA LISTED
Local name: Kajis (Oku) – Munyenyembe 879.
Erect annual herb 30–45 cm; stem wiry, unbranched in basal half; leaves alternate, sessile, linear-lanceolate, c. 25 × 1–2 mm; inflorescence a cluster of many-flowered racemes; flowers pale pink, c. 7mm long. Savanna.
Kilum: Elak, 2200m, fl., 28 Oct. 1996, Munyenyembe 747; Path from Project HQ to KA, 2200m, fl., 30 Oct. 1996, Cheek 8499; Shambai, 2400m, fl., 5 Nov. 1996, Munyenyembe 879; Summit, near savanna zone, 2500m, fl., 5 Nov. 1996, Zapfack 1193.
Ijim: Gikwang road towards Nyasosso forest, Ijim Mountain Forest Reserve, 2300m, fl., 19 Nov. 1996, Etuge 3485; Above Fon's Palace, about 30 mins. walk along path to Fulani settlement, 2050m, fl., 22 Nov. 1996, Cheek 8732; Muteff, 2000m, fl., 22 Nov. 1996, Satabie 1088; Ijim plateau above Fon's Palace, Laikom, 1900m, fl., 3 Dec. 1998, Cheek 9739.

POLYGONACEAE

Fagopyrum snowdenii (Hutch. & Dandy) S.P.Hong

Herb, weakly erect or scandent, to 1 m; leaves sagittate, lower ones ovate-triangular up to 8 × 3 cm, upper linear-lanceolate; inflorescences in slender terminal and axillary racemes, up to 15 cm long; flowers green; fruits with red or purple barbed setae. Forest.
Syn. ***Harpagocarpus snowdenii*** Hutch. & Dandy, F.W.T.A. 1: 140 (1954).
Kilum: Oku, by river below village in well-developed forest, 1850m, fl., 17 Feb. 1958, Hepper 2033; Oku, evergreeen forest, 2000m, fl., 21 Jun. 1962, Brunt 598; Oku, montane forest, fl. fr, 14 Jul. 1967, Letouzey 8911; Mt. Oku, 2200m, fl. fr., 1 Feb. 1970, C. N.A.D. 1836; Elak, 2000m, fl., 8 Jun. 1996, Zapfack 759; KD, 2800m, fr., 10 Jun. 1996, Etuge 2259; KA, 2600m, fl. fr., 27 Oct. 1996, Buzgo 590; Elak, 2600m, fl., 27 Oct. 1996, Munyenyembe 710; KJ, 2400m, fl., 29 Oct. 1996, Zapfack 1062.
Ijim: Laikom, edge of forest, 2000m, fl., Apr. 1931, Maitland 1491; Mbesa, 1660m, fl., Jun. 1931, Maitland 1503; Gikwang, towards Nyasosso bush, 2260m, fr., 18 Nov. 1996, Etuge 3442.

Polygonum cf. glomeratum Dammer
Local name: Mbas ngi (Kom at Laikom) – Etuge 4511.

Herb 60 cm; stems erect or semi-decumbent with recurved hairs or glabrous; leaves linear-lanceolate, up to 15 × 2 cm, glabrous; ochrea glabrous, 3–4 cm long, amarginal cilia 1 mm; inflorescence a glandular-hairy raceme up to 5 cm long; flowers white. Swamp.

Differs from *Polygonum glomeratum* in the glandular inflorescence, broad leaves and ochreal cilia of c. 1 mm in length. Reference material unavailable at K, matches F.T.E.A. description well.

Ijim: Above Fon's palace, Laikom, about 30 mins. walk on path to Fulani settlement, 2050m, fl., 21 Nov. 1996, Cheek 8701; Laikom, swamp forest in water, 1860m, 6.17N, 10.20E, fl., 2 Dec. 1998, Etuge 4511.

Polygonum nepalense Meisn.

Local name: Qua Qua – Cheek 8687.

Annual herb; stems prostrate; leaves ovate, 1.5–2 cm long; ochrea entire, brownish; flowers in pedunculate clusters; nutlets closely pitted. Fallow and grassland.

Kilum: Lake Oku, 2200m, fl., 11 Jun. 1996, Zapfack 871; Elak, KJ, 2300m, fl., 28 Oct. 1996, Buzgo 596; Oku-Elak, lower parts of KA, 2200m, fl., 28 Oct. 1996, Cheek 8461.
Ijim: Above Aboh village, 1.5 hours walk along Gikwang road, track towards TA, 2400m, fl., 20 Nov. 1996, Cheek 8687; Aboh, Ijim mountain forest, the ridge toward a point where Lake Oku can be viewed, fl., 22 Nov. 1996, Kamundi 690; Anyajua Forest, montane rain forest along the ridge and disturbed farmland and abandoned garden on lower slopes, 1800m, fl., 27 Nov. 1996, Kamundi 723; Laikom, forest patch towards foot of waterfall above Akwamofu sacred forest, 1860m, fl., 11 Dec. 1998, Etuge 4603.

Polygonum salicifolium Brouss. ex Willd.

Annual herb 1–1.2m; stem erect or semi-decumbent; leaves glabrous beneath, except on margin and midrib, linear-lanceolate, up to 15 cm long; inflorescence a lax raceme over 2 cm long; flowers pink and white. Marshy places.

Ijim: Above Fon's palace, Laikom, about 30 mins. on path to Fulani settlement, 2050m, fl., 21 Nov. 1996, Cheek 8701; Afua swamp, 1950m, 6.08N, 10.24E, fl., 7 Dec. 1998, Cheek 9849.

Polygonum cf. salicifolium Brouss. ex Willd.

Differs from *Polygonum salicifolium* in: ochreal cilia only 6 mm long; leaves with lower midrib appressed strigose; nutlets triquetrous. Swamp.

Kilum: N side of the lake, lake shore, water level, 2100m, fl., 5 Nov. 1996, Buzgo 703.

Polygonum cf. strigosum R.Br.

Straggling herb 60 cm; stems red; leaves sessile, ligulate-oblong, 5 × 0.4 cm; ochrea 6mm long, entire; flowers in branched glandular hairy panicles c. 8 cm long, pink. Swamp.

Differs from *Polygonum strigosum* in: glabrous stems (not strigose); leaf-bases attenuate (not hastate), bearing narrow leaf blades (not broad); ochrea only 1 cm. long (not 2–3 cm), pointed (not truncate).

Mbi Crater: 1950m, 6.05N, 10.21E, fl., 9 Dec. 1998, Cheek 9902.

Rumex abyssinicus Jacq.

Local name: Kintongtong (Oku) – Cheek 8443.

Herb to 1 m; stout perennial rootstock, 5 cm diam., stems annual; leaves hastate, triangular, 6-15 cm long; flowers in large much-branched panicles. Forest edge.

Kilum: Oku-Elak to the forest at KA, 2200m, fl., 27 Oct. 1996, Cheek 8443; Elak, KC, 2400m, fl., 30 Oct. 1996, Buzgo 650.

Rumex nepalensis Spreng.

Local name: Klanluk (Oku) – Munyenyembe 760.

Perennial herb, 1–2m; leaves lanceolate, to 30 × 5 cm; flowers in much-branched panicles. Forest edge.

Syn. *Rumex bequartii* De Wild., F.W.T.A. 1: 139 (1954).

Kilum: Lake Oku, high forest around the Lake, 2400m, fl. fr., 7 Jan. 1951, Keay & Lightbody FHI 28494; Oku, inside Fon's Palace, area in waste ground, 2000m, fl. fr., 17 Feb. 1958, Hepper 2023; Road to Oku, 2100m, fl. fr., 19 Jun. 1962, Brunt 576; Elak, 2100m, fl., 8 Jun. 1996, Zapfack 792; Path above water-tower, KJ, about 1 km into forest, 2200m, fl., 9 Jun. 1996, Cable 2963; Lake Oku, 2300m, fl., 11 Jun. 1996, Zapfack 861; Elak, KJ, 2400m, fl., 28 Oct. 1996, Buzgo 616; Elak, 2900m, fl., 29 Oct. 1996, Munyenyembe 760; KA, close to top of the mountain, 2600m, fl., 31 Oct. 1996, Buzgo 676.
Ijim: Ijim Mountain Forest, 2160m, fl., 19 Nov. 1996, Kamundi 625; Gikwang main road towards Nyasosso forest, 2250m, fl., 20 Nov. 1996, Etuge 3513; Laikom, grassland near the Fulani village, 2000m, 6.17N, 10.20E, fl., 3 Dec. 1998, Etuge 4536.

PRIMULACEAE

Anagallis djalonis A.Chev.

Erect annual herb 5–15 cm; stem branched from base; leaves petiolate, obovate-spathulate; flowers axillary almost to base of stem, white. Damp places.

Ijim: Laikom, Ijim plateau above Fon's Palace, 1900m, 6.16N, 10.19E, fl., 3 Dec. 1998, Cheek 9722.

Anagallis minima (L.) E.H.L.Krause

Local name: Unknown – Cheek 8672.

Annual herb, stems prostrate, to 5 cm long; leaves alternate, 5 × 2 mm; flowers white or pink, 2 mm diam., sessile. Damp open places.

Ijim: Above Aboh village, 1.5 hours walk along Gikwang road, 2450m, fr., 19 Nov. 1996, Cheek 8672.

Anagallis tenuicaulis Baker

Herb; stems 30 cm long, rooting at base, ascending; leaves sessile, ovate, 6 × 4 mm; flowers single, in apical axils, white, c. 6 mm diam.; pedicel c. 8 mm long. Swamp.

Ijim: Afua swamp, 1950m, 6.08N, 10.24E, fl., 7 Dec. 1998, Cheek 9833.

Ardisiandra sibthorpioides Hook.f.

Creeping pubescent herb; stems slender, up to 60 cm long; leaves suborbicular, 3–4 cm diam., deeply dentate-incised, lobes coarsely toothed; petiole as long as or longer than lamina; inflorescences of few-flowered axillary racemes;

flowers white, 5 mm long; calyx-segments ovate-lanceolate; corolla campanulate. Montane forest.
Kilum: Around Lake Oku, 2200m, fl., 5 Nov. 1996, Etuge 3338.

Lysimachia ruhmeriana Vatke
Perennial herb to 1.2 m; stems branched, erect or decumbent; leaves narrowly elliptic, c. 5 × 1 cm, sessile; inflorescences of terminal racemes; flowers numerous, white or pink, 3 mm; fruit globose, 3 mm. Forest and swamp.
Kilum: KJ to KD, 3000m, fr., 13 Jun. 1996, Etuge 2354; Elak, 2400m, fr., 1 Nov. 1996, Munyenyembe 853; Main road from Ijikijem to Oku, 2350m, fr., 7 Nov. 1996, Etuge 3353.
Ijim: Afua swamp, 1950m, 6.08N, 10.24E, fr., 7 Dec. 1998, Cheek 9836.

PROTEACEAE
Det. R.K.Brummitt (K)

Protea madiensis Oliv. subsp. *madiensis*
Fire adapted shrub to 2 m; inflorescence a capitulum; bracts yellow, tinged pink at tips, densely tomentose; flowers numerous, whitish-yellow; flower 'limb' glabrous. Savanna.
Syn. *Protea elliotii* C.H.Wright var. *angustifolia* Keay, F.W.T.A. 1: 179 (1954).
Ijim: Aboh, Ijim Mountain Forest, Grassland savanna, 2280m, 6.1116N, 10.2543E, fl., 20 Nov. 1996, Pollard 60.

Protea madiensis Oliv. subsp. *madiensis* var *nov. ?*
This possible new variety has pinkish-red flowers, quite distinct from the typical yellowish-white. Bracts crimson in both varieties. Savanna.
Ijim: Ijim Mountain Forest, 2010m, fl., 20 Nov. 1996, Kamundi 659; Belo to Afua, roadside in cultivated area, 1350m, 6.08N, 10.22E, fl., 7 Dec. 1998, Cheek 9825.
Bambili: Near Ring Road, 8 km E of Bambili, 2000m, fl., 7 Nov. 1970, Bauer 129.

RANUNCULACEAE

Clematis grandiflora DC.
Climber; leaflets ovate with short acute teeth, pilose to pubescent beneath, 10 cm long; flowers 10 cm diameter. Low bush and forest margins.
Bambui: 1660m, fr., Apr. 1931, Maitland 1632 & 1633.
Ndop: fr., 20 Dec. 1952, Boughey GC 11137.

Clematis hirsuta Guill. & Perr. var. *hirsuta*
Climber; leaves opposite, pinnate; leaflets ovate, slightly 3-lobed, c. 4 × 3 cm, dentate, white-velvety beneath; inflorescence a dense panicle; flowers white, 2.5 cm diameter. Forest edge.
Ijim: Laikom, plateau above Fon's Palace, 2000m, 6.1642N, 10.1950E, fl., 2 Dec. 1998, Pollard 257.

Clematis simensis Fresen.
Local name: N'dang (Oku) – Munyenyembe 765.

Climber; leaves opposite, pinnate, leaflets ovate, c. 7 × 4 cm, not lobed, serrate, glabrous or sparsely hairy below; inflorescence a panicle; flowers white, 1.5 cm diameter. Forest edge.
Kilum: Lower parts of KA, 2200m, 28 Oct. 1996, Cheek 8472; KJ, 2500m, 29 Oct. 1996, Buzgo 627; KJ, 2620m, fl., 29 Oct. 1996, Buzgo 636; Elak, 2600m, fl., 29 Oct. 1996, Munyenyembe 765; KC, 2600m, fl., 30 Oct. 1996, Zapfack 1108; Around Lake Oku, 2200m, fl., 5 Nov. 1996, Etuge 3340.
Ijim: Main road from Atubeaboh (Gikwang Foe) towards Oku, 2250m, fl., 15 Nov. 1996, Etuge 3378; Ijim Mountain Forest, 2160m, fl., 19 Nov. 1996, Kamundi 604 & fr., Kamundi 630.
Plot voucher: Forest above Oku-Elak, Jun. 1996, Oku 90.

Clematis villosa DC. subsp. *oliveri* (Hutch.) Brummitt
Kew Bull. 55: 104 (2000).
Erect perennial herb with stout rhizome; stems 0.7–1.5 m, strongly striate, 1–5-flowered; leaves pinnate to bipinnate or trifoliate, lobes very irregular in shape; flowers solitary, 3.5–5 cm diam., white, pink or mauve; achenes up to 10 cm in diameter. Savanna.
Syn. *Clematopsis villosa* (DC.) Hutch. subsp. *oliveri* (Hutch.) J.Raynal & Brummitt, Adansonia, II, 18: 18 (1978).
Syn. *Clematopsis scabiosifolia* (DC.) Hutch., F.W.T.A. 1: 64 (1954).
Bambui: In grassland, 1330m, fr., Apr. 1931, Maitland 1628.
Sabga pass: 2000m, fl., Feb. 1951, Lightbody s.n.
Jakiri: Grassy hillside, 1850m, fl. fr., 18 Feb. 1958, Hepper 2714; Jakiri livestock improvement centre, paddock, high lava, 1800m, 6N, 10.30E, fr., 18 Jun. 1962, Brunt 524.

Delphinium dasycaulon Fresen.
Local name: Ntingji (Oku) – Munyenyembe 763.
Herb c. 1 m; leaves alternate, digitately lobed, to 8 cm diam.; flowers bright blue, 2–3 cm diam. Forest edge.
Kilum: Grassland, 3000m, fl., 29 Oct. 1996, Munyenyembe 763.

Ranunculus multifidus Forssk.
Herb, stems prostrate to 45 cm long; leaves bipinnate, to c. 5 cm diam.; flowers yellow, 1.5 cm diam. Forest edge.
Kilum: Summit, 2900m, fl., 10 Jun. 1996, Cable 3062; KD, close to the top of the mountain, 2600m, fl., 31 October 1996, Buzgo 678.
Ijim: Afua swamp, 1950m, 6.09N, 10.24E, fl. fr., 7 Dec. 1998, Etuge 4564; Ngengal swamp, 10 km E of Laikom, en route to Embezza, 2400m, 6.1615N, 10.2435E, fl. fr., 9 Dec. 1998, Pollard 345.
Mbi Crater: Ndawara area, Mbi swamp, slopes around the crater, 1900m, 6.05N, 10.21E, fl. fr., 9 Dec. 1998, Etuge 4581.

Thalictrum rhynchocarpum Quart.-Dill. & A.Rich. subsp. *rhynchocarpum*
Herb c. 1 m; leaves c. 40 cm long, tripinnate; leaflets ovate, 1 cm long, trilobed; inflorescence a diffuse panicle, c. 20 cm

diameter. Forest.
Kilum: Elak, 2100m, fr., 8 Jun 1996, Zapfack 793; Path above water-tower, KJ, about 1 km into forest, 2200m, fr., 9 Jun. 1996, Cable 2954; KD, fl. fr., 9 Jun. 1996, Etuge 2230; KJ, above water tank from Elak, 2 km into forest, 2400m, fl., 10 Jun. 1996, Cable 3011; Shore of Lake Oku, 2250m, fr., 12 Jun. 1996, Cable 3139; KJ, 2500m, fr., 29 Oct. 1996, Buzgo 626; Kitjim (Kitsem) Waterfall, 2200m, fr., 1 Nov. 1996, Buzgo 694.
Plot voucher: Forest above Oku-Elak, Jun. 1996, Oku 69.

RHAMNACEAE

Gouania longispicata Engl.
Local name: Ndang (Oku) – Munyenyembe 806.
Climber, tendrillate; leaves alternate, ovate, c. 4.5 × 3 cm, base truncate, serrulate; petiole 1.5 cm; racemes axillary, erect, c. 10 cm long, flowers white, 3 mm diameter. Forest edge.
Kilum: Lake Oku, 2200m, fl., 12 Jun. 1996, Etuge 2339; Elak, 2300m, fl., 30 Oct. 1996, Munyenyembe 806; Between Mankok, BirdLife Project HQ and Lumatu Traditional Forest (above the palace of the Fon), top end of Lumatu forest, 2200m, fl., 1 Nov. 1996, Buzgo 697.
Ijim: Gikwang towards Nyasosso bush, 2280m, fl., 18 Nov. 1996, Etuge 3462; Aboh village, 6.15N, 10.26E, fl., 18 Nov 1996, Onana 527; Ijim Mountain Forest, 6.11N, 10.25E, 2160m, fl., 19 Nov. 1996, Kamundi 619.
Plot voucher: Forest above Oku-Elak, Jun. 1996, Oku 94.

Maesopsis eminii Engl.
Tree c. 10 m, foul-smelling when cut; leaves alternate, narrowly elliptic, c. 9 × 3 cm, acuminate, serrate-dentate; petiole 1 cm long; drupes ellipsoid, succulent, 2.5 × 1.5 cm. Forest.
Ijim: Fundong near Touristic House, near waterfall, 6.15N, 10.26E, fr., 23 Nov. 1996, Onana 591.

Rhamnus prinoides L'Hér.
Local name: Findindi (Oku) – Munyenyembe 726.
Shrub 1–3 m; leaves glossy, alternate, ovate, c. 6 × 3.5 cm, acuminate, serrulate, petiole 1 cm long; flowers single, green, 3 mm diam.; fruits red, 7 mm across. Forest edge.
Kilum: Elak, KD, 2700m, fr., 10 Jun. 1996, Cable 3054; KD, 2800m, fl. fr., 10 Jun. 1996, Etuge 2270; Shore of Lake Oku, 2250m, fr., 12 Jun. 1996, Cable 3120; Path from Kilum Project HQ to summit, 2600m, fr., 12 Oct. 1996, Cheek 8417; Elak, 2420m, fr., 28 Oct. 1996, Munyenyembe 726; Summit, 2800m, fl. fr., 31 Oct. 1996, Zapfack 1113.
Ijim: Laikom, Maitland 1666; Above Fon's palace, Laikom, about 30 mins. on path to Fulani settlement, 6.1642N 10.1952E, 2050m, fr., 21 Nov. 1996, Cheek 8710; Zitum road, 1900m, 6.1116N 10.2543E, fr., 21 Nov. 1996, Etuge 3560; Forest patch, beside path from Fon's Palace, Laikom to grassland plateau above, 2040m, 6.1642N, 10.1950E, fl. fr., 2 Dec. 1998, Pollard 275.

RHIZOPHORACEAE

Cassipourea gummiflua Tul. var. *ugandense* (Stapf) J.Lewis
Tree c. 8 m; leaves glossy, opposite, elliptic, c. 12 × 4 cm, shortly acuminate, entire; petiole 1 cm long. Forest.
Kilum: Above Fon's Palace, about 30 mins. walk along path to Fulani settlement, 2050m, 22 Nov. 1996, Cheek 8733.

Cassipourea malosana (Baker) Alston
Tree to c. 15 m; stems hairy; leaves elliptic, c. 10 × 4 cm, serrate; petiole 5 mm long; stipule triangular, 7 mm long; fruits c. 8 cm diameter. Forest.
Kilum: KD, fl., 9 Jun. 1996, Etuge 2235; Elak, 2400m, fr., 28 Oct. 1996, Munyenyembe 740; KC, 2300m, fr., 30 Oct. 1996, Zapfack 1100; Lumeto Forest, near Oku-Elak, 2200m, 7 Nov. 1996, Cheek 8610.
Ijim: From Aboh, path towards Tum, 2200m, 6.1116N, 10.2543E, st., 22 Nov. 1996, Etuge 3596; Inside crater of Lake Oku, 2300m, 25 Nov. 1996, Cheek 8764.
Plot voucher: Forest above Oku-Elak, Jun. 1996, Oku 39.

ROSACEAE
Det. B.J.Pollard (K)

Alchemilla cryptantha Steud. ex A.Rich.
Herb; stems prostrate; leaves erect, alternate, blades reniform, c. 1.5 × 2.5 cm, 5–7-lobed, dentate, sparsely hairy; flowers white, inconspicuous c. 2 mm across. Forest edge.
Kilum: Road to Oku, 2100m, fl., 19 Jun. 1962, Brunt 569; Lake Oku, Keay FHI 28467; Elak, 2000m, fl., 8 Jun. 1996, Zapfack 775; KD above Oku village, fl., 9 Jun. 1996, Etuge 2205.
Belo: Belo, 1500m, fl., Apr. 1931, Maitland 1760.
Bambui: Bambui, by stream, fl., 13 Mar. 1960, Jackson 2514.
Jakiri: Near 'Journey's End' House, 1660m, fl., 18 Jun. 1962, Brunt 555.

Alchemilla fischeri Engl. subsp. *camerunensis* Letouzey
Local name: Mbaklum (Oku) – Munyenyembe 776; Mbakilum – Pers. comm. (Mar. 2000) DeMarco.
RED DATA LISTED
Robust herb resembling *Alchemilla cryptantha*, but leaf-blades 7–9-lobed, to 8 cm broad, thickly silvery hairy, velvety-soft to the touch. Growing in large patches. Forest edge.
Kilum: Mt. Oku, prairie, 2800m, st., 1 Feb. 1970, C. N.A.D. 1820; Summit, 2950m, fl., 10 Jun. 1996, Cable 3085; KD, 2800m, fl., 10 Jun. 1996, Etuge 2282; Summit, 3000m, st., 29 Oct. 1996, Munyenyembe 776; Summit, 2800m, st., 31 Oct. 1996, Zapfack 1115; Summit, in subalpine prairie, above a threshold of 2850–2900m, fl., 2 May 1997, Maisels 41.

Alchemilla kiwuensis Engl.

Herb resembling *Alchemilla cryptantha*, but leaves 7-lobed, 5–8 cm wide. Forest edge.

Kilum: KJ, 2620m, st., 29 Oct. 1996, <u>Buzgo</u> 639; Elak, high-mountain prairie, 2200m, fl., 1 Nov. 1996, <u>Satabie</u> 1064.

Prunus africana (Hook.f.) Kalkman
RED DATA LISTED

Tree to c. 20 m; leaves alternate, lanceolate, 7 × 3 cm, serrate; petiole 2 cm long, bearning 2 glands near apex; inflorescence a desne panicle; flowers white, 5 mm diam.; fruit a drupe, succulent, red, c. 1 cm diameter. Forest.

Kilum: KD, 2800m, fr., 10 Jun 1996, <u>Etuge</u> 2281.

Ijim: Gikwang road towards Nyasosso forest, 2300m, fl., 19 Nov. 1996, <u>Etuge</u> 3484.

Plot voucher: Forest above Oku-Elak, Jun. 1996, <u>Oku</u> 23.

Rubus pinnatus Willd. var. *afrotropicus* (Engl.) Gust.

Local name: Bakom (Oku) – <u>Cheek</u> 8442A.

Spiny shrub to 1 m; leaves 5-foliolate, leaflets elliptic, c. 6 cm long, glabrous below. Forest edge.

Kilum: Oku-Elak to the forest at KA, 2200m, 6.1349N, 10.3112E, fl. fr., 27 Oct. 1996, <u>Cheek</u> 8442A.

Ijim: Aboh, Gikwang road towards Oku, 2260m, fl. fr., 15 Nov. 1996, <u>Etuge</u> 3388.

Plot voucher: Forest above Oku-Elak, Jun. 1996, <u>Oku</u> 57.

Rubus pinnatus Willd. var. *ledermannii* Engl.

Local name: Bakom (Oku) – <u>Cheek</u> 8442B; Cheinle (Bekom) – <u>Kamundi</u> 621.

Spiny shrub resembling *Rubus pinnatus* var. *afrotropicus*, but leaflets white-hairy below. Forest edge.

Kilum: Elak, 2300m, fl. fr., 8 Jun. 1996, <u>Zapfack</u> 740; Path above water-tower, KJ, 1 km into forest, 2250m, fl. fr., 9 Jun. 1996, <u>Cable</u> 2980; KD, 2800m, fl. fr., 10 Jun. 1996, <u>Etuge</u> 2266; Oku-Elak to the forest at KA, 2200m, 6.1349N, 10.3112E, fl. fr., 27 Oct. 1996, <u>Cheek</u> 8442B; KJ, 2400m, fl. fr., 28 Oct. 1996, <u>Buzgo</u> 624; KJ, 2500m, fl. fr., 29 Oct. 1996, <u>Buzgo</u> 630.

Ijim: Above Aboh village, 1.5 hours along Gikwang road towards Lake Oku, 2380m, 6.1112N, 10.2529E, fl. fr., 18 Nov. 1996, <u>Cheek</u> 8629; Ijim Montain Forest, 2160m, 6.1125N, 10.2525E, fl. fr., 19 Nov. 1996, <u>Kamundi</u> 621.

Rubus rigidus Sm. var. *camerunensis* Letouzey

Spiny shrub; stems densely pubescent; leaves 3-foliolate, pubescent below; Petals 2–3, small, caducous; fruits ripening black. Forest edge.

Jakiri: Jakiri, 1750m, fl. fr., 13 Feb. 1958, <u>Hepper</u> 1951; 'Journey's End' House, 1660m, fl. fr., 18 Jun. 1962, <u>Brunt</u> 528.

Bambui: Bambui Agricultural Farm, 1660m, fl., 1 Jan. 1951, <u>Keay</u> FHI 28338; Bambui, fl. fr., Oct. 1960, <u>Cowell</u> M34.

RUBIACEAE
Det. M.Cheek (K), S.Dawson (K) & D.Bridson (K)

Anthospermum asperuloides Hook.f.
RED DATA LISTED

Subhrub to 30 cm; leaves narrowly oblanceolate, c. 8 mm long, acute, revolute; flowers yellowish green, inconspicuous. Grassland.

Syn. *Anthospermum cameroonense* Hutch. & Dalziel, F.W.T.A. 2: 223 (1963).

Kilum: Cliff-face, 2800m, fl., 31 Oct. 1996, <u>Munyenyembe</u> 835.

Chassalia laikomensis Cheek ined.
RED DATA LISTED

Shrub 2–3 m; leaf-blade narrowly elliptic; 4–12 × 1.5–4 cm, acuminate; stipules chaffy, not lobed; inflorescence terminal, loosely branched; flowers white, 6–10 mm long; fruits black, ovoid, 6–9 mm long. Forest.

Syn. *Chassalia sp.* nr. *umbraticola* sensu F.W.T.A. pro parte, non Vatke.

Kilum: KD, fl., 9 Jun. 1996, <u>Etuge</u> 2232; Lake Oku, 2300m, fl., 11 Jun. 1996, <u>Zapfack</u> 864; Shore of Lake Oku, 2250m, fl., 12 Jun. 1996, <u>Cable</u> 3114; KD-10/KR-1, 2400m, fr., 1 Nov. 1996, <u>Pollard</u> 34; Oku, 2010m, fl., 2 Nov. 1996, <u>Etuge</u> 3335.

Ijim: Laikom, forest, 2000m, fl., Apr. 1931, <u>Maitland</u> 1665; Ijim mountain forest, 1950m, 6.1116N, 10.2543E, fr., 21 Nov. 1996, <u>Pollard</u> 66; Inside crater of Lake Oku, 2300m, fr., 25 Nov. 1996, <u>Cheek</u> 8765; Laikom, 2000m, 6.17N, 10.20E, fr., 3 Dec. 1998, <u>Etuge</u> 4532.

Coffea liberica Bull. ex Hiern var. *nov* ?

Evergreen shrub to 3 m; leaves broadly elliptic, 7–13 × 3.5–6(–10) cm; fruit red when ripe, ellpisoid, c. 14–18 × 8–10 mm. Forest.

Ijim: Laikom, forest, 2000m, fr., May 1931, <u>Maitland</u> 1587; Main path from Ijim to Tum, 2100m, fr., 25 Nov. 1996, <u>Etuge</u> 3614; <u>Etuge</u> 3682 (No label); Laikom, forest slope above Fulani village, 2000m, 6.17N, 10.20E, fr., 3 Dec. 1998, <u>Etuge</u> 4520; Laikom, medicinal forest below basaltic grassland plateau, 1900m, 6.1630N, 10.1951E, fr., 12 Dec. 1998, <u>Pollard</u> 380.

Cremaspora triflora (Thonn.) K.Schum. subsp. *triflora*

Shrub up to 4 m; leaves broadly acuminate, 3–10 × 1.8–5 cm; stipules triangular-lanceolate; flowers fragrant, white, 15 mm long, in dense axillary clusters; fruits scarlet, about 1 cm long. Forest edge.

Ijim: Main path from Ijim to Tum, 2100m, fr., 25 Nov. 1996, <u>Etuge</u> 3612.

Cuviera longiflora Hiern

Local name: Kitilu (Oku) – <u>Munyenyembe</u> 892.

Tree to 10 m, spiny; leaves oblong to elliptic, 15–27 × 5–10

cm; bracts pale yellowish; flowers green; fruit fleshy, 3 × 2 cm, with several stones. Forest.
Kilum: Lumetu traditional forest edge, 2300m, fr., 6 Nov. 1996, Munyenyembe 892.
Ijim: Laikom, forest edge, 2000m, fl., May 1931, Maitland 1361; Gikwang road towards Oku, 2260m, fr., 15 Nov. 1996, Etuge 3410; Laikom, 6.17N, 10.20E, 2000m, fr., 3 Dec. 1998, Etuge 4538.

Galium simense Fresen.
Local name: Imbane (Oku) – Munyenyembe 721.
Scandent herb; stems to 2 m long, adhering to clothing; leaves linear-oblanceolate, 10–40 × 2–4 mm; flowers yellowish-green or white, solitary, axillary; fruits black, about 4 mm diameter. Grassland and forest edge.
Kilum: Elak, 2100m, fl., 8 Jun. 1996, Zapfack 763; Elak, KD, fl., 9 Jun. 1996, Etuge 2211; Elak, KA, 2760m, fl., 10 Jun. 1996, Asonganyi 1310; Elak, KD, 2800m, fl., 10 Jun. 1996, Etuge 2283; Elak, 2400m, fl., 28 Oct. 1996, Munyenyembe 721; Path from Project HQ to KA, 6.1349N, 10.3112E, 2800m, 30 Oct. 1996, Cheek 8512; Summit, 2800m, fl., 31 Oct. 1996, Zapfack 1129.
Ijim: Gikwang towards Nyasosso, 2280m, fr., 18 Nov. 1996, Etuge 3417; Abandoned farms near Laikom village, 1950m, 6.16N, 10.20E, fr., 8 Dec. 1998, Etuge 4575; Ngengal swamp, 10 km E of Laikom, en route to Mbesa, 2400m, 6.1615N, 10.2435E, fl., 9 Dec. 1998, Pollard 346.

Ixora foliosa Hiern
Local name: Feifyang (Oku) – Munyenyembe 764; Feguoh (Oku) – Cheek 8598; Fenkuyikush (Oku) – Munyenyembe 885.
Tree to 10 m; leaves obovate or elliptic, 7–14 × 3–6 cm; stipules awned; flowers white, tinged pink; corolla lobes 4; fruits red, c. 9 mm wide. Forest and woodland.
Kilum: Forest around Lake Oku, NE side, 2230m, 6.13N, 10.28E, fl., 17 Feb. 1958, Thomas 4376; Forest, 2400m, fl., Mar. 1987, Thomas 7146; Elak, 2000m, 8 Jun. 1996, Zapfack 761; Elak, 2100m, 8 Jun. 1996, Zapfack 788; KD, 2800m, 9 Jun. 1996, Etuge 2220; Shore of Lake Oku, 2300m, 12 Jun. 1996, Cable 3112 & 3118; Elak, 2600m, 29 Oct. 1996, Munyenyembe 764; KC, 2300m, 30 Oct. 1996, Buzgo 649; KD-4, 2290m, 1 Nov. 1996, Pollard 32; Lake Oku, below
Baptist Rest House, 2200m, 6.1214N, 10.2733E, 5 Nov. 1996, Cheek 8598; Shambai, 2500m, 5 Nov. 1996, Munyenyembe 885.
Ijim: Laikom, forest, 2000m, fl., Apr. 1931, Maitland 1380; Gikwang road towards Nyasosso, 2280m, 18 Nov. 1996, Etuge 3420; Mountain forest, near Chufekhe stream, 2010m, 6.1124N, 10.2525E, 20 Nov. 1996, Kamundi 637; Along the top of the ridge, second gate from Aboh village, 2370m, 6.11N,10.26E, 21 Nov. 1996, Kamundi 666; Laikom, 2100m, 6.15N, 10.26E, 21 Nov. 1996, Onana 566; Inside crater of Lake Oku, 6.13N, 10.27E, 25 Nov. 1996, Cheek 8763; Mbalabo-Hill, 2100m, 26 Nov. 1996, Etuge 3634.

Keetia venosa (Oliv.) Bridson
Local name: Eshie (Oku) – Cheek 8609.
Woody climber to at least 4 m; leaves elliptic, c. 9 × 5 cm with hairy domatia; petiole 1 cm; stipules narrowly

triangular to aristate, 1 cm long; fruit orange, fleshy with 2 stones, 7 mm across, in dense clusters. Forest.
Syn. *Canthium venosum* (Oliv.) Hiern, F.W.T.A. 2: 184 (1963).
Kilum: Lumeto Forest, near Oku-Elak, 2200m, st., 7 Nov. 1996, Cheek 8609;
Ijim: Sappel, below Ardo's uncle's compound, 2300m, 6.16N, 10.24E, fr., 5 Dec. 1998, Cheek 9823.

Mussaenda arcuata Lam. ex Poir.
Shrub or climber to 2 m or more; leaves opposite, elliptic, c. 10 × 5 cm; inflorescences terminal, paniculate; flowers pale yellow with centre orange, turning red, 2 cm wide. Savanna.
Babungo-Mbi Crater: 1600m, 8 Nov. 1999, Cheek 9993.

Mussaenda erythrophylla Schum. & Thonn.
Climbing shrub to 15 m; leaves broadly ovate to elliptic, 7–15 × 5–11 cm; stipules bilobed; flowers cream, yellow or orange; corolla 1.5 cm diam., tube 2 cm long; one calyx-lobe enlarged, leaf-like, pink or red; fruit fleshy with many small seeds. Forest and beside streams.
Kilum: Lake Oku, 2200m, fl., 12 Jun. 1996, Etuge 2345.

Oldenlandia goreensis (DC.) Summerh. var. *goreensis*
Prostrate herb up to 45 cm long; leaves elliptic or ovate-elliptic, 1.5–2.5 × 0.5–1.3 cm; flowers white with mauve corolla-lobes or pink. Moist places.
Ijim: Afua swamp, 1950m, 6.08N, 10.24E, fl., 7 Dec. 1998, Cheek 9837.

Oldenlandia herbacea (L.) Roxb. var. *herbacea*
Local name: Igaigai (Oku) Munyenyembe 844.
Herb up to 30 cm; leaves narrowly linear 1.5–4 × 0.1–0.2 cm; flowers axillary; corolla 1–2mm long. Weed.
Kilum: Elak-Oku, KA path, 6.15N, 10.26E, fl., 2600–2800m, 30 Oct. 1996, Onana 475; Elak, 2200m, fl., 1 Nov. 1996, Munyenyembe 844; Ewook Etele, Oku-Mbae, 2350m, fl. fr., 6 Nov. 1996, Etuge 3342.

Oldenlandia lancifolia (Schumach.) DC.
Herb 30–60 cm; leaves narrowly lanceolate, 2–6 × 0.2–0.7 cm; flowers white, sometimes pale pink or mauve. Moist places.
Mbi Crater: 1950m, 6.0536N, 10.21E, fr., 9 Dec. 1998, Cheek 9900.

Oldenlandia rosulata K.Schum. var. *rosulata*
Annual erect herb 2–4 cm; leaves in basal rosette, c. 7 mm long; partial peduncles often 2-flowered; flowers white, c. 2 mm diameter. Grassland, between tussocks.
Ijim: Aboh, village, 2300m, 6.15N, 10.26E, fl. fr., 18 Nov. 1996, Onana 515; Gikwang road, Aboh to Akeh, about 1 km after first track, 2400m, 6.15N, 10.26E, fl. fr., 25 Nov. 1996, Cheek 8747; Laikom, Ijim plateau above Fon's Palace, 1900m, 6.16N, 10.19E, fl., 3 Dec. 1998, Cheek 9735.

Otomeria cameronica (Bremek) Hepper
Herb; stems 0.3–1 m long, usually prostrate; leaves ovate-elliptic or ovate-lanceolate, 2–8 × 0.4–4 cm; flowers white;

corolla-tube 3–5 mm long; fruits ovoid. Grassland.
Ijim: Aboh Village, 6.15N, 10.26E, fl., 18 Nov. 1996,
Onana 531; Aboh Village, Gikwang road towards Nyasosso
forest, 2080m, fl., 19 Nov. 1996, Etuge 3466; Aboh village,
1.5 hours walk along Gikwang road to Lake Oku, track
towards TA, fr., 2400m, 6.1112N, 10.2529E, 20 Nov. 1996,
Cheek 8696; Laikom, above Fon's palace, about 30 mins.
walk on path to Fulani settlement, 6.1642N 10.1952E,
2050m, fl., 21 Nov. 1996, Cheek 8704; Laikom, Ijim plateau
above Fon's Palace, 1900m, 6.16N, 10.19E, fl., 3 Dec. 1998,
Cheek 9743.

Oxyanthus okuensis Cheek & Sonké ined.
RED DATA LISTED
Shrub or small tree 3–8 m; leaf-blade elliptic or elliptic
oblong, 9–13 × 2.5–6 cm; petiole 5–8mm long; stipule
oblong, apical third triangular, 12–19 x 3–9 mm;
inflorescences in only one axil, at alternate sides on
successive nodes, erect, 20–50-flowered; flowers white,
corolla tube 3.3 cm long. Forest.
Kilum: Forest around Lake Oku, NE side, 2230m, 6.13N,
10.28E, fr., 17 Feb. 1985, Thomas 4377; Lake Oku, 2300m,
fl., 11 Jun. 1996, Zapfack 858; Lake Oku, near Baptist Rest
House, 2300m, fr., 4 Nov. 1996, Cheek 8583.
Ijim: Etuge 3664.

Pauridiantha paucinervis (Hiern) Bremek. subsp.
paucinervis
Local name: Feban (Oku) – Munyenyembe 868.
Shrub 3–6 m; leaves lanceolate to oblong-elliptic, 7–11 × 2–
3 cm; stipules lanceolate; flowers axillary, white; fruits red.
Forest.
Kilum: Upkim Traditional Forest, 2060m, fr., 4 Nov. 1996,
Munyenyembe 868; Ethiale, near the stream, 2000m, fr., 4
Nov. 1996, Zapfack 1173.

Pavetta hookeriana Hiern var. *hookeriana*
Shrub to 3m; leaves oblanceolate to obovate, 7–12 × 2.5–5
cm with scattered bacterial nodules; stipules awned; flowers
white; corolla 8–9 mm long; fruit black shiny, 2-seeded.
Forest.
Kilum: Ridge above Lake Oku, margin of forest, 2600m, fl.
fr., 7 Jan. 1951, Keay FHI 28514; Evergreen forest, on road
to Oku from the 'Ring Road', 2000m, fl., 21 Jun. 1962, Brunt
607; Slope of Oku village, 2800m, fr., 14 Jul. 1967,
Letouzey 8937; Mount Oku, forest, fl., 1 Feb. 1970, C.
N.A.D. 1805; Achatabaw slope to Lake Oku, 1700m to
2230m, fl. fr., 5 Dec. 1974, Letouzey 13448; Junction of KJ
and KN above Elak, 2550m, 10 Jun. 1996, Cable 3018; KJ,
2620m, 29 Oct. 1996, Zapfack 1090 & 1092; Elak, 2800m,
30 Oct. 1996, Satabie 1058.
Ijim: Aboh, Ijim mountain forest, 2400m, 6.1116N,
10.2543E,fl., 20 Nov. 1996, Pollard 62; Near Aboh-Anyajua,
2500m, fl. fr., 20 Nov. 1996, Satabie 1079; Top of the ridge,
second gate from Aboh village, 2370m, 6.11N, 10.26E, 21
Nov. 1996, Kamundi 664; Laikom, 2100m, fr., 21 Nov.
1996, Onana 575; Laikom, Ijim ridge, about 30 mins. walk
above Fon of Kom's Palace, 2050m, 6.16N, 10.19E, fr., 2
Dec. 1998, Cheek 9771; Laikom, near Akwamofu sacred
forest, 1800m, 6.1633N, 10.1940E, fr., 5 Dec. 1998, Pollard
323.

Plot voucher: Forest above Oku-Elak, Jun. 1996, Oku 30.

Pentas ledermannii Krause
RED DATA CANDIDATE
Shrub 1.5 m, red hairy; leaves lanceolate-ellpitic, 4–6 × 1–2
cm; flowers white; corolla-tube c. 5 mm long. Forest edge.
Syn. *Pentas pubiflora* S.Moore subsp. *bamendensis* Verdc.,
F.W.T.A. 2: 216 (1963).
Kilum: Forest, scrublands, grasslands around Lake Oku and
Mount Oku, 2200–3000m, 6.11N, 10.28E, Thomas &
McLeod H.L. 6012; Oku, cattle-trail, 2330m, 2660m, fl., 3
Sep. 1952, Savory UCI 464.
Ijim: Laikom, in grassland on plateau, 2000m, fl., Apr.
1931, Maitland 1422; Laikom, edge of forest, fl., May 1931,
Maitland 1743.

Pentas pubiflora S.Moore subsp. *pubiflora*
Herb or subshrub 1.5 m; leaves narrowly elliptic, c. 7 × 2
cm; petiole 1 cm; flowers white; corolla tube c. 7 mm long.
Forest edge.
Kilum: Elak, 2200m, fl., 8 Jun. 1996, Zapfack 768; KJ,
above water tank from Elak, 2 km into forest, 2400m, fl., 10
Jun. 1996, Cable 3012; Shore of Lake Oku, 2250m, fl., 12
Jun. 1996, Cable 3137; Elak, between Fon's Palace and
Manchok, 2000m, 6.1447N, 10.3058E, fl., 31 Oct. 1996,
Cheek 8543; Elak, 2200m, fl., 1 Nov. 1996, Munyenyembe
851; Emoghwo traditional forest, 1900m, fl., 3 Nov. 1996,
Zapfack 1161.
Ijim: Zitum road, fl., 21 Nov. 1996, Etuge 3544; Ijim
Mountain Forest, montane forest along the ridge towards a
point where Lake Oku can be veiwed, 2500m, fl., 22 Nov.
1996, Kamundi 686.

Pentas purpurea Oliv. subsp. *purpurea*
Herb, 30–60 cm; leaves lanceolate, ovate-lanceolate or
oblong-lanceolate, 4.5–11.5 × 1–3.2 cm; inflorescence
densely capitate, 1.3–2.5 cm diam.; flowers mauve.
Grassland.
Ijim: Laikom, 2000m, fl., May 1931, Maitland 1371.
Jakiri: Near 'Journey's End' House, 1660m, fl., 18 Jun. 1962,
Brunt 550.

Pentas schimperiana (A.Rich.) Vatke subsp.
occidentalis (Hook.f.) Verdc.
Herb or shubshrub 1.5 m; leaves ovate-elliptic, 9 × 4 cm,
pubescent below, petiole 1 cm; flowers yellow white, corolla
tube 17 mm long. Forest edge.
Kilum: Elak, 2460m, fl., 9 Jun. 1996, Zapfack 802; Junction
of KJ and KN above Elak, fl., 10 Jun. 1996, Cable 3025;
KD, 2900m, fl., 10 Jun. 1996, Cable 3057; KD, 2800m, fl.,
10 Jun. 1996, Etuge 2284; KA to summit, 2800m, fl., 30
Oct. 1996, Cheek 8530; KD, 2500m, fl., 31 Oct. 1996,
Pollard 25; Road from Rest House E of lake, to N of
mountain range, 2300m, fl., 5 Nov. 1996, Cheek 8590.
Ijim: Aboh village, 2450m, fl., 19 Nov. 1996, Onana 542;
Zitum road, 1900m, fl., 21 Nov. 1996, Etuge 3535; Akeh,
2400m, fl., 25 Nov. 1996, Satabie 1093; Laikom, grassland
plateau above Fon's Palace, 2000m, 6.1642N, 10.1950E, fl.,
2 Dec. 1998, Pollard 265.

Psychotria peduncularis (Salisb.) Steyerm. var. *hypsophila* (K.Schum. & K.Krause) Verdc.
Local name: Fubom (Bekom) – Kamundi 603.
Shrub c. 1 m; leaves elliptic, 9–18 × 3.5–9 cm; stipules bifid, 20–35 × 10–15 mm; flowers white in a tight terminal head, 3.5 cm diam.; involucral bracts free or deeply lobed. Forest.
Syn. *Cephaelis peduncularis* Salisb. var. *hypsophila* (K.Schum. & K.Krause) Hepper, F.W.T.A. 2: 204 (1963).
Kilum: Elak, 2000m, 8 Jun. 1996, Zapfack 795; Path above water-tower, KJ, about 1 km into forest, 2250m, 9 Jun. 1996, Cable 2972; KD, 2800m, 9 Jun. 1996, Etuge 2219; Shore of Lake Oku, 2300m, 12 Jun. 1996, Cable 3109; KA, 2600m, 27 Oct. 1996, Buzgo 587; KA, 2400m, 27 Oct. 1996, Buzgo 592; Montane forest, 2600m, 27 Oct. 1996, Munyenyembe 705; KJ, 2400m, fr., 29 Oct. 1996, Zapfack 1058; KJ, 2500m, 29 Oct. 1996, Zapfack 1074; KC-9, 2500m, 30 Oct. 1996, Pollard 13.
Ijim: Gikwang road towards Oku, 2260m, 15 Nov. 1996, Etuge 3392; Gikwang towards Nyasosso bush, 2260m, 18 Nov. 1996, Etuge 3454; Ijim mountain forest, 2160m, 6.11N, 10.25E, 19 Nov. 1996, Kamundi 603; Aboh village, 2450m, 6.15N, 10.26E, 20 Nov. 1996, Onana 552; Near Aboh-Anyajua, 2500m, 20 Nov. 1996, Satabie 1081; Aboh, 6.15N, 10.26E, 28 Nov. 1996, Onana 640; Laikom, Akwamofu, 1760m, 6.17N, 10.20E, fr., 5 Dec. 1998, Etuge 4557.
Plot voucher: Forest above Oku-Elak, Jun. 1996, Oku 2; 31; 45.

Psychotria succulenta (Hiern) Petit
Shrub or small tree to 10 m; leaves oblong-elliptic, coriaceous, reticulately veined, 12 × 5 cm; stipules bifid at apex; flowers white; fruits 2-seeded, yellow turning red. Beside upland streams.
Kilum: Elak, secondary forest near the village, 6.15N, 10.26E, fr., 6 Nov. 1996, Onana 507.
Ijim: Main path from Ijim to Tum, 2100m, 6.1116N, 10.2543E, st., 25 Nov. 1996, Etuge 3619.

Psychotria sp. nov. ?
Local name: Inchya (Oku) – Munyenyembe 873.
RED DATA LISTED
Shrub to 3 m; leaves elliptic, c. 12 × 5 cm, shortly acuminate, acute; nerves red below; domatial cavities glabrous c. 0.5 mm diam.; stipules triangular, 3–4 × 3–4 mm; inflorescences in dense terminal panicles; flowers white c. 4 mm diameter. Forest.
Syn. *Psychotria chalconeura* (K.Schum.) Petit var. *chalconeura* sensu F.W.T.A. 2: 215 (1963).
Kilum: Upkim Traditional forest, 2060m, fr., 4 Nov. 1996, Munyenyembe 873.
Ijim: Laikom, 2000m, fl., May 1931, Maitland 1744; Zitum road, 1950m, fr., 21 Nov. 1996, Etuge 3548; Laikom, 2000m, 6.17N, 10.20E, fr., 3 Dec. 1998, Etuge 4518.

Psydrax dunlapii (Hutch. & Dalziel) Bridson
Local name: Mbangaformbese (Oku) – Munyenyembe 727.
Medium-sized tree; branches horizontal; leaves oblong or oblong-elliptic, 15–20 × 6 cm; flowers white, in axillary 'corymbs'; fruit 2-lobed. Forest.

Syn. *Canthium dunlapii* Hutch. & Dalziel, F.W.T.A., 2: 184 (1963).
Kilum: Elak, 2200m, fr., 8 Jun. 1996, Zapfack 766; KD, fr., 9 Jun. 1996, Etuge 2227; Elak, 2400m, fr., 28 Oct. 1996, Munyenyembe 727; Elak, KC, 2300m, fr., 30 Oct. 1996, Zapfack 1094.
Ijim: Aboh, Gikwang, towards Nyasosso, 2280m, fr., 18 Nov. 1996, Etuge 3423.

Psydrax kraussioides (Hiern) Bridson
Woody climber; leaves elliptic, shiny coriaceous, 9–13 × 4–6 cm; stipules with a keeled lobe; flowers yellow-green; fruit 2-lobed (each almost spherical). Beside upland streams.
Syn. *Canthium henriquezianum* sensu Hepper, F.W.T.A., 2: 181 (1963).
Kilum: Path from Mboh village, about 6 km from Elak-Oku, 2200–2500m, fr., 3 Nov. 1996, Onana 497.
Ijim: Main path from Ijim to Tum, 2100m, 6.1116N, 10.2543E, fl., 25 Nov. 1996, Etuge 3613.

Rothmannia urcelliformis (Hiern) Robyns
Shrub or small tree to 9 m or more; leaves elliptic, 8–13 × 2.8–5.5 cm; corolla white, mottled purple within, corolla-tube 4.5–6.5 cm long; corolla lobes 1.5–3 cm long; fruits 3.5–7.5 x 3.3–5.5 cm. Forest.
Ijim: Laikom, Akwamofu, 1760m, 6.17N, 10.20E, fl., 5 Dec. 1998, Etuge 4543; Medicinal forest below basaltic grassland plateau, Laikom, 1900m, 6.1630N, 10.1951E, fl., 12 Dec. 1998, Pollard 382.

Rytigynia neglecta sensu F.W.T.A., non (Hiern) Robyns
Local name: Febhan (Oku) – Cheek 8444; Feban (Oku) – Munyenyembe 728; Fekole (Kom) – Cheek 8640.
Shrub or small tree about 5 m; leaves long acuminate at apex; flowers greenish white; corolla 8–10 mm long, tube about 2.5 mm diameter. Forest.
Kilum: Lake Oku, Keay & Lightbody FHI 28490; Elak, 2100m, fr., 8 Jun. 1996, Zapfack 765; KD, fr., 9 Jun. 1996, Etuge 2251; Elak, 2450m, fr., 9 Jun. 1996, Zapfack 803;Lake Oku, 2300m, fr., 11 Jun. 1996, Zapfack 855; Shore of Lake Oku, 2250m, fr., 12 Jun. 1996, Cable 3135; Oku-Elak to the forest at KA, 2200m, fr., 27 Oct. 1996, Cheek 8444; Elak, 2420m, fr., 28 Oct. 1996, Munyenyembe 728; KJ, 2500m, fr., 29 Oct. 1996, Buzgo 634; Summit, near savanna zone, 2500m, fr., 5 Nov. 1996, Zapfack 1192.
Ijim: Laikom, Maitland 1359; Above Aboh village, 1.5 hours walk along Gikwang road, 2350m, fl., 18 Nov. 1996, Cheek 8640; Gikwang towards Nyasosso, 2250m, fl., 18 Nov. 1996, Etuge 3422; Aboh, Gikwang towards Nyasosso bush, 2280m, fr., 18 Nov. 1996, Etuge 3457.
Plot voucher: Forest above Oku-Elak, Jun. 1996, Oku 25; 28.

Sacosperma parviflorum (Benth.) G.Taylor
Climbing shrub; leaves ovate-lanceolate, 5–7 × 2–2.5 cm; flowers bluish-purple; corolla-tube 5–6 mm long, lobes 1–2 mm long. Thickets.
Ijim: Belo, by stream, 1330m, fl., Apr. to May 1931, Maitland s.n.

Spermacoce natalensis (Hochst.) K.Schum. ex S.Moore

Herb; leaves linear-lanceolate, 1–4 x 0.3–1 cm; flowers white, tinged mauve; fruits glabrous. Wayside weed.
Kilum: Elak, KA, fl., 11 Jun. 1996, 2200m, Etuge 2313.
Ijim: Laikom, Ijim plateau above Fon's Palace, 1900m, 6.16N, 10.19E, fl., 3 Dec. 1998, Cheek 9725.

Spermacoce princeae K.Schum. var. princeae

Local name: Eptingekeyu (Oku) – Munyenyembe 771.
Herb to about 60 cm; leaves ovate to ovate-lanceolate, 2–5 x 1–2 cm; flowers whitish. Forest.
Kilum: Path above water-tower, KJ, about 1 km into forest, 2200m, 9 Jun.1996, Cable 2952; KD, 2800m, 9 Jun. 1996, Etuge 2228; Lake Oku, 2300m, 11 Jun. 1996, Zapfack 866; KA, 2600m, 27 Oct. 1996, Buzgo 588; Oku-Elak, lower parts of KA, 2200m, 6.1349N, 10.3112E, 28 Oct. 1996, Cheek 8474; Elak, 2400m, 29 Oct. 1996, Munyenyembe 771.
Ijim: Above Aboh village, 1.5 hours walk along Gikwang road, track towards TA, 2400m, 6.1112N, 10.2529E, 20 Nov. 1996, Cheek 8682; Anyajua Forest, montane rain forest along the ridge and disurbed farmland and abandoned garden on lower slopes, 2200m, 27 Nov. 1996, Kamundi 725; Aboh, 6.15N, 10.26E, 28 Nov. 1996, Onana 641.
Plot voucher: Forest above Oku-Elak, Jun. 1996, Oku 6; 16.

Spermacoce pusilla Wall.

Herb 15–40 cm; leaves linear, 2–4 x 0.3–0.6 cm; flowers whitish; fruits more or less pubescent at apex, 1 mm long. Sandy places.
Kilum: Elak, Mbokengfish, fl., 7 Nov. 1996, Munyenyembe 903.

Spermacoce sphaerostigma (A.Rich.) Vatke

Local name: Kiyung (Oku) – Munyenyembe 902.
Herb c. 30 cm; stem pubescent; leaves linear-elliptic, c. 2.5 × 0.5 cm; inflorescence a head-like cluster; flowers white, long-tubed, 3 mm wide. Grassland.
Kilum: Jikijem, 2200m, fl., 5 Nov. 1996, Pollard 47; Ewook Etele Mbae, 10 km from Oku, 2350m, fr., 6 Nov. 1996, Zapfack 1216; Mbokengfish, 1900m, fl., 7 Nov. 1996, Munyenyembe 902.

Tarenna pavettoides sensu F.W.T.A., non (Harv.) Sim

RED DATA LISTED
Shrub; leaves obovate, 7–16 x 3–6.5 cm; stipules darker in centre; flowers white in terminal corymbs; corolla-tube 4–5 mm, lobes 4–5.5 mm. Forest.
Kilum: Elak, 2600m, fr., 27 Oct. 1996, Munyenyembe 706; KC, 2300m, fr. 30 Oct. 1996, Buzgo 648; KDH, 2300m, fr., 1 Nov. 1996, Zapfack 1148.
Ijim: Aboh, Gikwang, towards Nyasosso, 2280m, fr., 18 Nov. 1996, Etuge 3421; Laikom, 2000m, 6.17N, 10.20E, fr., 3 Dec. 1998, Etuge 4527.

Virectaria major (K.Schum.) Verdc. var. major

Herb to 2 m; leaves lanceolate to ovate-lanceolate, 2.3–9.5 × 0.7–4.5 cm; corolla 15–20 mm long, pink outside, paler or white inside. Grassland.
Ijim: Anyajua forest, 1800m, fl., 27 Nov. 1996, Kamundi 716.

RUTACEAE

Clausena anisata (Willd.) Hook.f. ex Benth.

Shrub or small tree to 7 m; leaves imparipinnate; leaflets 17–32, alternate, obliquely ovate or ovate-lanceolate, odorous, very variable in size, up to 11 × 5 cm; inflorescence a lax narrow panicle about half as long as leaves; flowers cream-white; fruit a drupe, ellipsoid, black, shiny, 9 × 7 mm. Forest.
Kilum: Lake Oku, Keay & Lightbody FHI 28491; Elak, 2200m, fr., 8 Jun. 1996, Zapfack 787; KD, fr., 9 Jun. 1996, Etuge 2239; Shore of Lake Oku, 2250m, fl., 12 Jun. 1996, Cable 3102; Elak, 2400m, 30 Oct. 1996, Munyenyembe 803.
Ijim: Ijim Mountain Forest, along the ridge towards a point where Lake Oku can be viewed, 2500m, fr., 22 Nov. 1996, Kamundi 695.
Bambui: Maitland 1464.

Zanthoxylum rubescens Planch. ex Hook.f.

Tree to 10 m, armed with black thorns; leaves to 45 cm long; leaflets 9–13, oblong-elliptic or ovate elliptic, 5–20 × 2.5–9 cm, upper leaflets larger than lower, with translucent yellowish gland-dots; inflorescence paniculate; flowers greenish or whitish; fruits 5–7 mm long. Forest.
Syn. *Fagara rubescens* (Planch. ex Hook.f.) Engl., F.W.T.A. 1: 685 (1958).
Kilum: Lumeto Forest near Oku-Elak, 2200m, 7 Nov. 1996, Cheek 8611; Main road from Ijikijem to Oku, 2350m, fr., 7 Nov. 1996, Etuge 3359.

SALICACEAE

Salix ledermannii Seemen

Local name: Efumejes (Oku) – Cheek 8582.
Tree or shrub to 8 m, leaves oblong-lanceolate, c. 6 cm long, obscurely veined when mature; male catkins c. 4 cm long; capsules broadly ovoid, 4–5 mm long. Near streams and rivers.
Kilum: Lake Oku, near Baptist Rest House, 6.1214N 10.2733E, 2300m, 4 Nov. 1996, Cheek 8582; Lake Oku, 2020m, fr., 5 Nov 1996, Munyenyembe 887.
Ijim: Laikom, Maitland 1495; Fundong, Johnstone 9; 224 & 225; Fundong, near touristic house near the waterfall, 6.15N, 10.26E, fl., 22 Nov. 1996, Onana 586; Afua swamp, 1950m, 6.08N, 10.24E, fr., 7 Dec. 1998, Cheek 9850.
Kumbo: Banso, Keay FHI 28453.

SAPINDACEAE

Allophylus bullatus Radlk.

Local name: Ingwey (Oku) – Munyenyembe 746.
Tree 10–17 m; leaves 3-foliolate; leaflets bullate, central ones obovate, 8–19 × 3.7–9.5 cm; inflorescence erect, terminal, in upper leaf-axils, with several well-developed lateral branches, mostly longer than leaves. Forest.
Kilum: Lake Oku, Keay FHI 28487; Forest above Oku-Elak,

2300m, fl., 8 Jun. 1996, Zapfack 745; KJ, 2400m, fl., 28
Oct. 1996, Buzgo 611; Oku-Elak, 2200m, fr., 28 Oct. 1996,
Cheek 8471; Elak, 2400m, fr., 28 Oct. 1996, Munyenyembe
746; Elak, KJ, 2620m, fl., 29 Oct. 1996, Zapfack 1089;
KDH, 2300m, fr., 1 Nov. 1996, Zapfack 1149.
Ijim: Gikwang road towards Oku, 2270m, fr., 15 Nov. 1996,
Etuge 3401; Laikom, forest above Fulani village, 2000m,
6.17N, 10.20E, 3 Dec. 1998, Etuge 4515.
Plot voucher: Forest above Oku-Elak, Jun. 1996, Oku 34.

Deinbollia cf. pinnata Schum. & Thonn.
Erect shrub or small tree 1.2–5m; leaflets elliptic to oblong,
5–18 × 2.3–7.5 cm; inflorescence racemes or panicles to 60
cm, erect towards top of shoots; flowers creamy-white, 5–7
mm long; fruits orange, subglobose, c. 1.5 mm diameter.
Forest. Unusual in being sparingly hirsute, nerves up to 16
each side. Leaflets 9 pairs.
Ijim: Above Fon's palace, Laikom, about 30 mins. on path to
Fulani settlement, 2050m, st., 21 Nov. 1996, Cheek 8709;
From Aboh, path towards Tum, 2200m, 6.1116N, 10.2543E,
fl., 22 Nov. 1996, Etuge 3600.

Paullinia pinnata L.
Woody or subwoody climber with tendrils; branchlets softly
pubescent, ribbed; leaves imparipinnate, 5-foliolate, rhachis
winged, leaflets oblong to obovate; flowers small, white, in
racemes; fruit scarlet, capsular, 3-angled, c. 3mm long.
Forest.
Jakiri: Tan or Ntan, Al Hadji Gey's land, 1600m, fl., 6 Nov.
1998, Maisels 173.

SCROPHULARIACEAE
Det. E.Fischer (KOBLENZ), B.J.Pollard (K)
& J-P.Ghogue (YA)

Alectra sessiliflora (Vahl) Kuntze var. *monticola* (Engl.) Melch.
Erect scabrid herb 30–90 cm, slightly hispid or glabrous;
leaves petiolate, lanceolate to elliptic, serrate, 10–50 × 10–
20 mm; flowers pale yellow, seed in transparent envelope c.
1 mm long. Grassland.
Bambui: Bambui Agric. Dept. Farm., near office, 1730m, fl.
fr., 13 Nov. 1962, Brunt 889.

Alectra sessiliflora (Vahl) Kuntze var. *senegalensis* (Benth.) Hepper
Local name: Unknown (Ernest Kiming) – Cheek 8427.
Slender, erect, roughly pilose herb c. 30 cm; leaves sessile,
ovate, serrate, 10–50 × 10–20 mm; flowers yellow, seed in
transparent envelope c. 1 mm long, brown. Grassland.
Kilum: Kumbo to Oku, Hepper 2005; Oku-Elak to the forest
at KA, 2200m, fl., 27 Oct. 1996, Cheek 8427; Ewook Etele
Mbae, 10 km from Oku, 2350m, fl., 6 Nov. 1996, Zapfack
1222.
Ijim: Gikwang road to Nyasosso forest,, 2090m, fl., 19 Nov.
1996, Etuge 3470; Ijim Mountain Forest, near Chufekhe
stream, 2010m, fl., 20 Nov. 1996, Kamundi 661; Isebu
Mountain Forest, 1700m, fl., 26 Nov. 1996, Kamundi 696;
Ijim plateau above Fon's Palace, Laikom, 1900m, 6.16N,
10.19E, fl. fr., 3 Dec. 1998, Cheek 9757.

Buchnera leptostachya Benth.
Local name: Fu Fukafuk (Kom at Laikom – Yama Peter) –
Gosline 203.
Erect slender-branched herb, almost glabrous; stem 30–60
cm; leaves in a basal rosette; leaf-veins sometimes tinged
purple beheath; calyx glabrous with ciliate teeth; flowers
mauve, occasionally white, 5 mm. Roadsides and moist
places.
Kilum: Kumbo to Oku, Hepper 2015.
Ijim: Afua swamp, in volcanic caldera, 1950m, 6.09N,
10.24E, fl. fr., 7 Dec. 1998, Etuge 4562; In bog below Ardo's
compound, in grass beside swamp, 2100m, 6.16N, 10.21E,
fl., 7 Dec. 1998, Gosline 203.

Craterostigma sp. nov. ?
RED DATA CANDIDATE
Epilithic herb, perhaps a 'resurrection plant' to 1–3 cm;
leaves 1–2 cm × 0.5–1 cm; flowers solitary; upper 'lip' with
one small hooded petal, mauve; lower 'lip' with one central
and two lateral petals, white with mauve margins.
Rockfaces.
Ijim: Seasonally wet N-facing rockface in 'gallery forest'
along path from Fon's Palace, Laikom, to Ijim ridge
grassland plateau above, 1960m, 6.1642N, 10.1950E, fl. fr.,
2 Dec. 1998, Pollard 273.

Cycnium adonense (E.Mey.) ex Benth. subsp. *camporum* (Engl.) O.J.Hansen
Erect herb 15–30 cm, stems subscabrid; leaves opposite,
sessile; elliptic, c. 4–5 × 2 cm, flowers single in upper axils,
4 cm long, 3 cm wide, mauve. Savanna.
Syn. *Cycnium camporum* Engl., F.W.T.A. 2: 373 (1963).
Syn. *Cycnium petunioides* Hutch., F.W.T.A. 2: 374 (1963).
Ijim: Laikom, in grassland on plateau, fl. fr., Apr. 1931,
Maitland 1733.
Kumbo: Fuel plantations by the side of extraction road, fl.,
25 Apr. 1951, Ejiofor FHI 30080.
Ndop: Near Ndop on Kumbo road, grassland, 1330m, fl., 4
Jan. 1951, Keay FHI 28427.

Hedbergia abyssinica (Hochst. ex Benth.) Molau
Local name: Ngei (Oku) – Munyenyembe 759.
Erect herb from underground rootstock, 10–45 cm; stems
several; leaves opposite, sessile, ovate-lanceolate, 1–3 × 0.5–
1 cm, serrate, pubescent; flowers solitary, from upper third
of stem, to c. 2 cm long, 1 cm diam., pale pink to deep red.
Grassland.
Syn. *Bartsia mannii* Hemsl. , F.W.T.A. 2: 367 (1963).
Syn. *Bartsia petitiana* (A.Rich.) Hemsl. , F.W.T.A. 2: 367
(1963).
Kilum: Kumbo to Oku, fl. fr., 15 Feb. 1958, Hepper 2843;
Summit, 2900m, fl., 29 Oct. 1996, Munyenyembe 759;
Epiphyte on a tree trunk, KA, top plateau of the mountain,
2800m, fl., 31 Oct. 1996, Buzgo 669.
Ijim: Above Aboh village, 1.5 hours walk along Gikwang
road, 2000m, fr., 18 Nov. 1996, Cheek 8657; Gikwang road
towards Nyasosso forest, Ijim Mountain Forest Reserve,
2300m, fl. fr., 19 Nov. 1996, Etuge 3488; Chuhuku River
near Anyajua Forest, 1600m, fr., 28 Nov. 1996, Kamundi
731; Steep grassland slopes near first waterfall from Ijim

ridge grassland plateau, Laikom, 1850m, 6.17N. 10.20E, fl. fr., 10 Dec. 1998, Pollard 364.

Lindernia abyssinica Engl.
Small perennial, erect or ascending, 5–10 cm; leaves elliptic-oblong, 3–5 × 7–13 mm; calyx-lobes lanceolate, flowers blue; fruits c. 1 cm, shortly beaked. Grassland.
Ijim: Laikom, 2000m, fl., May 1931, Maitland 1790.

Lindernia rotundifolia (L.) Alston
Creeping herb; stems rooting at the nodes; leaves suborbicular, 6–12 mm; flowers few, white or pale mauve, 5 mm long; fruits c. 5 mm long. Boggy ground.
Syn. *Ilysanthes rotundifolia* (L.) Benth., F.W.T.A. 2: 365 (1963).
Mbi Crater: Ndawara area, swamp, 1760m, 6.06N, 10.23E, fl., 9 Dec. 1998, Etuge 4578.
Jakiri: Muddy and rocky ground with water trickling over, and odd tussocks of grass, above farm, 1780m, fl. fr., 13 Feb. 1958, Hepper 1956.

Rhabdotosperma densifolia (Hook.f.) Hartl
Local name: Anf saaghi (Kom at Laikom) – Etuge 4510.
Robust erect herb to 90 cm; stems pithy, woody at base, tomentose; leaves tomentose beneath, lanceolate, closely serrate, 2–7 × 0.7–2 cm; inflorescence a terminal raceme; flowers yellow, c. 2 cm diam.; fruits 6–8 mm long. Montane grassland and thickets.
Syn. *Celsia densifolia* Hook.f., F.W.T.A. 2: 355 (1963).
Kilum: Kumbo to Oku, dry hillside 2 miles along Oku road, 1850m, fl., 15 Feb. 1958, Hepper 1996; KJ to KD, 3000m, fl., 13 Jun. 1996, Etuge 2349; Between Mankok, BirdLife Project HQ and Lumatu Traditinal Forest, 2100m, fl., 1 Nov. 1996, Buzgo 693; Summit, near savanna zone, 2480m, fl., 5 Nov. 1996, Zapfack 1194.
Ijim: Turbo path, 1 km after its splitting from the main road at the border of the Ijim Mountain Forest Reserve, N from there, 1800–2100m, fl., 18 Nov. 1996, Buzgo 767; Gikwang towards Nyasosso bush, 2270m, fl., 18 Nov. 1996, Etuge 3433; Grassland and gallery forest on volcanic grassland plateau above Laikom, 2000m, 6.17N, 10.20E, fl. fr., 2 Dec. 1998, Etuge 4510.

Rhabdotosperma ledermannii (Murb.) Hartl
Local name: Lenuwan (Oku) – Munyenyembe 739.
Stout erect herb 1.2–1.6 m; stems sparsely tomentose, woody at base; leaves oblong-lanceolate, sparsely tomentose on nerves beneath, up to 16 cm long; flowers yellow, c. 2.5 cm diam.; fruits 8–10 mm long. Grassland.
Syn. *Celsia ledermannii* Schltr. ex Murb., F.W.T.A. 2: 355 (1963).
Kilum: Elak, 2200m, fl., 28 Oct. 1996, Munyenyembe 739.
Ijim: Laikom, in grassland on plateau, 2000m, fl. fr., Jun. 1931, Maitland 1547.
Bambili: Bambili lakes, side of S crater, 2500m, short grassland, Bauer 109.

Sibthorpia europaea L.
Prostrate herb; stems thinly pubescent, rooting at nodes; leaves reniform-orbicular, crenate, pilose, 1.5–2 cm diam.; flowers purple. Forest edges.

Kilum: Kumbo to Oku, 2 miles along Oku road, 1850m, fl., 15 Feb. 1958, Hepper 2009.

Sopubia mannii Skan var. *mannii*
Erect undershrub, c. 30 cm; rootstock woody; stems branched, internodes short; leaves densely congested on stem; flowers purplish or magenta, numerous, often densely arranged. Grassland.
Kilum: Lake Oku, Lightbody FHI 26255; Elak, 2200m, fl., 31 Oct. 1996, Munyenyembe 821.
Ijim: Zitum road, 1950m, fl., 21 Nov. 1996, Etuge 3528; Inside crater of Lake Oku, 2300m, fl., 27 Nov. 1996, Cheek 8762; Grassland plateau above Fon's Palace, Laikom, 2020m, 6.1642N, 10.1950E, fl., 3 Dec. 1998, Pollard 286.

Veronica abyssinica Fresen.
Prostrate creeping herb; stem branched from the base, pilose; leaves ovate, serrate except towards base, 2–4 × 1–2 cm; inflorescence a slender axillary peduncle; flowers blue or pinkish, paired or a few together, 8–10 mm diam.; fruit bilobed, pubescent, 3–4 mm long. Grassland and scrub beside small streams.
Kilum: Elak, 2900m, fl. fr., 29 Oct. 1996, Munyenyembe 798.
Ijim: Gikwang towards Nyasosso bush, 2280m, fl., 18 Nov. 1996, Etuge 3450; Anyajua Forest, 2200m, fl., 27 Nov. 1996, Kamundi 724.

Veronica mannii Hook.f.
RED DATA LISTED
Erect or ascending perennial herb 15–45 cm; stems with 2 lines of hairs, internodes short; leaves sessile, lanceolate, 1–4 × 0.2–1.2 cm; inflorescence a dense terminal spicate raceme, densely glandular-pubescent; flowers blue, 10–14 mm diam.; fruits retuse, pubescent. Grassland.
Kilum: Cliff-face, 2800m, fl., 31 Oct. 1996, Munyenyembe 820; Summit, 2800m, fl., 31 Oct. 1996, Zapfack 1133.

SIMAROUBACEAE

Brucea antidysenterica J.F.Mill.
Shrub or tree to 10 m; leaves imparipinnate, 10–35 cm long; leaflets 4–5, oblong-ovate to ovate-lanceolate, rusty tomentose beneath, 4.5–14 × 2–7 cm, margins undulate; inflorescence an elongated panicle to 35 cm long; flowers clustered, subsessile, green. Fruits ellipsoid, brown when ripe. Montane forest.
Kilum: Elak, KD, 3000m, fr., 10 Jun. 1996, Etuge 2294.
Ijim: Laikom, Maitland 1588; Aboh village, 6.15N, 10.26E, Elak, 2450m, fl., 20 Nov. 1996, Onana 555.

SOLANACEAE
Det. E.Biye (YA) & B.J.Pollard (K)

Brugmansia × *candida* Pers.
Shrub or small tree to 3 m; leaves ovate-acuminate to 24 × 12 cm; flowers white, fragrant, pendulous funnel-shaped, 25–30 cm long; fruit lemon-shaped. Cultivated.
Kilum: KA, 2200m, fl., 11 Jun. 1996, Etuge 2314.

Ijim: Laikom, boundary of Akwamofu sacred forest, 1760m, 6.1633N, 10.1940E, fl., 5 Dec. 1998, Pollard 314.

Cyphomandra betacea (Cav.) Sendtn.

Local name: Garden Egg (English) – Etuge 3603.
Tree c. 4 m; leaves alternate, ovate, to c. 20 × 15 cm, base cordate; petiole c. 10 cm long; inflorescence few-flowered, axillary, pendulous; flowers campanulate, c. 1.5 cm × 0.7 cm wide; petals pink; stamens yellow. Fruit ellipsoid, c. 7 x 4 cm, orange. Cultivated.
Ijim: From Aboh, path towards Tum, 2100m, fr., 22 Nov. 1996, Etuge 3603.

Discopodium penninervium Hochst.

Local name: Ajoua-aku (Kom) – Etuge 4560.
Small tree or shrub 5–7 m; leaves elliptic to oblong-elliptic, mostly glabrous, lateral nerves in 10–12 pairs, 10–25 × 3–10 cm; flowers white or yellowish fading to brown, fasciculate, axillary; corolla cylindrical, lobes reflexed or spreading, c. 8 mm long; berry globose, 6–8 mm diam. Forest.
Kilum: Elak, 2300m, fl., 8 Jun. 1996, Zapfack 749; Path above water-tower, KJ, about 1 km into forest, 2250m, fl., 9 Jun. 1996, Cable 2995; KD, 2800m, fr., 10 Jun. 1996, Etuge 2257; KJ, 2400m, fl., 29 Oct. 1996, Zapfack 1061; Mbijame forest edge, 2000m, fl. fr., 4 Nov. 1996, Munyenyembe 866.
Ijim: Gikwang road towards Nyasosso forest, 2300m, fl. fr., 19 Nov. 1996, Etuge 3475; Ijim Moutain Forest, near Chufekhe stream, 2010m, fl. fr., 20 Nov. 1996, Kamundi 639; Second gate from Aboh village, 2360m, fl., 21 Nov. 1996, Kamundi 668; Laikom, Akwa-mofu sacred forest, 1760m, 6.17N, 10.20E, fl. fr., 5 Dec. 1998, Etuge 4560.
Plot voucher: Forest above Oku-Elak, Jun. 1996, Oku 93.

Nicotiana tabacum L.

Local name: Ndabe (Kom) – Etuge 3443.
Robust annual to 2 m; upper leaves oblong-lanceolate to elliptic, 8–15 × 1.5–6 cm; inflorescence a terminal cyme, flowers viscid-glandular outside, tubular; corolla c. 4 cm long, lobes very acute, white, cream or pinkish. Cultivated.
Ijim: Gikwang road towards Nyasosso bush-forest, 2260m, fl. fr., 18 Nov. 1996, Etuge 3443.

Physalis peruviana L.

Erect perennial to 1 m, densely hairy, from creeping rootstock; leaves rhomboid to deltoid, 8–10 × 6–7.5 cm; flowers yellow with purple centre, 15 mm long. Fallow.
Kilum: Elak, 2100m, fl., 8 Jun. 1996, Zapfack 783; KD, fl., 9 Jun. 1996, Etuge 2209; Between Mankok, BirdLife Project HQ & Lumito Traditional Forest, 2100m, fl., 1 Nov. 1996, Buzgo 692.
Ijim: Gikwang towards Nyasosso forest, 2260m, fl. fr., 18 Nov. 1996, Etuge 3459; Gikwang road, Aboh to Akeh, about 1 km after first track, 2400m, 6.15N, 10.26E, fl., 25 Nov. 1996, Cheek 8751; Laikom to Fundong, 1700m, 6.17N, 10.20E, fl., 10 Dec. 1998, Etuge 4589.

Solanum aculeastrum Dunal var. *albifolium* (C.H.Wright) Bitter

Tree or shrub to 7 m; unarmed; white tomentum on all parts except surface of leaves; leaves to 15 mm long, compressed, base to 5 mm long, sharply recurved; inflorescence lateral, axillary. Cultivated.
Kilum: Secondary forest near Elak, 6.15N, 10.26E, fr., 6 Nov. 1996, Onana 505.
Ijim: Laikom, about villages, 2000m, fl., Apr. 1931, Maitland 1657; Gikwang road towards Nyasosso bush, 2270m, fr., 18 Nov. 1996, Etuge 3451.

Solanum aculeatissimum Jacq.

Undershrub 30–60 cm armed with almost straight spines; leaves distinctly petiolate, pinnately lobed, pubescent; flowers white, subsolitary, c. 1 cm diam.; fruits brownish-yellow. Forest edge.
Ijim: Laikom, about villages and on edge of bush and forest, 2000m, fl., May 1931, Maitland 1745.

Solanum distichum Thonn.

Local name: Kinyaju (Oku) – Munyenyembe 783; Ajuafekghe (Bekom) – Kamundi 629.
Coarse tomentose undershrub to 2 m, spiny or not; leaves elliptic, very shortly pubescent above, subtomentose beneath, up to 10–16 cm long; inflorescence a racemose-like cyme; flowers white, c. 5 mm long; fruits erect, globose, red, 1–1.5 cm diam. Forest edge.
Syn. *Solanum indicum* L. subsp. *distichum* (Thonn.) Bitter var. *distichum* Bitter, F.W.T.A. 2: 333 (1963).
Syn. *Solanum indicum* L. subsp. *distichum* (Thonn.) Bitter var. *grandemunitum* Bitter, F.W.T.A. 2: 334 (1963).
Syn. *Solanum indicum* L. subsp. *distichum* (Thonn.) Bitter var. *modicearmatum* Bitter, F.W.T.A. 2: 334 (1963).
Kumbo: Banso, grassland on the Banso-Bamenda Motor Road, fl. fr., 20 Oct. 1947, Tamajong FHI 23484.
Kilum: Oku evergreen forest, 2000m, fl., 21 Jun. 1962, Brunt 616; Elak, 2200m, fl. fr., 8 Jun. 1996, Zapfack 779; KJ, above water-tower, 2 km into forest from Elak, 2400m, fl. fr., 10 Jun. 1996, Cable 3010; KD, 2800m, fl. fr., 10 Jun. 1996, Etuge 2258; Shore of Lake Oku, Elak, 2250m, fl. fr., 12 Jun. 1996, Cable 3119; KJ, Elak, 2300m, fl. fr., 28 Oct. 1996, Buzgo 603; KJ, 2620m, fl. 29 Oct. 1996, Buzgo 640; Elak, 2800m, fl. fr., 29 Oct. 1996, Munyenyembe 783;
Ijim: Gikwang road towards Oku, 2270m, fl. fr., 15 Nov. 1996, Etuge 3409; Aboh, 6.11N, 10.25E, 2160m, fl. fr., 19 Nov. 1996, Kamundi 629; Laikom, 2000m, 6.17N, 10.20E, fr., 3 Dec. 1998, Etuge 4537.

Solanum nigrum L. *sens. lat.*

Local name: Kre' Fom (Oku) – Cheek 8455; Kifon (Oku) – Munyenyembe 790.
Herb, 30–60 cm; leaves mostly glabrous, shortly ciliate; inflorescence umbellate; flowers small, white, on a common peduncle; berries 0.6–2 cm diameter. Weed, sometimes cultivated.
Kumbo: Hepper 1981.
Kilum: KA, 2200m, fl., 11 Jun. 1996, Etuge 2305; Oku-Elak to the forest at KA, fl. fr., 27 Oct. 1996, Cheek 8455; Elak, 2500m, st., 29 Oct. 1996, Munyenyembe 790.
Ijim: Gikwang road towards Nyasosso, 2250m, fl. fr., 18 Nov. 1996, Etuge 3416; Second gate from Aboh village, 2360m, fl., 21 Nov. 1996, Kamundi 673.

Solanum terminale Forssk.
Climber, leaves alternate, elliptic, c. 12 x 5 cm, acuminate, glabrous, petiole 1–2 cm long; inflorescence terminal, spicate, 10–20 cm long; flowers c. 5 mm long, c. 8mm wide, petals purple, staminal tube yellow. Forest.
Ijim: Path from Aboh towards Tum, 2100m, 6.1116N, 10.2543E, fl., 22 Nov. 1996, Etuge 3576; Laikom, start of path from Fon's Place to grassland plateau, 1950m, 6.1642N, 10.1950E, fl., 4 Dec. 1998, Pollard 308.

STERCULIACEAE

Cola anomala K.Schum.
Local name: Ehbe (Oku – Ernest Kiming) – Cheek 8426; Ebii (Kom at Laikom – Yama Peter) – Etuge 4540.
Tree to 20 m; crown dense; leaves in whorls of 3, elliptic to obovate-elliptic, to 17 × 7.5 cm; flowers yellow without red markings, fruiting carpels green with knobs and ridges, to 12 cm long; seeds 2 per carpel, each with 4–5 cotyledons. Cultivated.
Kilum: Oku-Elak to the forest at KA, 2200m, fl., 27 Oct. 1996, Cheek 8426.
Ijim: Akwamofu medicinal forest, Laikom, 1760m, 6.17N, 10.20E, fl., 5 Dec. 1998, Etuge 4540.
Bambui: Russell FHI 14964; 14984.
Kumbo: Russell FHI 14974.

Dombeya ledermannii Engl.
RED DATA LISTED
Tree 3.5–15 m; leaf-blade suborbicular to ovate, slightly 5-lobed, 11–19 × 5.3–15 cm; cordate, petiole 4–7 cm long; inflorescence an axillary cyme, peduncle 3–7 cm long; petals white, 0.8–1.3 cm long. Forest and woodland.
Ijim: Ajung, 1300m, Apr. 1998, DeMarco in Maisels 113.

Pterygota mildbraedii Engl.
Tree c. 50 m; leaves alternate, ovate, c 15 × 15 cm, base cordate; domatia conspicuous; petiole 10 cm. Forest.
Bambui: 1200m, 11 Aug. 1963, Brunt 1259.

THEACEAE

Ternstroemia polypetala Melch.
RED DATA LISTED
Tree c. 18 m, glabrous; branching terminalia-type; leaves leathery, oblanceolate, 6–7 × 2–2.5 cm, apex rounded or notched; inflorescence of single flowers, 4–5 per stem; pedicels 12 mm long; petals 8, white, c. 7 mm long. Forest.
Ijim: Zitum road, 1900m, fl., 21 Nov. 1996, Etuge 3557.

THYMELAEACEAE

Gnidia glauca (Fresen.) Gilg
Local name: Ding (Oku) – Munyenyembe 767; Ling (Kom at Laikom – Yama Peter) – Pollard 318.
Tree to 15 m; trunk much-branched; leaves oblanceolate, very acute, glabrous, 5–8 cm long; flowerheads numerous, subsessile, c. 5 cm diam.; petals spathulate, surrounded by large ovate glabrascent bracts. Montane open woodland.

Syn. *Lasiosiphon glaucus* Fresen., F.W.T.A. 1: 176 (1954).
Kilum: Summit, 2900m, fl., 10 Jun. 1996, Cable 3063; Summit, 3000m, fl., 29 Oct. 1996, Munyenyembe 767; Lake Oku, 2200m, fl., 4 Nov. 1996, Pollard 36.
Ijim: Ijim Mountain Forest, Lat 6.11N, 10.25E, 2160m, fl., 19 Nov. 1996, Kamundi 606; Laikom, near Akwamofu sacred forest, farmbush, 1900m, 6.1633N, 10.1940E, fl., 5 Dec. 1998, Pollard 318.

TILIACEAE

Triumfetta annua L.
Annual herb up to 90 cm; leaves membranous, ovate to elliptic, up to 9 × 6 cm, serrate, inconspicuously simple-hairy, basal teeth glands absent; flowers yellow, c. 5 mm wide; fruits 1–1.5 cm diam., including bristles; bristles as long as fruit body, hooked at apex. Forest edge.
Ijim: Ijim Mountain Forest, 1800m, fr., 21 Nov. 1996, Pollard 72.

Triumfetta cordifolia A.Rich.
Local name: Ofie (Oku) – Munyenyembe 901.
Shrub to 4–5 m; stem pubescent, hairy, or nearly glabrous; leaves undivided, trilobed, or sub-5-lobed, ovate, cordate, long-petiolate, c. 10 × 10 cm; flowers yellow; fruits globose, dehiscent, 1–1.5 cm diam., including prickles. Forest edge.
Kilum: Lowland forest patches at lower altitude than the BirdLife HQ, 1800m, fl., 6 Nov. 1996, Buzgo 715; Mbokengfish, 1900m, fl., 7 Nov. 1996, Munyenyembe 901.
Ijim: Ijim Mountain Forest, 1800m, fl. fr., 21 Nov. 1996, Pollard 76.
Njinikom: Maitland 1766.

Triumfetta rhomboidea Jacq.
Local name: None – Cheek 8475.
Weedy undershrub, variable in habit and foliage; flowers yellow; fruits ovoid or globose, indehiscent, 4–10 mm diam., including prickles, tomentose. Weed.
Kilum: Oku-Elak, lower parts of KA, 2200m, fl., 28 Oct. 1996, Cheek 8475; Oku-Elak, 2200m, fl., 30 Oct. 1996, Munyenyembe 809.
Ijim: Ijim Mountain Forest, 2000m, fl., 21 Nov. 1996, Pollard 77; Laikom to Fundong, 1700m, 6.17N, 10.20E, fl. fr., 10 Dec. 1998, Etuge 4591; Path from Fon's palace downhill towards Akwamofu scared forest, 1900m, 6.1630N, 10.1951E, fl., 12 Dec. 1998, Pollard 374.

Triumfetta tomentosa Bojer var. *tomentosa*
Undershrub 0.7–1.7 m; tomentose all over; leaves mostly undivided, mostly ovate, 5–12 × 2.5–7 cm; flowers yellow; fruit 1–1.5 cm diam.; prickles of ovary and fruit slender, nearly glabrous. Forest edge.
Njinikom: Njinikom, in grassland, 1660m, fl., Jun. 1931, Maitland 1767.

ULMACEAE

Trema orientalis (L.) Blume
Local name: Fang (Oku) – Munyenyembe 804; Fang (Oku) – Cheek 8608.

Shrub or tree to 13 m; branchlets elongated, pubescent; leaves distichous, ovate, pubescent, 6–12 × 2.5–5 cm; flowers polygamous, very small, glomerate in leaf-axils; drupe subglobose, glabrous, c. 3 mm long. Forest edge.
Kilum: Elak, 2100m, fl. fr., 8 Jun. 1996, Zapfack 762; Path above water-tower, KJ, about 1 km into forest, 2250m, 9 Jun. 1996, Cable 2999; Elak, KD, fl. fr., 9 Jun. 1996, Etuge 2222; Elak, forest edge, 2300m, fl., 30 Oct. 1996, Munyenyembe 804; Lumeto Forest near Oku-Elak, 2200m, 7 Nov. 1996, Cheek 8608.

UMBELLIFERAE

Afrocarum imbricatum (Schinz) Rauschert
First record of genus from Cameroon.
Perennial rhizomatous herb 0.2–1 m; stem hollow, glabrous; leaves mostly basal, pinnate; leaflets 16–32, orbicular to oblong, 5–20 × 3–15 mm, denticulate; petiole 0.5–5 cm long; umbels 5–10 cm diam., peduncles 2–14 cm; rays 5–10, involucre of 4–10 bracts, conspicuous; flowers white. Swamp.
Mbi Crater: 1950m, 6.05N, 10.21E, fl. fr., 9 Dec. 1998, Cheek 9889.

Agrocharis melanantha Hochst.
Local name: Nsah (Oku) – Etuge 2207; Nsah (Oku) – Cheek 8447; Nsa (Oku) – Munyenyembe 757.
Herb c. 30 cm; leaves bipinnately divided into small coarsely serrate leaflets; umbels congested on long, slender, tomentose peduncles; flowers white; fruits c. 5 mm long, with numerous hooked spines. Grassland.
Syn. *Caucalis melanantha* (Hochst.) Hiern, F.W.T.A. 1: 754 (1958).
Kilum: KD, fl., 9 Jun. 1996, Etuge 2207 & fr., Etuge 2253; Elak, 2450m, fr., 9 Jun. 1996, Zapfack 805; KD, 2400m, fr., 10 Jun. 1996, Cable 3050; Oku-Elak to the forest at KA, 2200m, fr., 27 Oct. 1996, Cheek 8447; Montane *Podocarpus* forest, 2800m, fr., 29 Oct. 1996, Munyenyembe 757; KA, top plateau to the mountain, 2800m, fr., 31 Oct. 1996, Buzgo 674.
Ijim: Nyasosso, Gikwang towards Nyasosso, 2290m, fr., 18 Nov. 1996, Etuge 3415; Ijim Mountain Forest, near Chufekhe stream, 2010m, fr., 20 Nov. 1996, Kamundi 646; Laikom, Ijim ridge, about 30 mins. walk above Fon of Kom's Palace, 2050m, 6.16N, 10.19E, fr., 2 Dec. 1998, Cheek 9777.

Centella asiatica (L.) Urb.
Local name: Kesi (Oku) – Munyenyembe 899.
Perennial creeping herb; stems rooting at nodes; leaves reniform, glabrous or sub-glabrous, 3–5 cm diam.; flowers purplish, c. 1 mm diam.; peduncle 5–8mm long, pubescent; fruit widely orbicular, c. 2 mm diameter. Damp grassy places.
Kilum: Oku-Elak, between Fon's Palace and Manchok, bank along road in village, 2000m, fl., 31 Oct. 1996, Cheek 8542; Mbokengfish, 1900m, fr., 7 Nov. 1996, Munyenyembe 899.

Cryptotaenia africana (Hook.f.) Drude
Herb 0.6–1.2 m; lower part of stem and petioles rusty-pilose; radical leaves ternate, leaflets broadly ovate, cauline leaflets linear-lanceolate; umbels with 3–6 primary and 3–5 secondary rays; flowers very small, narrowly ovoid, green, 5 mm long. Forest.
Kilum: KJ, above water tank from Elak, 2 km into forest, 2400m, fl., 10 Jun. 1996, Cable 3006; KJ, 2300m, fl., 28 Oct. 1996, Buzgo 609; KA path, 2300m, fl., 28 Oct. 1996, Onana 444.
Ijim: Path from Aboh towards Tum, 2100m, fl., 22 Nov. 1996, Etuge 3604; Gikwang road, Aboh to Akeh, about 1 km after first track, 2500m, fr., 25 Nov. 1996, Cheek 8745; Laikom, boundary of Akwamofu sacred forest, under waterfall, very wet rocks, 1820m, 6.1633N, 10.1940E, fl. fr., 5 Dec. 1998, Pollard 317; Laikom, forest patch towards foot of waterfall above Akwamofu sacred forest, 1860m, 6.17N, 10.20E, fl. fr., 11 Dec. 1998, Etuge 4608.

Hydrocotyle hirta R.Brown ex A.Rich. sensu Fl. Cameroun, non F.T.E.A.
Creeping herb; leaves suborbicular-reniform, deeply lobed and distinctly pilose-setulose, 5–10 mm diam.; flowers 10–30 in each head, peduncle 1–3 cm long. Damp habitats.
Syn. *Hydrocotyle mannii* Hook.f., F.W.T.A. 1: 753 (1958).
Kumbo: Banso, Tamajong FHI 23460.

Hydrocotyle sibthorpioides Lam.
Creeping herb; leaves suborbicular, shallowly lobed, subglabrous, 5–10 mm diam.; flowers 2–4 in each head, peduncle up to 1 cm long. Damp habitats.
Ijim: Laikom, Maitland 1772; Laikom, swamp E of Ardo's compound, 2000m, 6.16N, 10.22E, st., 5 Dec. 1998, Cheek 9793.

Lefebvrea abyssinica A.Rich.
Perennial herb 0.9–3 m; stem smooth, 5–10 mm diam. at base; lower leaves deltoid in outline, bipinnatisect with 2–3 pairs pinnae, to 50 cm long; lowest pinnae long petiolulate (to c. 7 cm), with 1–2 pairs leaflets; leaflets linear to broadly lanceolate, 4–33 × 0.2–3 cm; umbels c. 7 cm diam.; partial umbels 11–27-flowered; flowers green or purplish green. Forest edge.
Syn. *L. stuhlmannii* Engl., Fl. Cameroun 10: 96 (1970).
Ijim: Aboh, Ijim Mountain Forest Reserve, 2100m, 6.116N, 10.2543E, fl., 21 Nov. 1996, Pollard 67.

Peucedanum angustisectum (Engl.) Norman
RED DATA LISTED
Annual herb to 1.2 m; stems smooth or slightly striate; leaves biternate to bipinnate, up to 25 × 20 cm; pinnae subpalmatisect; pinnules linear-lanceolate, to 8 × 1.5 cm, coarsely serrate; umbels lacking involucre; rays 6–8, 1.5–1.8 cm long, flowers white. Grassland.
Ijim: Laikom, path from Fon's palace to Mbororo settlement in grassland, 1900m, fr., 21 Nov. 1996, Cheek 8698.

Peucedanum cf. angustisectum (Engl.) Norman
As *Peucedanum angustisectum*, but leaflets longer and more sparsely toothed. Grassland.
Ijim: Laikom, near cliff-face dropping 100m at S edge of 'headland', 200m W of top of path from village, 2000m, 6.1642N, 10.1950E, fr., 2 Dec. 1998, Pollard 264.

Peucedanum camerunensis Jacq.-Fél.
RED DATA LISTED
Annual (?) herb 30–40 cm; stems glabrous, smooth; basal leaves 18–20 cm long, 2–3-jugate; pinnae ternate; leaflets sessile, linear, 7–8 × 0.2–0.3 cm; umbels lacking involucre; rays 5–7, 1.5 cm long; umbellules poorly developed. Grassland.
Kilum: Cliff-face, 2800m, fr., 31 Oct. 1996, Munyenyembe 819; Ewook Etele Mbae, 10 km from Oku, 2350m, fr., 6 Nov. 1996, Zapfack 1229.
Ijim: Border of Lake Oku, along the path to the lake, fr., 26 Nov. 1996, Onana 639.

Peucedanum winkleri H.Wolff
Herb; leaves bipinnatisect; leaflets ovate or narrowly ovate, crenate-serrate, to 3 × 1 cm; fruit elliptic, broadly winged, 10 × 6.5 mm. Forest edge.
Kilum: KA to summit, 2800m, fr., 30 Oct. 1996, Cheek 8528.

Pimpinella oreophila Hook.f.
Herb, 5–30 cm; leaves pinnate, ovate-orbicular, crenate-dentate, terminal one c. 1.5 cm diam., lateral ones smaller, sessile; umbels c. 6-rayed; pedicels short and stout; fruit ovoid-oblong, glabrous, 2.5 mm long. Grassland
Kilum: Summit, 2800m, fl., 31 Oct. 1996, Zapfack 1134.

Pimpinella sp. aff. praeventa Norman
RED DATA CANDIDATE
Possibly a new species.
Kilum: Above Mboh, path above village to forest through Wambeng's farm, 2200m, fl., 1 Nov. 1996, Cheek 8561.
Ijim: Turbo path, 1 km after its splitting from the main road at the border of the Ijim Mountain Forest Reserve, N from there, 1800–2100m, fl., 18 Nov. 1996, Buzgo 765; Above Aboh village, 1.5 hours walk along Gikwang road, 2380m, fl., 18 Nov. 1996; Cheek 8630; Gikwang towards Nyasosso, 2280m, fl., 18 Nov. 1996, Etuge 3430; Ijim Mountain Forest, 2100m, fl., 25 Nov. 1996, Pollard 80; Laikom, Ijim ridge, about 30 mins. walk above Fon of Kom's Palace, 2050m, 6.16N, 10.19E, fl. fr., 2 Dec. 1998, Cheek 9769; Afua swamp, grassland, 1950m, 6.09N, 10.24E, fl. fr., 7 Dec. 1998, Etuge 4568.

Sanicula elata D.Don
Herb; radical leaves long-petiolate, palmate with 5 obovate lobes, c. 4 cm long, sharply crenate, glabrous; cauline leaves progressively smaller, bract-like; heads small, few-flowered; bracts lanceolate; fruits densely covered with hooked bristles. Forest.
Kilum: Path above water-tower, KJ, about 1 km into forest, 2250m, fl., 9 Jun. 1996, Cable 2988 & 2991; Junction of KJ and KN, 2550m, fl., 10 Jun. 1996, Cable 3031; KD, 2800m, fl., 10 Jun. 1996, Etuge 2274; Elak, 2400m, fl., 10 Jun. 1996, Zapfack 836; Shore of Lake Oku, 2250m, fl., 12 Jun. 1996, Cable 3131; KJ to KD, 3000m, fl., 13 Jun. 1996, Etuge 2350; KJ, 2500m, fl., 29 Oct. 1996, Pollard 6; KC, 2600m, fl., 30 Oct. 1996, Buzgo 654.
Ijim: Aboh village, 2450m, fl., 20 Nov. 1996, Onana 551.

Torilis arvensis (Huds.) Link
Erect herb to 1 m; leaves pinnate to bipinnate, ultimate segments lanceolate; leaves and inflorescence clothed with appressed strigose bristles; inflorescence of compound umbels; fruit with prominent hooked spines. Forest edge.
Kilum: KA path, 2600–2800m, fl., 30 Oct. 1996, Onana 466; KA, 2600m, fl., 31 Oct. 1996, Buzgo 687.
Ijim: Ijim Mountain Forest, 2250m, fr., 19 Nov. 1996, Pollard 56; Ijim Mountain Forest, along the ridge towards a point where Lake Oku can be viewed, 2600m, fl., 22 Nov. 1996, Kamundi 689.

URTICACEAE
Det. J-M.Onana (YA), M.Wilmot-Dear (K), M.Cheek (K) & B.J.Pollard (K)

Boehmeria macrophylla Hornem.
Local name: Di (Oku) – Munyenyembe 742.
Shrub 1–5 m; branches soft-wooded; leaves opposite, variable from broadly ovate to elliptic-lanceolate, margin coarsely triangular-dentate or serrate with inconspicuous dot-like cystoliths, up to 20 × 18 cm; inflorescences axillary, spicate, to 30 cm long; flowers in separated clusters along rhachis. Forest edge.
Syn. **Boehmeria platyphylla** D.Don, F.W.T.A. 1: 622 (1958).
Kilum: Elak, 2100m, fl., 8 Jun. 1996, 2100m, Zapfack 790; Elak, KD, fl., 9 Jun. 1996, Etuge 2218; Elak, KD, 2500m, fl., 10 Jun. 1996, Cable 3044; Elak, 2400m, fl., 28 Oct. 1996, Munyenyembe 742.
Ijim: Laikom, Maitland 1362.
Ndop: Adams GC 11061.
Plot voucher: Forest above Oku-Elak, Jun. 1996, Oku 47; 77.

Droguetia iners (Forssk.) Schweinf. subsp. iners
Local name: None known – Cheek 8692.
Perennial herb or undershrub to 1.8 m; stems trailing and ascending; leaves opposite, ovate, 3-nerved, with dot-like cystoliths above, sparsely pubescent, 1.8–2.5 cm; inflorescences axillary, androgynous, bowl-shaped or campanulate involucres. Forest.
Ijim: Above Aboh village, 1.5 hours along Gikwang road to Lake Oku, track towards TA, 2400m, 6.1112N, 10.2529E, fl., 20 Nov. 1996, Cheek 8692; Aboh, grassland, montane forest edge, 2400m, 6.1116N, 10.2543E, fl., 20 Nov. 1996, Pollard 63.
Bambui: Adams GC 11218.

Elatostema monticola Hook.f.
Weak herb; stems trailing; leaves obliquely ovate, coarsely dentate with 5–10 teeth per margin, 1.5–3.5 × 0.8–2 cm; bracts small, shortly acuminate; male inflorescence sessile. Forest.
Kilum: Path from Project HQ to KA, Elak, 2800m, 6.1349N, 10.3112E, fl., 30 Oct. 1996, Cheek 8518; Elak, above Mboh, path above village to forest through Wambeng's farm, 2200m, 6.1146N, 10.2856E, fl., 1 Nov. 1996, Cheek 8571.
Ijim: Aboh village, 6.15N, 10.26E, fl., 25 Nov. 1996, Onana 613.

Elatostema paiveanum Wedd.

Perennial herb to 60 cm; stem erect, fleshy; leaves obliquely elongate-elliptic, 6–24 × 2–8.5 cm, numerous conspicuous cystoliths on upper surface, closely serrate, c. 20 teeth on each margin; male inflorescence sessile, flat, subtriangular or lobed, pubescent. Forest.

Ijim: Top of Ijim Mountain ridge, second gate from Aboh village, 2250m, fl., 21 Nov. 1996, <u>Kamundi</u> 679.

Plot voucher: Forest above Oku-Elak, Jun. 1996, <u>Oku</u> 36.

Elatostema welwitschii Engl.

Perennial herb to 50 cm; stem fleshy, erect; leaves obliquely ovate to elliptic, 5–15 × 0.5–3.5 cm; serrate, 25–50 per margin; male inflorescence densely clustered, sessile, 0.5–0.8 cm diameter. Forest.

Kilum: Stream flowing into Lake Oku, 2800m, fl., 5 Nov. 1996, <u>Pollard</u> 40; Inside crater of Lake Oku, 2300m, 6.13N, 10.27E, fl., 25 Nov. 1996, <u>Cheek</u> 8768.

Ijim: Aboh village, 6.15N, 10.26E, fl., 25 Nov. 1996, <u>Onana</u> 619.

Girardinia diversifolia (Link) Friis

Erect herb 0.4–2 m with stiff stinging hairs on stem, leaves and inflorescence; stem subsucculent; leaves alternate, pinnately or subdigitately lobed to below the middle lobes, coarsely serrate, up to 20 × 20 cm; male inflorescence axillary, often branched; female inflorescence short, dense. Fruits warted. Forest.

Syn. **Girardinia condensata** (Hochst. ex Steud.) Wedd., F.W.T.A. 1: 618 (1958).

Kilum: Elak, KJ, 2300m, fl., 28 Oct. 1996, <u>Buzgo</u> 607.

Ijim: Laikom, <u>Maitland</u> 1774; Top of Ijim Mountain ridge, second gate from Aboh village, 2250m, fl., 21 Nov. 1996, <u>Kamundi</u> 681.

Bambui: <u>Adams</u> GC 11227.

Laportea alatipes Hook.f.

Local name: Kinshiem (Oku) – <u>Cheek</u> 8438; Kinshir (Oku) – <u>Munyenyembe</u> 717.

Robust perennial herb to 1 m; stems with sharp reflexed stinging bristles; leaves ovate-elliptic, coarsely dentate-serrate, setose beneath, 10–15 × 5–7 cm; leaf-surface with dot-like cystoliths; male inflorescence axillary; female inflorescence much longer than leaves, subsecund; fruits flat, obliquely ovate, slightly rugose. Forest.

Kilum: Elak, 2300m, fr., 8 Jun. 1996, <u>Zapfack</u> 755; Path above water-tower, KJ, about 1 km into forest, 2250m, fl., 9 Jun. 1996, <u>Cable</u> 2979; Elak, KD, fl., 9 Jun. 1996, <u>Etuge</u> 2231; Shore of Lake Oku, 2250m, fl., 12 Jun. 1996, <u>Cable</u> 3093; Elak to the forest at KA, 2200m, 6.1349N, 10.3112E, fl., 27 Oct. 1996, <u>Cheek</u> 8436 & 8438; Elak, 2400m, fr., 28 Oct. 1996, <u>Munyenyembe</u> 717.

Ijim: Laikom, <u>Maitland</u> 1368; Aboh, 6.11N, 10.25E, fr., 19 Nov. 1996, <u>Kamundi</u> 633.

Plot voucher: Forest above Oku-Elak, Jun. 1996, <u>Oku</u> 7.

Laportea ovalifolia (Schum.) Chew.

Local name: Kenshiey (Oku) – <u>Cheek</u> 8614.

Creeping herb, sometimes half-shrubby, usually with stinging hairs; leaves ovate, dentate, 10 × 8 cm with short linear cystoliths on lower surface; male inflorescence spicate or catkin-like; peduncles long, axillary or arising from underground stem; flowers in dense clusters, greenish-white; female inflorescence smaller, loosely few-flowered; fruits compressed, ovate, slightly rugose, sometimes subterranean, 3–5 mm long. Forest.

Kilum: Shore of Lake Oku, 2250m, fl., 12 Jun. 1996, <u>Cable</u> 3122; Lake Oku, 2200m, fl. 12 Jun. 1996, <u>Etuge</u> 2340; Lumeto forest, 2200m, 6.1412N, 10.2943E, fl., 7 Nov. 1996, <u>Cheek</u> 8614.

Lecanthus peduncularis (Royle) Wedd.

Local name: Mbol (Oku) – <u>Cheek</u> 8432; Kambas (Oku) – <u>Munyenyembe</u> 716.

Weak herb 2.5–25 cm, simple or branched, glabrous; stem densely rooting, creeping underground; leaves ovate-lanceolate, coarsely dentate, numerous linear cystoliths prominent on upper surface, 2–3.5 × 0.8–1.5 cm; inflorescence orbicular, up to 1 cm diam., peduncle slender 2–2.5 cm long. Forest.

Kilum: Elak to the forest at KA, 2200m, 6.1349N, 10.3112E, fl., 27 Oct. 1996, <u>Cheek</u> 8432; Elak, KJ, 2300m, fl., 28 Oct. 1996, <u>Buzgo</u> 598; Elak, 2400m, fl., 28 Oct. 1996, <u>Munyenyembe</u> 716; KJ, Elak, 2500m, fl., 29 Oct. 1996, <u>Zapfack</u> 1080.

Ijim: Zitum road, 1950m, 6.1116N, 10.2543E, fl., 21 Nov. 1996, <u>Etuge</u> 3566; Top of Ijim mountain ridge, second gate from Aboh village, 2380m, fl., 21 Nov. 1996, <u>Kamundi</u> 674; Main path from Ijim to Tum, 2100m, 6.1116N, 10.2543E, fl., 25 Nov. 1996, <u>Etuge</u> 3610.

Parietaria debilis G.Forst.

Local name: No known name – <u>Cheek</u> 8691; Afzijai iwu (Kom at Laikom – Yama Peter) – <u>Pollard</u> 316.

Weak decumbent herb, 6–30 cm; stems slender; leaves broadly ovate, very thin, flaccid with conspicuous dot-like cystoliths, 1–2 × 1–2 cm; flowers few, subsolitary or in sessile axillary clusters. Forest.

Kilum: Lake Oku, 2200m, fl., 11 Jun. 1996, <u>Zapfack</u> 868; Shore of Lake Oku, 2250m, fl., 12 Jun. 1996, <u>Cable</u> 3136; Elak, KJ, 2400m, fl., 28 Oct. 1996, <u>Buzgo</u> 610 & 612; Path from HQ to KA, 2800m, 6.1349N, 10.3112E, fl., 30 Oct. 1996, <u>Cheek</u> 8509 & 8516.

Ijim: Gikwang towards Nyasosso bush, 2280m, fl., 18 Nov. 1996, <u>Etuge</u> 3447; Above Aboh village, 1.5 hours on Gikwang road towards Lake Oku, track towards TA, 2400m, 6.1112N, 10.2529E, fl., 20 Nov.1996, <u>Cheek</u> 8691; Top of Ijim Mountain ridge, second gate from Aboh village, 2380m, fl., 21 Nov. 1996, <u>Kamundi</u> 675; Main path from Ijim to Tum, 2100m, 6.1116N, 10.2543E, fl., 25 Nov. 1996, <u>Etuge</u> 3621; Laikom, boundary of Akwamofu sacred forest, 1820m, 6.1633N, 10.1940E, fl., 5 Dec. 1998, <u>Pollard</u> 316.

Pilea rivularis Wedd.

Terrestrial or epiphytic herb to 60 cm; stems lax, creeping and rooting; branches erect, smooth; leaves broadly ovate, coarsely dentate, with cystoliths scattered on both surfaces, 1.7 × 1–5.8 cm; inflorescence a nodal, whorl-like sessile cluster or dense terminal cyme. Forest.

Syn. **Pilea ceratomera** Wedd., F.W.T.A. 1: 621 (1958).

Kilum: Above Lake Oku, <u>Keay</u> FHI 28505; Elak, KD,

2200m, fl., 10 Jun. 1996, Cable 3047; Shore of Lake Oku, 2250m, fl., 12 Jun. 1996, Cable 3091; Elak, KJ, 2400m, fl., 28 Oct. 1996, Buzgo 620; Elak, KJ, 2500m, fl., 29 Oct. 1996, Zapfack 1077.

Ijim: Aboh, near Chufekhe stream, 2010m, 6.1124N, 10.2525E, fl., 20 Nov. 1996, Kamundi 645; Top of Ijim Mountain ridge, second gate from Aboh village, 2250m, fl., 21 Nov. 1996, Kamundi 680; Aboh, Ijim Mountain Forest, 2000m, 6.1116N, 10.2543E, fl., 21 Nov. 1996, Pollard 70; Main path from Ijim to Tum, 2100m, 6.1116N, 10.2543E, fl., 25 Nov. 1996, Etuge 3609; Aboh village, 6.15N, 10.26E, fl., 25 Nov. 1996, Onana 620; Laikom, forest patch towards foot of waterfall, above Akwamofu sacred forest, 1860m, 6.17N, 10.20E, fl., 11 Dec. 1998, Etuge 4602.

Pilea tetraphylla (Steud.) Blume

Local name: Mbol (Oku) – Cheek 8454; Mbolakambash (Oku) – Munyenyembe 715.

Herb to 40 cm; stem glabrous, erect, slender, simple or branched, sometimes from creeping base; leaves broadly ovate, crenate-serrate, laxly clothed with cystoliths, 2–4 × 1.5–3 cm; petioles to 2 cm long; inflorescence a flat terminal sessile corymb involucrate by 4 uppermost leaves. Forest.

Kilum: Elak to the forest at KA, 2200m, 6.1349N, 10.3112E, fl., 27 Oct. 1996, Cheek 8454; Elak, 2400m, fl., 28 Oct. 1996, Munyenyembe 715; Elak, KA path, 2300m, 6.15N, 10.26E, fl., 28 Oct. 1996, Onana 446.

Ijim: Gikwang road towards Nyasosso forest, 2250m, 6.1116N, 10.2543E, fl., 20 Nov. 1996, Etuge 3519; Aboh village, 2450m, 6.15N, 10.26E, fl., 20 Nov. 1996, Onana 550.

Pouzolzia parasitica (Forssk.) Schweinf.

Perennial shrubby herb to 2 m; leaves serrate. Forest edge.

Kilum: Shore of Lake Oku, 2250m, fl., 12 Jun. 1996, Cable 3123; Inside crater of Lake Oku, 2300m, 6.13N, 10.27E, fl., 25 Nov. 1996, Cheek 8769.

Urera sp.

Climber, leaves alternate, glossy, elliptic, c. 10 × 4 cm long. Forest.

Ijim: Ajung, 1900m, 12 Nov. 1999, Cheek 10091.

VERBENACEAE

Clerodendrum silvanum Henriq. var. *buchholzii* (Gürke) Verdc.

Local name: Feban (Oku) – Munyenyembe 865; Alam-nse (Kom at Laikom) – Etuge 4546.

Woody climber to 10 m; leaves elliptic or ovate, glabrous, 8–20 × 3–10 cm; petiolar thorns; inflorescence an elongate, leafless panicle, frequently cauliflorous, rhachis 5–30 cm long; flowers white, fragrant; corolla tube up to 2.5 cm long; fruits red. Forest.

Syn. *Clerodendrum buchholzii* Gürke, F.W.T.A. 2: 443 (1963).

Kilum: Path from Mbo village (about 6 km from Elak-Oku), 2200–2500m, fl. fr., 3 Nov. 1996, Onana 496; Mbijame forest, 2000m, fl., 4 Nov. 1996, Munyenyembe 865.

Ijim: Main road from Atubeaboh (Gikwang Foe) towards Oku, 2260m, fl., 15 Nov. 1996, Etuge 3379; Laikom, behind basecamp, Fon's Palace, 1950m, 6.1642N, 10.1950E, fl., 4 Dec. 1998, Pollard 310; Laikom, Akwamofu, medicinal forest, 1760m, 6.17N, 10.20E, fl. 5 Dec. 1998, Etuge 4546.

Lantana camara L. var. *camara*

Shrub, stems erect or spreading, much-branched, quadrangular, usually armed with short recurved prickles; leaves, petioles and peduncles pilose or strigose; flowers in convex heads; corolla much longer than subtending bract, white, yellow, red, orange or pink. Introduced weed.

Kilum: Elak, 2000m, fl., 8 Jun. 1996, Zapfack 760.

Vitex cf. doniana Sweet

Tree 10–20 m; branches glabrous; leaves coriaceous, 5-foliolate; leaflets obovate to elliptic, middle ones 5–16 × 4–10 cm, petiolule 1–2.5 cm; inflorescence axillary or axillary and with terminal cymes, congested; peduncle 2–8 cm long; fruits obovoid to sub-globose, c. 1 cm long. Savanna woodland.

Jakiri: Tan, Al Hadji Gey's land, grasslands, 1600m, st., 6 Nov. 1998, Maisels 179.

VITACEAE

Cyphostemma rubrosetosum (Gilg & Brandt) Desc.

Herbaceous climber resembling *Cyphostemma mannii*, but glabrous apart from numerous long red glandular hairs. Forest edge.

Ijim: Laikom, near stream, 1660m, fl., May 1931, Maitland 1432.

Cyphostemma mannii (Baker) Desc.

Local name: Kintiseh (Oku) – Cheek 8449; Kinkeble - Munyenyembe 723.

Herbacous climber; stems, petioles, inflorescence and flower-buds densely tomentellous; leaflets elliptic or obovate-elliptic, crenate-serrate, up to 10 × 4 cm. Forest.

Kilum: Lake route, 2100m, fl., 11 Jun. 1996, Zapfack 842; Oku-Elak to the forest at KA, 2200m, fl., 27 Oct. 1996, Cheek 8449; Elak, 2400m, fl., 28 Oct. 1996, Munyenyembe 723.

Ijim: Gikwang road towards Oku, 2260m, fl., 15 Nov. 1996, Etuge 3406; Ijim Mountain Forest, near Chufekhe stream, 2010m, fl., 20 Nov. 1996, Kamundi 640; Above Fon's Palace, about 30 mins. walk along path to Fulani settlement, 2050m, st., 22 Nov. 1996, Cheek 8734.

ANGIOSPERMAE

MONOCOTYLEDONAE

AMARYLLIDACEAE
Det. I.Nordal (O)

Scadoxus multiflorus (Martyn) Raf.
Slender bulbous herb; bulb cylindrical, about 2 × 1.5 cm; leaves small, expanding after the flowers; inflorescence lateral, 7–25 cm; flowers scarlet. Forest edge.
Syn. *Haemanthus rupestris* Baker, F.W.T.A. 3: 132 (1968).
Ijim: Aboh, Isebu Mountain Forest, along the ridge, 1700m, fl., 26 Nov. 1996, Kamundi 706; Laikom, between medicinal forest and upper waterfall (of two), 1960m, 6.1630N, 10.1951E, st., 11 Dec. 1998, Pollard 369.

ANTHERICACEAE
Det. I.Nordal (O)

Chlorophytum comosum (Thunb.) Jacq. *sens. lat.*
A rather fleshy herb, to about 50 cm; leaves erect, ovate to ovate-lanceolate, 15–35 cm; petiole long and distinct; flowers white or greenish, sometimes viviparous. Forest.
Syn. *Chlorophytum sparsiflorum* Baker, F.W.T.A. 3: 100 (1968).
Kilum: Main road from Ijikijem to Oku, 2350m, fl., 7 Nov. 1996, Etuge 3354.
Ijim: Zitum road, 1800m, 21 Nov. 1996, Etuge 3549.

ARACEAE

Amorphophallus staudtii (Engl.) N.E.Br.
Herb; spathe ovate-orbicular, about 12 cm diameter when spread out, with numerous stiff hairs in the basal portion; spadix slender, 18 cm. Forest.
Kilum: Lowland forest patches at lower altitude than the BirdLife HQ, 1900m, 6 Nov. 1996, Buzgo 716 & 717.

Amorphophallus sp.
Leaves needed for identification, further material required. Forest.
Ijim: Laikom, gallery forest alongside path from Fon's Palace to grassland plateau, 2020m, 6.1642N, 10.1950E, fr., 4 Dec. 1998, Pollard 311.

Anchomanes difformis (Blume) Engl.
Rain forest herb; leaf developing after and becoming taller than the inflorescence; lamina 1.5 m diameter; petiole to 3 m; spathe dark purplish; ovary pink or purple with a rather small white stigma; peduncle 0.9–2 m. Forest; forest edge.
Babungo-Mbi Crater: 1600m, 8 Nov. 1999, Cheek 10018.

Nepthytis sp.
Herb; leaves sagittate or 3-lobed. Forest.
Babungo-Mbi Crater: 1600 m, 8 Nov. 1999, Cheek 10004.

ASPHODELACEAE

Kniphofia reflexa Codd
RED DATA LISTED
An orthographic error has led to this taxon being known as *Kniphofia reflexum* in F.W.T.A. 3: 94 (1968).
A tufted perennial herb; leaves arranged in a basal rosette, 20–60 cm; inflorescence a conspicuous spike to 2 m or more; flowers small, campanulate, yellow. Swamp.
Ijim: Plateau, Laikom, 2000m, fl., Apr. 1931, Maitland 1624; Cattle watering hole by Fulani settlement, 11 km E of Laikom, 2400m, 6.1608N, 10.2500E, fr., 9 Dec 1998, Pollard 359.

COLCHICACEAE
B.J.Pollard (K)

Wurmbea tenuis (Hook.f.) Baker subsp. *tenuis*
Cormous herb; corm ovoid, about 1 cm long, tunicate, bulb-like; stem with basal sheaths membranous, truncate, 1–3 cm; stem leaves 2–3, decreasing in size upwards; leaves 1 or 2 arising from the corm, linear, acute, 6–16 × 0.15–4 cm, glabrous; spike 2–6-flowered; perianth segments white and purple, fading in older flowers, 6 mm. Rocky grassland.
Ijim: Laikom, in grassland and about stony places, 2000m, fl., Apr. 1931, Maitland 1407; Laikom, in grassland and about rocks, fl. fr., Apr. 1931, Maitland 1511.

COMMELINACEAE
Det. R.B.Faden (US)

Aneilema umbrosum (Vahl) Kunth subsp. *umbrosum*
Local name: Kii yung (Kom) – Etuge 4551.
Herb, to 1m; leaves lanceolate to elliptic or ovate, to 13 × 4 cm; inflorescence branches about 8–30; flowers white to mauve or purple. Forest; forest edge.
Ijim: Akwo-mufu, Laikom, 1760m, 6.17N, 10.20E, fl., 5 Dec. 1998, Etuge 4551.

Commelina africana L. var. *africana*
Prostrate herb; stems c. 90 cm, rooting at nodes; flowers yellow, c. 6 mm wide. Forest edge.
Jakiri: 1500m, Brunt 553.

Commelina benghalensis L. var. *hirsuta*
C.B.Clarke
Sprawling gregarious herb about 45 cm; leaves mostly ovate to lanceolate; leaf-sheaths ± densely rusty-hairy all over outside; flowers bright blue, open in morning. Forest; grassland.
Ijim: Tum, 1800m, 4 Nov. 1999, Cheek 9952.

Commelina cameroonensis J.K.Morton
Erect herb to 60 cm, gregarious; leaf-sheaths rusty-hairy; spathes 1–3; flowers white, c. 1.5 cm wide. Forest.
Kilum: Lake Oku, 2200m, Jan. 1951, Keay FHI 28475.

Commelina diffusa Burm.f. subsp. *diffusa*
Local name: Ekekeyun (Kom) – Cheek 8645.
Herb to 1.2 m; leaves lanceolate to elliptic 8 × 0.8–2.5 cm; spathes 1.5–3 cm; flowers deep blue to violet or lilac-blue; capsule 5-seeded. Open wet places.
Ijim: Above Aboh village, 1.5 hours along Gikwang road to Lake Oku, 2350m, 6.1112N, 10.2529E, fl., 18 Nov. 1996, Cheek 8645.

Cyanotis barbata D.Don
Local name: Kyung Kekwing (Kom at Laikom -Yama Peter) – Pollard 330.
Herb, 5–50 cm; leaves linear, to about 8 cm; individual cincinni normally in terminal or axillary clusters of 2 or more together; outer bracts leafy, ± falcate; flowers normally blue, rarely mauve or white. Grassland.
Ijim: Gikwang road towards Oku, Aboh, 2260m, fl., 15 Nov. 1996, Etuge 3389; Ijim plateau above Fon's Palace, Laikom, 1900m, 6.16N, 10.19E, fl., 3 Dec. 1998, Cheek 9732; Swamp, half a mile W of Ardo's Fulani compound, Ijim ridge grassland plateau, Laikom, 2060m, 6.1609N, 10.2051E, fl., 7 Dec. 1998, Pollard 330.

Floscopa glomerata (Willd. ex Schult. & Schult.f.) Hassk. subsp. *glomerata*
Annual or perennial herb, erect or straggling, sometimes rhizomatous at base; stems about 0.15–2 m, rooting at base or from lower nodes; inflorescence terminal, hairy, rather compact; flowers mauve. Swamp.
Ijim: Swamp E of Ardo's compound, Laikom, 2000m, 6.16N, 10.22E, fr., 5 Dec. 1998, Cheek 9788; Afua swamp, 1950m, 6.08N, 10.24E, fr., 7 Dec. 1998, Cheek 9845.

Palisota schweinfurthii C.B.Clarke
Robust herb; stems to 2 m; leaves mostly in terminal rosettes; inflorescence spicate; flowers dirty white to pink. Forest.
Ijim: Laikom to Fundong, 1700m, 6.17N, 10.20E, fl., 10 Dec. 1998, Etuge 4588.

COSTACEAE

Costus afer Ker Gawl.
Perennial herb to 3.5 m; stems leafy; ligule entire, glabrous; inflorescence terminal, succulent; flowers white and yellow with pink tip. Forest.
Ijim: Anyajua, 1400m, 14 Nov. 1999, Cheek 10109.

CYPERACEAE
Det. K.A.Lye (NLH)

Ascolepis brasiliensis (Kunth.) Benth. ex C.B.Clarke
Annual or perennial herb, shortly rhizomatous; sheaths reddish, not fibrous; inflorescence lobed, consisting of 4 spikelets in a pyramidal head. Swamp.
Ijim: Sappel, below Ardo's compound, 2300m, 6.16N, 10.24E, fl., 5 Dec. 1998, Cheek 9803 & 9807; Ngengal swamp, 10 km E of Laikom, en route to Mbesa, 2460m,

6.1615N, 10.2435E, fl. fr., 9 Dec. 1998, Pollard 347.

Ascolepis protea Welw. subsp. *protea*
Annual or perennial herb; stems slender, about 0.5 mm diameter, often thickened at the base with pale fibrous sheaths; inflorescence consisting of a single spikelet with white, often curved, scales in a radiate head about 6–8 mm across. Grassland; granite outcrops.
Kilum: Ewook Etele Mbae, 10 km from Oku, 2350m, fl., 6 Nov. 1996, Zapfack 1212.

Bulbostylis densa (Wall.) Hand.-Mazz. var. *densa*
Annual herb, very variable in size; spikelets ovate to broadly elliptic, always pedicillate; glumes broadly ovate, clearly keeled. Grassland.
Kilum: Summit, 2800m, 31 Oct. 1996, Zapfack 1126.
Ijim: Above Fon's palace, Laikom, about 30 minutes walk on path to Fulani settlement, 2050m, 6.1642N, 10.1952E, 21 Nov. 1996, Cheek 8715 & 8721; Ijim Mountain Forest, 2100m, 6.1116N, 10.2543E, fl., 25 Nov. 1996, Pollard 81; Laikom, N-facing rockface, in 'gallery' forest along path from Fon's palace to Ijim ridge, 2000m, 6.1642N, 10.1950E, fl., 2 Dec. 1998, Pollard 280, Ijim plateau above Fon's palace, 1900m, 6.16N, 10.19E, fl., 3 Dec. 1998, Cheek 9754.

Bulbostylis hispidula (Vahl) R.W.Haines subsp. *hispidula*
Densely tufted annual herb, sometimes persisting; inflorescence consisting of few spikelets; glumes mucronate, glabrous or rarely puberulous to shortly pubescent, dark chestnut on the sides and with a green keel; achenes yellowish white. Grassland.
Syn. *Fimbristylis hispidula* (Vahl) Kunth subsp. *hispidula*, F.W.T.A. 3: 324 (1968).
Kilum: Ewook Etele Mbae, 10 km from Oku, 2350m, 6 Nov. 1996, Zapfack 1226.

Bulbostylis sp. nov. ?
RED DATA CANDIDATE
Ijim: Grassland plateau above Fon's Palace, Laikom, 2020m, 6.1642N, 10.1950E, fr., 3 Dec. 1998, Pollard 293.

Carex chlorosaccus C.B.Clarke
Tufted perennial, 0.6–1 m; basal leaf-sheaths pale brown or greenish; inflorescence of much-branched panicles with short green spikes. Forest.
Kilum: Oku-Elak, lower parts of KA, 2200m, 28 Oct. 1996, Cheek 8490; KC, 2500m, fl., 30 Oct. 1996, Buzgo 653; Above Mboh, path above village to forest through Wambeng's farm, 2200m, 6.1146N, 10.2856E, fr., 1 Nov. 1996, Cheek 8565.

Carex echinochloë Kunze
Tufted perennial about 1 m; inflorescence of simple or compound panicles of short greenish spikes; utricles ciliolate. Forest.
Kilum: Near maternity hospital, 2000m, 10 Jun. 1996, Asonganyi 1322.

Carex mannii E.A.Bruce

Tufted perennial to 1.4 m; inflorescence of 6–10 solitary dark reddish brown spikes. Forest.
Kilum: Path from BirdLife HQ to KA, 2500m, 6.1349N, 10.3112E, fl., 30 Oct. 1996, Cheek 8504.

Carex neo-chevalieri Kük.

Tufted perennial 0.6–1 m; inflorescence of scanty panicles of light brown spikes; glumes glabrous or shortly hairy. Forest.
Ijim: From Aboh, path towards Tum, 2100m, 6.1116N, 10.2543E, fl., 22 Nov. 1996, Etuge 3594; Forest patch near foot of waterfall above Akwamofu sacred forest, 1860m, 6.17N, 10.20E, fl., 11 Dec. 1998, Etuge 4600.

Carex preussii K.Schum.
RED DATA LISTED
Tufted perennial, 0.3–1 m; inflorescence scanty; spikelets usually rust and green, but sometimes redder. Forest.
Kilum: Lake Oku, Jan. 1951, Keay & Lightbody FHI 28465.
Ijim: Ijim Mountain Forest, near Chufekhe stream, 2010m, 6.1124N, 10.2525E, 20 Nov. 1996, Kamundi 657.

Cyperus atrorubidus (Nelmes) Raymond
Local name: Insansakitele (Oku) – Munyenyembe 876.
Herb to 10 cm, ± caespitose; spikelets in a sessile pseudo-lateral cluster, dark red-black. Damp places.
Syn. *Pycreus atrorubidus* Nelmes, F.W.T.A. 3: 302 (1968).
Kilum: Shambai, 2500m, fl., 5 Nov. 1996, Munyenyembe 876; Iwook Etele Mbae, 2500m, fl. fr., 5 Nov. 1996, Zapfack 1188; Ewook Etele Mbae, 10 km from Oku, 2350m, 6 Nov. 1996, Zapfack 1218.

Cyperus atroviridis C.B.Clarke
Local name: Saghe-Saghe (Kom at Aboh) – Cheek 8669.
Tufted herb to 1.3 m; inflorescence of subcylindric spikes, secondary rays (when present) spreading or deflexed; spikelets slender, elongated, black and green. Damp places; grassland.
Kilum: Near market in Oku village, 2000m, 6N, 10.30E, fr., 21 Jun. 1962, Brunt 634; Shore of Lake Oku, 2250m, 12 Jun. 1996, Cable 3111; Path from Mboh village, 2200–2500m, 6.15N, 10.26E, fl., 1 Nov. 1996, Onana 487; Manchok, 2200m, fl., 1 Nov. 1996, Pollard 31; Ewook Etele Mbae, 2500m, 5 Nov. 1996, Zapfack 1188.
Ijim: Above Aboh village, 1.5 hours walk along the Gikwang road to Lake Oku, 2450m, 6.1112N, 10.2529E, fl., 19 Nov. 1996, Cheek 8669; Grassland plateau above Fon's Palace, 2020m, 6.1642N, 10.1950E, fl., 3 Dec. 1998, Pollard 291; On rock in stream, 2040m, 6.1642N, 10.1950E, fl., 3 Dec. 1998, Pollard 304.
Bambili: Bambili, 1660m, fl. fr., 5 Jun. 1970, Bauer 85.

Cyperus cyperoides (L.) Kuntze sens. lat.

Robust perennial herb; culms 20–80 cm; inflorescence a 4–25 cm wide umbel-like anthela of 6–18 spikes; major inflorescence bracts 5–15, leafy, largest 6–30 cm; spikelets oval, 1–3-flowered, greenish yellow and often with a brownish tinge. Damp grassland.
Ijim: Grassland plateau above Fon's palace, Laikom, 2040m, 6.1642N, 10.1950E, fl., 3 Dec. 1998, Pollard 307.

Cyperus cyperoides (L.) Kuntze subsp. flavus Lye

Tufted plant; rhizome woody, ± composed of swollen stem bases; leaf sheaths purple; inflorescence variable, rays ± well-developed; spikes of ± crowded, small greenish or reddish, 1–2-flowered spikelets. Damp grassland.
Syn. *Mariscus alternifolius* Vahl, F.W.T.A. 3: 296 (1968).
Kilum: Oku, Brunt 565.
Bambili: Bambili College of Arts and Science, 1660m, fl., 14 Mar. 1970, Bauer 51.

Cyperus cyperoides (L.) Kuntze subsp. macrocarpus (Kunth) Lye

A robust perennial, stem-base swollen; rhizome woody; culms 15–60 cm, triangular; inflorescence a 2–6 cm wide anthela of 5–10 sessile or subsessile spikes, 7–20 mm, with 25–100 spreading spikelets. Damp places.
Ijim: Laikom, near Akwamofu sacred forest, by waterfall, 1870m, 6.1633N, 10.1940E, fl., 5 Dec. 1998, Pollard 320 & 322.

Cyperus densicaespitosus Mattf. & Kük.

Caespitose herb; inflorescence with numerous small, green and white, 3–4-spiked pyramidal heads. Damp places.
Syn. *Kyllinga pumila* Michx., F.W.T.A. 3: 305 (1972).
Ijim: Fundong to Belo road, 2050m, 6.17N, 10.20E, 21 Nov. 1996, Cheek 8726; Laikom, near Akwamofu sacred forest, by waterfall, 1870m, 6.1633N, 10.1940E, fl., 5 Dec. 1998, Pollard 321.
Ndop: Ndop Baptist School, 1160m, fl., 9 Nov. 1960, Gillett 11.

Cyperus denudatus L.f. var. denudatus

Rhizome creeping; leaves reduced; inflorescence rays slender, longer than bract. Marshy grassland.
Mbi Crater: 1950m, 6.0536N, 10.21E, fl., 9 Dec. 1998, Cheek 9890.

Cyperus dichroöstachyus Hochst. ex A.Rich.

Leafy herb; rhizome thin; spikelets very small, blackish and green; bracts many, subequal; achene slightly shorter than the glume. Wet ground.
Ijim: Laikom, swamp E of Ardo's compound, 2000m, 6.16N, 10.22E, fl., 5 Dec. 1998, Cheek 9790.
Mbi Crater: 1950m, 6.0536N, 10.21E, fl., 9 Dec. 1998, Cheek 9896.

Cyperus digitatus Roxb. subsp. auricomus (Spreng.) Kük. var. bruntii Hooper

Herb to 3 m; leaves to 3 cm wide; inflorescence bisumbellate; spikes arranged digitately, lax; spikelets spreading with a central golden stripe; rays scabrid. Shallow water and swamps.
Mbi Crater: Ndawara area, 1900m, 6.05N, 10.21E, fr., 9 Dec. 1998, Etuge 4583.
Ndop: Ndop plain, near Babungo, 1260m, st., 5 Mar. 1962, Brunt 115.

Cyperus distans L.f. subsp. *longibracteatus* (Cherm.) Lye var. *longibracteatus*

Caespitose perennial; rhizome short, thick; culms usually set in a row or solitary; involucral bracts to over 25 cm; spikelets greenish, probably always falling off entire; glumes 1.7–2.6 mm, slightly larger than the type subspecies. Wet places; fallow.
Syn. *Mariscus longibracteatus* Cherm., F.W.T.A. 3: 295 (1968).
Kilum: Elak, 2000m, fl. fr., 8 Jun. 1996, Zapfack 774 & 785; Path above water-tower, KJ, about 1 km into forest, 2250m, fl., 9 Jun. 1996, Cable 2973.
Ijim: Grassland plateau above Fon's palace, Laikom, 2040m, 6.1642N, 10.1950E, fl., 3 Dec. 1998, Pollard 306.

Cyperus distans L.f. subsp. *longibracteatus* (Cherm.) Lye var. *niger* C.B.Clarke

Similar to the other varieties in habit, differs in its very dark chestnut or black spikelets; glumes slightly longer. Wet places; fallow.
Syn. *Cyperus keniensis* Kük., F.W.T.A. 3: 295 (1968).
Kilum: Elak, 2000m, fl. fr., 8 Jun. 1996, Zapfack 773.
Ndop: Near Bamessi village, 1260m, fl., 3 Apr. 1962, Brunt 311.
Bambui: Bambui Farm, May 1943, Pedder 22.

Cyperus esculentus L. var. *esculentus*

Robust stoloniferous perennial; stolons to about 15 cm, covered with brown to blackish scales, ending in a blackish tuber, 3–8 mm in diameter; spikelets rather blunt, 5–20 mm long, brown or rust-coloured, 6–22-flowered. Wet grassland; swamp.
Bambui: Bambui Farm, May 1943, Pedder 18.

Cyperus haspan L. subsp. *haspan*

Annual or perennial herb; stems weak, usually crowded, 5–40 cm, triangular, glabrous; sheaths purple; inflorescence spreading; bracts 2, unequal; spikelets near-linear, in groups of 1–3. Open wet places.
Ijim: Fundong to Belo road, 2050m, 6.17N, 10.20E, 21 Nov. 1996, Cheek 8727.
Jakiri: Jakiri, 2000m, fl., 18 Feb. 1958, Hepper 2717.

Cyperus laxus Lam. subsp. *buchholzii* (Boeck.) Lye

Leafy, caespitose perennial, 25–50 cm; leaf sheaths purple; inflorescence 2–3 times umbellate with small clusters of few-flowered spikelets; glumes spreading. Grassland.
Syn. *Cyperus diffusus* Vahl subsp. *buchholzii* Kük., F.W.T.A. 3: 289 (1968).
Ijim: Grassland plateau above Fon's palace, Laikom, 2020m, 6.1642N, 10.1950E, fl., 3 Dec. 1998, Pollard 292.

Cyperus mannii C.B.Clarke

Leafy perennial to 1.5 m; inflorescence an irregular thrice-branched umbel bearing small clusters of brown spikelets; glumes not clearly apiculate, pale. Forest.
Kilum: Path above water-tower, KJ, 1 km into forest, 2250m, fl., 9 Jun. 1996, Cable 2985; Lake Oku, 2200m, fl., 12. Jun 1996, Etuge 2331; Oku-Elak, lower parts of KA,

2200m, 6.1349N, 10.3112E, 28 Oct. 1996, Cheek 8466; KJ, 2500m, fl. fr., 29 Oct. 1996, Zapfack 1078; KC, 2300m, fl., 30 Oct. 1996, Zapfack 1098; KC, 2500m, fl., 30 Oct. 1996, Zapfack 1104; Summit, 2800m, fl. fr., 31 Oct. 1996, Zapfack 1118; KD-4, 2300m, fl., 1 Nov. 1996, Pollard 30.
Ijim: Gikwang road towards Oku, 2270m, fl., 15 Nov. 1996, Etuge 3405; Ijim Mountain Forest, 2160m, 6.11N, 10.25E, 19 Nov. 1996, Kamundi 609; Above Aboh village, 1.5 hours along Gikwang road to lake Oku, track towards transect TA, 2400m, 6.1112N, 10.2529E, fl. fr., 20 Nov. 1996, Cheek 8683; Zitum road, 1950m, 6.1116N, 10.2543E, fl., 21 Nov. 1996, Etuge 3567; Ijim Mountain Forest, 2100m, 6.1116N, 10.2543E, fl., 25 Nov. 1996, Pollard 79; In 'gallery' forest, 15 minutes along path from Fon's palace at Laikom to grassland plateau above, 2000m, 6.1642N, 10.1950E, fl., 2 Dec. 1998, Pollard 259 & 260; Grassland plateau above Fon's palace, Laikom, 2020m, 6.1642N, 10.1950E, fl., 3 Dec. 1998, Pollard 285; Roadside in forest understorey, near Akwamofu sacred forest, Laikom, 1800m, 6.1633N, 10.1940E, fl., 5 Dec. 1998, Pollard 324; 327 & 328; Plot between Fon's palace at Laikom and Ardo's compound, 2000m, 6.16N, 10.21E, fl., 8 Dec. 1998, Cheek 9869; Grassland below Ijim ridge, montane forest understorey near fast running stream, 1950m, 6.1630N, 10.1951E, fl., 11 Dec. 1998, Pollard 371 & 372.
Bambui: Boughey GC 10778.

Cyperus margaritaceus Vahl var. *nduru* (Cherm.) Kük.

Robust perennial; leaves few, short; inflorescence small, with 2–4 spikelets; spikelets 5–10 mm; involucral bracts leafy, 1–3, frequently shorter than the inflorescence; glumes white to brownish. Grassland; woodland.
Syn. *Cyperus nduru* Cherm., F.W.T.A. 3: 292 (1968).
Bambili: Bambili, 1660m, fl., 3 Mar. 1970, Bauer 39.
Ndop: Ndop to Kumbo, Boughey GC 11158B; Small hill by Ndop village rest house, 1260m, fl., 1 Mar. 1962.

Cyperus niger Ruiz & Pav. subsp. *elegantulus* (Steud.) Lye

Caespitose perennial; inflorescence anthelate, compact; spikelets strongly compressed, small, black with green keels; involucral bracts 2–many, erect or spreading, at least the lower two much exceeding the inflorescence. Grassland; swamp.
Syn. *Pycreus elegantulus* (Steud.) C.B.Clarke, F.W.T.A. 3: 300 (1968).
Kilum: Elak, 2000m, fl. fr., 8 Jun. 1996, Zapfack 780; KD, fl., 9 Jun. 1996, Etuge 2250; Towards KJ, 2200m, fl., 30 Oct. 1996, Pollard 19; Elak, 2100m, fl., 1 Nov. 1996, Munyenyembe 856; Ewook Etele Mbae, 10 km from Oku, 2350m, 6 Nov. 1996, Zapfack 1215.
Ijim: Gikwang road towards Oku, 2260m, fl., 15 Nov. 1996, Etuge 3399; Above Aboh village, 1.5 hours walk along Gikwang road to Lake Oku, 2450m, 6.1112N, 10.2529E, fl., 19 Nov. 1996, Cheek 8668; Ijim Mountain Forest, along the ridge towards a point where Lake Oku can be viewed, 2300m, 22 Nov. 1996, Kamundi 684; Grassland plateau, Laikom, 2040m, 6.1642N, 10.1950E, fl., 3 Dec. 1998, Pollard 305; Afua swamp, 1950m, 6.08N, 10.24E, fl. 7 Dec. 1998, Cheek 9847A.

Cyperus niveus Retz. var. nov.
RED DATA CANDIDATE
Herb 3–8 cm; capitula white, bulbous. Grassland.
Ijim: Sappel, below Ardo's uncle's compound, 2300m, 6.16N, 10.24E, fl. fr., 5 Dec. 1998, Cheek 9810; Cattle watering hole by Fulani settlement, 11 km E of Laikom, en route to Mbesa, 2400m, 6.1615N, 10.2435E, fl., 9 Dec. 1998, Pollard 354.

Cyperus pectinatus Vahl
Tufted perennial; culms 0.3–1.2 m; stems slender, leafless; inflorescence capitate; spikelets few, broad, lanceolate; glumes numerous, closely imbricated, yellow-green to brown, sometimes proliferating; floating in mats. Swamp.
Syn. **Cyperus nudicaulis** Poir., F.W.T.A. 3: 293 (1968).
Ndop: Ndop, 3 miles S of Baba village, 1260m, fl., 5 Mar. 1962, Brunt 125.

Cyperus renschii Boeck. var. renschii
Robust perennial to 1.5 m; inflorescence lax, much-branched; involucral bracts leafy to 90 cm; spikelets few-flowered, dark, small, very numerous, tips recurved to the glumes. Wet places.
Ijim: Laikom to Fundong, roadside, 1700m, 6.17N, 10.20E, fl. fr., 10 Dec. 1998, Etuge 4595.
Kumbo: By River Wi, S of town, 1680m, fl., 14 Feb. 1958, Hepper 1977.

Cyperus sesquiflorus (Torr.) Mattf. & Kük. subsp. appendiculatus (K.Schum.) Lye
Culms 10–80 cm, the base usually swollen and covered by hardened scales which split; inflorescence a compound head of a larger ovoid to ovoid-cylindrical central spike and usually much smaller lateral spikes; involucral bracts 3–6, leafy, 3–15 cm. Forest edge; swamp.
Syn. **Kyllinga appendiculata** K.Schum., F.W.T.A. 3: 307 (1968).
Ijim: Aboh, Gikwang road towards Nyasosso forest, 2300m, fl., 19 Nov. 1996, Etuge 3486; Aboh village, 2450m, fl., 19 Nov. 1996, Onana 543; Afua swamp, 1950m, 6.08N, 10.24E, fl., 7 Dec. 1998, Cheek 9841.

Cyperus sesquiflorus (Torr.) Mattf. & Kük. subsp. cylindricus (Nees) Koyama
Culms 3–60 cm, triangular, ridged, glabrous; inflorescence either a single cylindrical or rarely globose spike or a compund head of one larger cylindrical spike and several smaller lateral spikes; spikelets 1.8–2.5 mm, 1-flowered. Grassland.
Bambui: Bambui Farm, fl., May 1943, Pedder 20.

Cyperus sesquiflorus (Torr.) Mattf. & Kük. subsp. sesquiflorus
Culms 0.2–0.5(–1) m, sharply triangular, deeply ridged, glabrous; inflorescence usually a single ovate spike, 6–15 mm, rarely with a few small additional lateral spikes, white or greyish and not fading pale-brownish; spikelets 3–4 mm, 1–2-flowered. Grassland; paths; roadsides.
Syn. **Kyllinga odorata** Vahl subsp. **odorata**, F.W.T.A. 3: 304 (1968).

Kilum: KD, fl., 9 Jun. 1996, Etuge 2206; Lake Oku, 2200m, fl., 11 Jun. 1996, Zapfack 877; Shore of Lake Oku, 2250m, fl., 12 Jun. 1996, Cable 3107; Path from BirdLife HQ to KA, 2200m, 6.1349N, 10.3112E, 30 Oct. 1996, Cheek 8502; KA, top plateau for the Mountain, 2800m, fl., 31 Oct. 1996, Buzgo 668; Elak, 2200m, fl., 1 Nov. 1996, Munyenyembe 843; KD-6, 2400m, fl., 1 Nov. 1996, Pollard 35.
Ijim: Gikwang road towards Nyasosso forest - Ijim Moutain Forest Reserve, 2300m, fl., 19 Nov. 1996, Etuge 3486; Ijim Mountain Forest, near Chufekhe stream, 2010m, 6.1124N, 10.2525E, fl., 20 Nov. 1996, Kamundi 662.

Cyperus tomaiophyllus K.Schum.
Very robust perennial; rhizome woody, branching, to 2 cm thick; inflorescence anthelate, 6–15 cm wide; major spikes 6–12; peduncles 0.5–4 (rarely to 15) cm. Swamp; grassland; forest.
Syn. **Mariscus tomaiophyllus** (K.Schum.) C.B.Clarke, F.W.T.A. 3: 295 (1968).
Kilum: Lake Oku, 2300m, fr., 11 Jun. 1996, Zapfack 876; KJ, 2400m, fl., 29 Oct. 1996, Zapfack 1060; Path from BirdLife HQ to KA, 2600m, 6.1349N, 10.3112E, fl., 30 Oct. 1996, Cheek 8526; Elak, forest edge, 2200m, fl., 30 Oct. 1996, Munyenyembe 812.
Ijim: Ijim Mountain Forest, 2160m, 6.11N, 10.25E, 19 Nov.1996, Kamundi 634; Aboh village, 2450m, 6.15N, 10.26E, fl., 20 Nov. 1996, Onana 557.

Cyperus triceps Endl.
Perennial, arhizomatous; base of stems bulbous; culms 5–30 cm; inflorescence a pale green to greyish white head, 4–9 mm, usually with one central spike; spikelets 2–2.5 mm, 1-flowered. Streambanks.
Syn. **Kyllinga tenuifolia** Steud., F.W.T.A. 3: 305 (1968).
Kilum: Manchok, 2100m, fl., 30 Oct. 1996, Pollard 20.

Cyperus unioloides R.Br.
Caespitose perennial; rhizome short, soon dying off, sometimes none; culms 0.3–1 m; leaves 10–60 cm, 1–4 per culm; leaf sheaths usually reddish brown; inflorescence an anthela of 1 sessile and 1–7 stalked heads of spikelet clusters; peduncle of stalked heads to 10 cm with a purple tubular basal prophyll. Swamp.
Syn. **Pycreus unioloides** (R.Br.) Urb., F.W.T.A. 3: 300 (1968).
Mbi Crater: Approach to Mbi crater, by short cut from Belo-Afua, 1750m, 6.05N, 10.23E, fl. fr., 9 Dec. 1998, Cheek 9886.

Cyperus sp. A
RED DATA CANDIDATE
Putative new species. Caespitose perennial, 30–50 cm; leaves filiform, almost as long as culm, 2–3 mm broad; inflorescence composed of subsessile spikelets; spikelets black; involucral bracts 2, unequal in length; glumes with a white margin. Swamp.
Ijim: Swamp, E of Ardo's compound, Laikom, 2000m, 6.16N, 10.22E, fl. fr., 5 Dec. 1998, Cheek 9789; Sappel, below Ardo's uncle's compound, 2300m, 6.16N, 10.24E, fl. fr., 5 Dec. 1998, Cheek 9809; Cattle watering hole by Fulani settlement, 11 km E of Laikom, near Mbesa, 2400m,

6.1615N, 10.2435E, fl. fr., 9 Dec. 1998, Pollard 355.

Cyperus sp. B
RED DATA CANDIDATE

Putative new species. A slender weak perennial to 30 cm; leaves few, yellowish; inflorescence of 4–20 pedunculate spikelets; spikelets with peduncles of differing length; glumes reddish brown, the margin yellowish green.

Ijim: Afua swamp, 1950m, 6.08N, 10.24E, fl., 7 Dec. 1998, Cheek 9835.

Cyperus sp. C
RED DATA CANDIDATE

Local name: Nseansea (Kom – Yama Peter) – Pollard 336. Putative new species. Erect, rhizomatous perennial, leaves to 35 × 0.7 cm, reddish green; inflorescence with one central subsessile spikelet and 5–20 pedunculate clusters of spikelets; spikelets reddish brown; glumes 2.5 mm, margin pale reddish-brown; style-branches 3; achene triangular, papillose. Swampy sites.

Ijim: Fundong, near Touristic House, by waterfall, 6.15N, 10.26E, fl., 23 Nov. 1996, Onana 600; Afua swamp, 1950m, 6.08N, 10.24E, fl., 7 Dec. 1998, Cheek 9839; Swamp, half a mile W of Ardo's Fulani compound, Ijim ridge, Laikom, 2060m, 6.1609N, 10.2051E, fl., 7 Dec. 1998, Pollard 336.

Cyperus sp.

Specimens immature, though probably *Cyperus mannii*.

Kilum: KA, 2760m, fl., 10 Jun. 1996, Asonganyi 1314; Summit, 2800m, fl., 10 Jun. 1996, Zapfack 831.
Ijim: From Aboh, path towards Tum, 2100m, 6.1116N, 10.2543E, fl., 22 Nov. 1996, Etuge 3577.

Fuirena stricta Steud. subsp. *chlorocarpa* (Ridl.) Lye

Loosely caespitose perennial; stems 20–70 cm; inflorescence of 1–several sessile or pedunculate clusters of 2–8 spikelets; spikelets terete; glumes arranged spirally, not in 5 rows. Swamp.

Ijim: Sappel, below Ardo's uncle's compound, 2300m, 6.16N, 10.24E, fl. fr., 5 Dec. 1998, Cheek 9820; Swamp, half a mile W of Ardo's compound, Ijim ridge grassland plateau, Laikom, 2060m, 6.1609N, 10.2051E, fl., 7 Dec. 1998, Pollard 337.
Mbi Crater: 1950m, 6.05N, 10.21E, fl., 9 Dec. 1998, Cheek 9897.
Jakiri: Jakiri, 2000m, fl., 18 Feb. 1958, Hepper 2062; Near 'Journey's End' House, 1660m, fl., 18 Jun. 1962, Brunt 534.

Isolepis fluitans (L.) R.Br. var. *fluitans*

A low-growing, carpet-forming perennial with creeping, much-branched stems, 5–50 cm; inflorescence a solitary pedunculate spikelet; glumes 5–8 per spikelet, greyish or brownish above and with green midrib. Swamp.
Syn. *Scirpus fluitans* L., F.W.T.A. 3: 309 (1968).
Kilum: Ntogemtuo swamp, Oku summit, 2900m, fl., 14 May 1997, Maisels 53.
Ijim: Swamp E of Ardo's compound, 6.16N, 10.23E, fl., 5 Dec. 1998, Cheek 9798.

Isolepis setacea (L.) R.Br. var. *setacea*

A low-growing, dark green, leafy, glabrous annual; inflorescence of 1–3 sessile, apparently lateral spikelets; glumes, the upper greyish, the lower dark reddish brown, to almost black. Wet places.
Syn. *Scirpus setaceus* L., F.W.T.A. 3: 309 (1968).
Kilum: Path from BirdLife HQ to KA, 2500m, 6.1349N, 10.3112E, fl., 30 Oct. 1996, Cheek 8508.

Lipocarpha chinensis (Osb.) Kern

Caespitose perennial; leaves persistant at the base; spikes pale and blunt, well-separated, 4–12, oval to rounded conical; lowest scales of spikes fall off to reveal notched spike axis. Wet places.
Jakiri: Jakiri, 2000m, fl., 18 Feb. 1958, Hepper 2064; Near 'Journey's End' House, 1660m, fl., 18 Nov. 1962, Brunt 535.

Rhynchospora brownii Roem. & Schult.

Slender, leafy perennial; stems 0.4–1 m; leaves 5–40 cm; inflorescence a slender panicle of terminal corymbose clusters, each of 5–15 spikelets; spikelets elliptical, 4–5 mm, brown. Swamp.
Syn. *Rhynchospora rugosa* (Vahl) Gale, F.W.T.A. 3: 333 (1968).
Ijim: Swamp below Ardo's compound, 2100m, 6.16N, 10.21E, fl. fr., 7 Dec. 1998, Gosline 205; Ngengal swamp, 10 km E of Laikom, en route to Mbesa, 2460m, 6.1615N, 10.2435E, fl. fr., 9 Dec. 1998, Pollard 348.

Rhynchospora corymbosa (L.) Britt.

A coarse, leafy, caespitose perennial, 0.6–2.5 m; leaves with minute spinose teeth on the margin and midrib; inflorescence of one terminal and several lateral corymbs; glumes reddish brown. Swamp; wet places.
Ijim: Afua swamp, 1950m, 6.08N, 10.24E, fl., 7 Dec. 1998, Cheek 9838.
Mbi Crater: Ndawara area, 1900m, 6.05N, 10.21E, fl., 9 Dec. 1998, Etuge 4582.
Jakiri: Jakiri, 2000m, fl., 18 Feb. 1958, Hepper 2075.

Schoenoplectus corymbosus (Roth ex Roem. & Schult.) J.Raynal var. *brachyceras* (A.Rich.) Lye

A stout, tough, leafless perennial to 2 m; inflorescence spreading, pseudolateral; main inflorescence bract 1–3 cm, stiff and leaf-like, boat-shaped; spikelets acute, reddish brown, clustered. Swamp.
Syn. *Scirpus brachyceras* Hochst. ex A.Rich., F.W.T.A. 3: 311 (1968).
Kilum: Ntogemtuo swamp, Oku summit, in standing water on tussocks of *Sphagnum* moss, 2950m, fl., 14 May 1997, Maisels 46; Mount Oku, 2900m, fl., 25 Jun. 1998, Maisels 140.
Ijim: Laikom, Maitland 1546; Swamp E of Ardo's compound, 2000m, 6.16N, 10.22E, fl., 5 Dec. 1998, Cheek 9786; Sappel, below Ardo's uncle's compound, 2300m, 6.16N, 10.24E, fl., 5 Dec. 1998, Cheek 9817; Cattle watering hole by Fulani settlement, 11 km E of Laikom, 2400m, 6.1615N, 10.2435E, fl., 9 Dec. 1998, Pollard 353.
Mbi Crater: Near outlet, fl., 11 Feb. 1999, Maisels 1001.
Jakiri: Hepper 1960.

Scleria achtenii De Wild.

Perennial 0.4–1.3 m; inflorescence of 1 (rarely 2), pendulous lateral panicles; peduncles long and slender; glumes glabrous, stramineous or somewhat reddish. Swamp.
Mbi Crater: 1950m, 6.0536N, 10.21E, fl. fr., 9 Dec 1998, Cheek 9893.

Scleria distans Poir. var. distans

A slender perennial with a creeping rhizome and numerous stems at 2–20 mm intervals; stems 20–90 cm (usually 30–60) cm; inflorescence a lax 'spike', 5–10 cm, reddish brown; glumes reddish-brown to blackish. Seasonally wet grassland.
Syn. *Scleria nutans* Willd. ex Kunth., F.W.T.A. 3: 344 (1968).
Ijim: Sappel, below Ardo's uncle's compound, 2300m, 6.16N, 10.24E, fl. fr., 5 Dec. 1998, Cheek 9800; Cattle watering hole by Fulani settlement, 11 km E of Laikom, en route to Mbesa, 2400m, 6.1615N, 10.2435E, fl., 9 Dec. 1998, Pollard 357.

Scleria hispidior (C.B.Clarke) Nelmes

Caespitose annual; stems crowded; root system minute, grey or purple; stems 5–25 cm, densely set with white hairs; leaves 2–3 on each stem; inflorescence spicate or a narrow panicle, 3–7 cm; bracts glume-like. Wet places.
Kilum: Shambai, 2500m, fl., 5 Nov. 1996, Munyenyembe 875; Ewook Etele Mbae, 10 km from Oku, 2350m, 6 Nov. 1996, Zapfack 1210.

Scleria interrupta Rich.

This species is closely related to *Scleria melanotricha*, but differs in having erect or spreading glomerules. Grassland.
Syn. *Scleria hirtella* Sw., F.W.T.A. 3: 344 (1968).
Ijim: Ijim plateau above Fon's palace, Laikom, 1900m, 6.16N, 10.19E, fl. fr., 3 Dec. 1998, Cheek 9747; Grassland plateau above Fon's palace, Laikom, 2020m, 6.1642N, 10.1950E, fl. fr., 3 Dec. 1998, Pollard 284.

Scleria melanotricha Hochst. ex A.Rich. var. grata (Nelmes) Lye

A slender annual; stems 10–50 cm; leaves 1–2 mm wide, hairy; inflorescence spicate, 3–20 cm; glomerules paired, shortly pedunculate; glumes straw-coloured to reddish brown, densely hairy. Wet places.
Syn. *Scleria grata* Nelmes, F.W.T.A. 3: 346 (1968).
Ijim: From Aboh, path towards Tum, 2200m, 6.1116N, 10.2543E, fl., 22 Nov. 1996, Etuge 3597.
Bambili: Bauer 122.

Cyperaceae sp. 1

Very unusual and interesting climbing herb to 6m, sterile voucher, fertile material most desirable.
Ijim: Aboh, Gikwang road towards Nyasosso forest, 2300m, st., 19 Nov. 1996, Etuge 3501.

Cyperaceae sp. 2

Another unusual collection, more material required.
Ijim: Ijim plateau above Fon's palace, Laikom, 1900m, 6.16N, 10.19E, fr., 3 Dec. 1998, Cheek 9756.

DIOSCOREACEAE
Det. P.Wilkin (K)

Dioscorea schimperiana Hochst. ex Kunth

Pubescent climber, 3–7 m. Upland areas.
Kilum: Mbijame forest, 2000m, fr., 4 Nov. 1996, Munyenyembe 864; Lumeto Forest near Oku-Elak, 6.1412N 10.2943E, 2200m, 7 Nov. 1996, Cheek 8613.
Ijim: Above Fon's palace, Laikom, about 30 minutes walk on path to Fulani settlement, 6.1642N 10.1952E, 2050m, 21 Nov. 1996, Cheek 8706; Zitum road, 6.1116N 10.2543E, 1950m, 21 Nov. 1996, Etuge 3530; Aboh village, along the main road, 6.15N 10.26E, 1700–1800m, fr., 24 Nov. 1996, Onana 605; Main path from Ijim to Tum, 6.1116N 10.2543E, 2100m, fr., 25 Nov. 1996, Etuge 3617; Laikom, forest, 2000m, fr., 3 Dec. 1998, Etuge 4523.

DRACAENACEAE

Dracaena fragrans (L.) Ker-Gawl

Shrub about 1.5 m; flowers white with pink lines, very fragrant. Forest
Syn. *Dracaena deisteliana* Engl., F.W.T.A. 3: 157 (1968).
Kilum: Elak, 2300m, 8 Jun. 1996, Zapfack 747; KA, 2200m, fr., 11 Jun. 1996, Etuge 2319; Manchok forest, above Elak-Oku, 2300m, fl., 13 Mar. 1997, Maisels 40.

ERIOCAULACEAE
Det. S.M.Phillips (K)

Eriocaulon asteroides S.M.Phillips
RED DATA LISTED
Annual rosulate herb, about 2–3 cm diameter; leaves linear-subulate, 0.8–1.5 cm, about 1 mm wide; scapes up to 10, 1–2.5 cm; capitula 5–7 mm wide, few flowered, star-like. Basalt pavement in grassland.
Kilum: Elak to Kumbo, Iwooketele Mbae, 2500m, 5 Nov. 1996, Zapfack 1204A.
Ijim: Laikom, Ijim plateau above Fon's Palace, 1900m, fr., 3 Dec. 1998, Cheek 9749.

Eriocaulon bamendae S.M.Phillips
RED DATA LISTED
Perennial rosulate herb, about 10 cm diameter when sterile to 30–40 cm diameter when fertile; leaves 10–22 cm (when fertile); scapes to 15, 30–90 cm; capitula 4–7 mm wide, globose, black and white, often viviparous. Swamp.
Syn. *Eriocaulon zambesiense* sensu F.W.T.A. 3: 62 (1968) non Ruhland.
Kilum: Summit, Kinkolong swamp, small running stream in subalpine prairie, 2950m, fl., 13 May 1998, Maisels 115; Tadu stream, open grassy bog near stream, 2000m, 11 Jul. 1998, Maisels 146.
Ijim: Laikom, in a pond, 1830m, Jun. 1931, Maitland 1400.
Kumbo: Mile 3 on Kumbo to Oku road, open very wet stagnant flush, 1850m, fl., 15 Feb. 1958, Hepper 2021.

Eriocaulon parvulum S.M.Phillips
RED DATA LISTED
Annual rosulate herb; 1.5–3 cm diameter; leaves linear
subulate, 0.8–1.5 cm × c. 1 mm wide, acute; scapes about 7,
1.5–3 cm; capitula subglobose, 4–4.5 mm wide, dirty white,
± glabrous. Basalt pavement in grassland.
Local name: Mbasigok (Oku) – Munyenyembe 877.
Kilum: Elak to Kumbo, Iwooketele Mbae, 2500m, 5 Nov.
1996, Zapfack 1204; Munyenyembe 877.
Ijim: Laikom, Ijim plateau above Fon's Palace, 1900m, fr., 3
Dec. 1998, Cheek 9748.

GRAMINEAE
Det. T.A.Cope (K)

Acritochaete volkensii Pilg.
Perennial, with weak trailing culms to 1 m; inflorescence of
1–4 slender racemes appressed to the main axis.
Ijim: Above Aboh village, 1.5 hours walk along Gikwang
road, 2350m, 6.1112N, 10.2529E, 18 Nov. 1996, Cheek
8644; Ijim Mountain Forest, 2610m, fl., 19 Nov. 1996,
Kamundi 610.

Agrostis mannii (Hook.f.) Stapf subsp. *mannii*
RED DATA LISTED
Caespitose perennial to 1 m; leaf-blade about 3 mm wide;
inflorescence a loose, open panicle, 35–40 cm, with flexuose
filiform branches; spikelets 4–4.5 mm, purple. Forest edge.
Kilum: Prairie, 2800m, fl., Aug. 1970, C.N.A.D. 1729.
Verkovi: Verkovi to summit, 2850m, fl., 8 Dec. 1974,
Letouzey 1729.

Agrostis quinqueseta (Steud.) Hochst.
Perennial to 1 m; panicle dense and spike-like; spikelets
3.5–4.5 mm. Wet places.
Ijim: Aboh village, 2450m, fr., 19 Nov. 1996, Onana 545.

Aira caryophyllea L.
Slender annual 5–30 cm; panicle delicate, with spreading
filiform branches; spikelets 2.7–3.2 mm. Grassland; fallow.
Kilum: Ewook Etele Mbae, 10 km from Oku, 2350m, 6 Nov.
1996, Zapfack 1225.
Ijim: Gikwang road, Aboh to Akeh, about 1 km after first
track, 2400m, 6.15N, 10.26E, 25 Nov. 1996, Cheek 8759.

Andropogon amethystinus Steud.
Perennial about 30 cm, tufted or with short underground
rhizomes; leaf-blades to 4 mm broad, herbaceous; spikelets
pilose, 6–7 mm; internodes and pedicels ciliate on both
margins, glabrous on the face; racemes paired. Forest edge.
Ijim: Aboh, 2450m, fl., 19 Nov. 1996, Onana 546; Ijim
Mountain Forest, along the ridge towards a point where Lake
Oku can be viewed, 2500m, 22 Nov. 1996, Kamundi 694.

Andropogon gabonensis Stapf
Robust perennial to 3.5 m; pedicelled spikelet 4–4.5 mm;
sessile spikelet 4–5 mm, usually with a line of hairs along
the median groove; external ligule absent. Roadside.
Ndop: In gully, 10 miles from Ndop along Bamenda road,
1800m, fl., 20 Dec. 1952, Boughey 10464.

Andropogon lacunosus J.G.Anderson
Perennial to 60 cm; inflorescence of 2–3 digitate racemes.
Swamp.
Mbi Crater: 1950m, fr., 9 Dec. 1998, Cheek 9892.

Andropogon lima (Hack.) Stapf
Densely tufted perennial to 1 m; leaf blades 30 cm or more ×
0.2 cm; spikelets sessile 6–8 mm. Grassland.
Kilum: KA, 2760m, 10 Jun. 1996, Asonganyi 1304; Summit,
2800m, fl., 31 Oct. 1996, Zapfack 1125.

Andropogon schirenis Hochst. ex A.Rich.
Local name: Hahaendehoh (Bororo at Bambui) – Pedder 2.
Erect caespitose perennial to 2 m; leaf blades linear, to 45 ×
1.4 cm, mostly cauline; racemes 6–12 cm; sessile spikelets
5–7 mm. Forest edge.
Syn. *Andropogon dummeri* Stapf, F.W.T.A. 3: 486 (1972).
Kilum: Oku-Elak, lower parts of KA, 2200m, 6.1349N,
10.3112E, fl., 28 Oct. 1996, Cheek 8486.
Bambui: Bambui, fl., Dec. 1945, Pedder 2; Bambui, fl., 21
Dec. 1952, Boughey GC 10384.

Arthraxon hispidus (Thunb.) Makino var.
hispidus
Decumbent annual; stems tough and wiry, rooting at the
lower nodes, hairs on the internodes to 0.5 mm; leaf blades
to 7 × 2 cm; spikelets 3–4 mm. Forest edge.
Syn. *Arthraxon quartinianus* (A.Rich.) Nash, F.W.T.A. 3:
470 (1972).
Ijim: Gikwang road towards Nyasosso forest, 2300m, fl., 19
Nov. 1996, Etuge 3490; Ijim Mountain Forest Reserve, near
Chufekhe stream, 2010m, 6.1124N, 10.2525E, fl., 20 Nov.
1996, Kamundi 650; Laikom, 1900m, fr., 3 Dec. 1998,
Cheek 9734.
Bambui: Bambui Farm, fl., Jul. 1945, Bumpus 13.

Arundinaria alpina K.Schum.
Large woody bamboo; culms to about 6 m or more , 3–5 cm
diameter, hollow, with thick walls. Forest.
Kilum: Oku, in open area where road cuts through tall forest
at mile 11, 2000m, fl., 17 Feb. 1958, Hepper 2045; KD,
3000m, 10 Jun. 1996, Etuge 2292; 2600m, 10 Jun. 1996,
Zapfack 834; Lake Oku, 2200m, 12 Jun. 1996, Etuge 2333;
Elak, 2200m, fl., 1 Nov. 1996, Satabie 1063.
Ijim: Anyajua Forest, along the ridge, 2300m, 27 Nov. 1996,
Kamundi 726.

Brachypodium flexum Nees
Weak-stemmed perennial, 30–90 cm; leaf-sheaths nearly
always scabrid; racemes 6–12 cm, usually with 5–8 narrowly
lanceolate spikelets. Forest edge; grassland.
Kilum: Summit, 2950m, 10 Jun. 1996, Cable 3087.

Chloris pycnothrix Trin.
Stoloniferous annual or rarely perennial to 30 cm;
inflorescence tinged with pink; fallow land and lawns.
Forest edge.
Kilum: Near maternity hospital, 2000m, 10 Jun. 1996,
Asonganyi 1316.
Ijim: Anyajua, 1800m, fl., 27 Nov. 1996, Kamundi 717.

Bambui: Bambui Experimental Station, 1660m, fl., 15 Jun. 1962, Brunt 507.

Coelorhachis afraurita (Stapf) Stapf
Perennial to nearly 2 m; basal sheaths strongly compressed; racemes 5–7 cm. Swamp.
Jakiri: Jakiri, overgrown pond, 2000m, fl., 18 Feb. 1958, Hepper 2073.

Ctenium ledermannii Pilg.
Perennial 60–90 cm with 2–5, dark green, paired (rarely solitary) digitate spikes. Forest edge.
Ijim: Gikwang road towards Nyasosso forest - Ijim Mountain Forest Reserve, 2300m, fl., 19 Nov. 1996, Etuge 3504.

Digitaria abyssinica (Hochst. ex A.Rich.) Stapf
A slender creeping perennial with culms to about 30 cm. Fallow weed; fallow.
Kilum: Near maternity hospital, 2000m, fl., 10 Jun. 1996, Asonganyi 1318 & 1332; Oku-Elak, lower parts of KA, 2200m, 6.1349N, 10.3112E, fl., 28 Oct. 1996, Cheek 8482.
Ijim: Ijim Mountain Forest Reserve, near Chufekhe stream, 2010m, 6.1124N, 10.2525E, 20 Nov. 1996, Kamundi 653.
Bambui: Bambui Farm, fr., May 1943, Pedder 24; Bambui, fr., 21 Dec. 1952, Boughey GC 10414; Hill behind Bambui, 2460m, fl., 21 Dec. 1952, Boughey GC 10859.

Digitaria diagonalis (Nees) Stapf var. diagonalis
Robust, erect perennial herb, 1–3 m, with handsome spreading inflorescence; spikelets glabrous. Grassland.
Ijim: Plateau above Fon's Palace, Laikom, 2020m, fr., 3 Dec. 1998, Pollard 289.

Digitaria diagonalis (Nees) Stapf var. hirsuta (De Wild. & T.Durand) Troupin
Local name: Gene Sabere (Bororo at Bambui) – Pedder 3.
As the above taxon, but with densely silky base. Grassland.
Bambui: Bambui, fr., Dec. 1945, Pedder 3; Bambui Farm, 2000m, fr., 21 Dec. 1952, Boughey GC 10751.

Digitaria debilis (Desf.) Willd.
Annual, geniculately ascending, to 60 cm; spikelets 2.5–3 mm, obscurely puberulous. Damp places.
Kilum: Near maternity hospital, 2000m, 10 Jun. 1996, Asonganyi 1319; KA, 2200m, fl., 11 Jun. 1996, Etuge 2311.
Bambui: Bambui Farm, fl., May 1943, Pedder 25; Bambui Experimental Station, 1600m, fl., 15 Jun. 1962, Brunt 509.
Babungo: Babungo Agriculture Department Farm, 1260m, fl. fr., 17 May 1962, Brunt 431.

Digitaria ternata (A.Rich.) Stapf
Caespitose annual to 60 cm; spikelets 1.8–2 mm; fruits black. Fallow weed.
Bambui: Bambui Exp. Station, 1600m, fl., 10 Aug. 1963, Brunt 1241.

Echinochloa crus-pavonis (Kunth) Schult.
Erect annual, 1–2 m; racemes compound, to 10 cm, spikelets 2.5–3 mm, densely clustered on short side branches. Near water.

Ijim: Fundong to Belo road, 2050m, fl., 21 Nov. 1996, Cheek 8722; Afua swamp, 1950m, fl., 7 Dec. 1998, Cheek 9831.
Babungo: Near Babungo, 1260m, fr., 5 Mar. 1962, Brunt 112; Babungo, 1260m, fr., 12 Feb. 1963, Brunt 962; Babungo sedge swamp, 1260m, fr., 10 Jul. 1962, Brunt 864.
Kumbo: By River Wi, to S of town, fl. fr., 14 Feb. 1958, Hepper 1983.

Eleusine indica (L.) Gaertn.
Erect annual to 60 cm; spikes 3–6 mm broad, slender, straight; spikelets 4–5.5 mm. Forest edge; wayside weed.
Kilum: Near maternity hospital, 2000m, 10 Jun. 1996, Asonganyi 1334; Path from BirdLife HQ to KA, 2100m, 6.1349N, 10.3112E, 30 Oct. 1996, Cheek 8527.
Ijim: Jikijem, 2200m, fl., 5 Nov. 1996, Pollard 48; Anyajua Forest, montane forest along the ridge and disturbed farmland and abandoned garden on lower slopes, 1900m, fl., 27 Nov. 1996, Kamundi 714.
Bambui: Bambui Farm, fl., May 1943, Pedder 27; Bambui Experimental Station, fl., 15 Jun. 1962, Brunt 506.
Bambili: College of Arts and Sciences, 1660m, fl., 11 Feb. 1970, Bauer 6.

Elymandra androphila (Stapf) Stapf
Coarse perennial to 2.5 m; racemes paired; pedicelled and homogamous spikelets glabrous. Fallow.
Bambui: Bambui Farm, 2100m, fr., 21 Dec. 1952, Boughey GC 10387.

Eragrostis atrovirens (Desf.) Trin. ex Steud.
A very variable perennial; culms 0.45–1 m, 1.5–3 mm diameter at base, leafy; leaf-blades 15–30 × 0.2–0.4 cm, flat or rolled; spikelets pallid to grey-purple. Swamp; wet places.
Kilum: Near maternity hospital, 2000m, 10 Jun. 1996, Asonganyi 1317 & 1326; Junction of KJ and KN above Elak, 2200m, 10 Jun. 1996, Cable 3022; Shore of Lake Oku, 2250m, 12 Jun. 1996, Cable 3108;
Ijim: Swamp E of Ardo's compound, Laikom, 2200m, fr., 5 Dec. 1998, Cheek 9799.

Eragrostis camerunensis Clayton
Densely tufted perennial, 30 cm; leaf-blades 3–7 × 0.1–0.2 cm, usually rolled; spikelets dark grey. Grassland; fallow weed.
Kilum: Elak, KD, fl., 9 Jun. 1996, Etuge 2203; Near maternity hospital, 2000m, 10 Jun. 1996, Asonganyi 1324; Ewook Etele Mbae, 10 km from Oku, 2350m, 6 Nov. 1996, Zapfack 1228.
Ijim: Nchan, fl. fr., May 1931, Maitland 10A; Ijim plateau above Fon's Palace, Laikom, 1900m, fr., 3 Dec. 1998, Cheek 9733.
Bambui: Bambui Farm, fl. fr., 1943, Pedder 23; Hill behind Bambui, fl. fr., 21 Dec. 1952, Boughey GC 10861.
Bambili: Bambili, 1 mile N of Lakes, 2160m, fl. fr., 28 Feb. 1970, Bauer 36.
Kumbo: Kumbo, 1600m, fl. fr., 1 Feb. 1970, C.N.A.D. 1788.

Eragrostis gangetica (Roxb.) Steud.
Loosely tufted annual; culms slender, geniculate, 15–45 cm;
a rather variable species. Roadside weed.
Bambui: Bambui, 1660m, fl. fr., 16 Feb. 1960, Cowell G20.

Eragrostis macilenta (A.Rich.) Steud.
Slender annual, 30–60 cm; spikelets very dark green to
almost black, 6–14-flowered. Fallow, roadsides.
Ijim: Nchan, 1660m, fl. fr., May 1931, Maitland 10C.
Bambui: Bambui Farm, Pasture 2, fl., May/Jun. 1943,
Pedder 15.

Eragrostis mokensis Pilg.
Local name: Ngei (Oku) – Munyenyembe 802.
Slender annual to 30 cm; much branched from the base;
spikelets 3–5.5 mm. Fallow.
Kilum: Oku-Elak, lower parts of KA, 2200m, 6.1349N,
10.3112E, fl. fr., 28 Oct. 1996 Cheek 8484; Elak, 2200m, 30
Oct. 1996, Munyenyembe 802.

Eragrostis pobeguinii C.E.Hubb.
Densely caespitose perennial, about 30 cm; panicle scantily
branched, bearing to 15 spikelets, pallid to olive-grey; leaf-
blades to 2 mm wide, usually rolled and setaceous; basal
sheaths bulbously swollen and hardened below. Grassland.
Ijim: Grassland plateau above Fon's Palace, Laikom,
2020m, fl., 3 Dec 1998, Pollard 287.

Eragrostis tenuifolia (A.Rich.) Hochst. ex Steud.
Slender caespitose perennial, 30–60 cm; leaves mostly
arising from tussocky base; spikelets dark green, the margins
conspicuously saw-toothed. Wayside weed.
Kilum: Near maternity hospital, 2000m, 10 Jun. 1996,
Asonganyi 1320 & 1333.
Ijim: Anyajua Forest, montane rain forest along the ridge
and disturbed farmland and abandoned garden on lower
slopes, 1800m, fl., 27 Nov. 1996, Kamundi 720.
Bambui: Bambui Experimental Station, 1660m, fr., 15 Jun.
1962, Brunt 508.
Bambili: College of Arts and Sciences, 1660m, fl. fr., 11
Feb. 1970, Bauer 7.

Eragrostis volkensii Pilg.
Straggly perennial with a dense tussock; culms 30–90 cm,
slender, wiry, many-noded; spikelets dark olive-green.
Grassland; wet places.
Kilum: Hill above Lake Oku, 2660m, fl. fr., 7 Jan. 1951,
Keay & Lightbody FHI 28500.
Ijim: Aboh village, 2450m, fr., 19 Nov. 1996, Onana 547;
Ijim Mountain Forest Reserve, near Chufekhe stream,
2010m, 6.1124N, 10.2525E, 20 Nov. 1996, Kamundi 652;
Ijim Mountain Forest Reserve, 2280m, fr., 20 Nov. 1996,
Pollard 64; Ijim, 2000m, fr., 2 Dec. 1998, Etuge 4514.
Bambui: Bambui Farm, 2000m, fr., 21 Dec. 1952, Boughey
GC 10750.

Festuca camusiana subsp. *chodatiana* St.-Yves
Loosely tufted perennial 60–90 cm; leaf-blades 1–2 mm
wide. Forest edge.
Ijim: Aboh, 2450m, fr., 20 Nov. 1996, Onana 549.

Festuca mekiste Clayton
Densely caespitose perennial, 15–60 cm; leaf-blades
narrowly linear, convolute. Forest edge.
Kilum: KA, 2760m, 10 Jun. 1996, Asonganyi 1307; Summit,
2950m, 10 Jun. 1996, Cable 3088; 3000m, 29 Oct. 1996,
Munyenyembe 795; KA to summit, 2800m, 6.1349N,
10.3112E, fl., 30 Oct. 1996, Cheek 8529; Summit, 2800m,
fr., 31 Oct. 1996, Zapfack 1116.
Ijim: Ijim Mountain Forest Reserve, along the ridge to a
point where Lake Oku can be viewed, 2400m, 22 Nov. 1996,
Kamundi 685.

Helictotrichon elongatum (Hochst. ex A.Rich.)
C.E.Hubb.
Perennial to 1.5 m; spikelets 0.8–1.6 cm; lower glume 5–12
mm, almost as long as the spikelet. Forest edge; grassland.
Ijim: Laikom, on grass covered plateau, fl., May 1931,
Maitland 9A; Gikwang road towards Nyasosso forest,
2300m, fl., 19 Nov. 1996, Etuge 3492.
Bambui: Bambui Farm, 2100m, fl., 21 Dec. 1952, Boughey
GC 10383; Agricultural Experimental Station, Bambui, fl.,
20 Sep. 1968, Nditapeh B47.
Jakiri: 21 km on road from Bamenda to Jakiri and then 2 km
along a track to the left, 1870m, 6.01N, 10.18E, fl., 4 Nov.
1975, de Wilde 8624.

Hyparrhenia bracteata (Humb. & Bonpl. ex
Willd.) Stapf
Densely tufted perennial; spikelet sessile, 4–6 mm.
Wayside; fallow.
Bambui: Bambui Farm, 1330–2000m, fl. fr., Jul. 1946,
Bumpus 26.
Jakiri: Footpath from Ring Road to Mbawver, 1660m, fl., 23
Jun. 1962, Brunt 682.

Hyparrhenia cymbaria (L.) Stapf
Robust perennial; culms 2–3.5 m, initially slender and
rambling, subsequently erect and sustained by stilted roots.
Forest edge; fallow.
Ijim: Gikwang road towards Nyasosso forest, 2300m, 19
Nov. 1996 Etuge 3493.
Bambui: Bambui Farm, 1330m, fl., 11 Aug. 1963, Brunt
1269.

Hyparrhenia diplandra (Hack.) Stapf
Coarse perennial, 2–3 m. Wet places.
Kilum: Near maternity hospital, 10 Jun. 1996, Asonganyi
1330.

Hyparrhenia filipendula (Hochst.) Stapf var.
filipendula
Caespitose perennial; culms 0.6–2 m; spikelets glabrous.
Open disturbed places.
Jakiri: Below Jakiri, 1500m, fl., 27 Jun. 1962, Brunt 766.

Hyparrhenia newtonii (Hack.) Stapf
Local name: Yemuwel Debbo (Bororo, Bambui) – Pedder 7.
Densely tufted perennial; culms 0.6–1.2 m; panicle loose,
often scanty. Grassland.
Bambui: Bambui, fl., Dec. 1945, Pedder 7.

Kumbo: Banso, grassland along Banso-Bamenda Motor Road, fl., 20 Oct. 1947, Tamajong FHI 23486.

Hyparrhenia poecilotricha (Hack.) Stapf

Perennial, 0.6–1.5 m; basal leaf sheaths glabrous. Savanna.
Bambui: Bambui Experimental Station, 1660m, fl., 25 Jun. 1963, Brunt 1185.

Hyparrhenia rufa (Nees) Stapf

Perennial or sometimes annual, 0.3–2.4 m; spikelets sessile, 3–5 mm; typically a savanna grass. Wet places; roadside.
Bambui: Bambui Farm, fl., May 1943, Pedder 11; Bambui Experimental Station, 1660m, fl., 12 Jul. 1963, Brunt 1191.
Bambili: Bambili, 1660m, fl., 14 Mar. 1970, Bauer 49.
Jakiri: Jakiri, footpath from Ring Road to Mbawver, near pit 134, 1660m, fl., 23 Jun. 1962, Brunt 680.

Hyparrhenia smithiana (Hook.f.) Stapf var. *major* Clayton

Local name: Gene Shabal (Bororo at Bambui) – Pedder 1.
Caespitose perennial, 1.5–2.4 m; spikelet indumentum rufous or fulvous. Grassland; savanna.
Kilum: Hill above Lake Oku, grassland, 2660m, fl., 7 Jan. 1951, Keay & Lightbody FHI 28499; Road to Oku, 2050m, fl., 22 Jun. 1962, Brunt 652; Oku-Elak, lower parts of KA, 2200m, 6.1349N, 10.3112E, 28 Oct. 1996, Cheek 8489.
Ijim: Gikwang towards Nyasosso, 2250m, fl. 18 Nov. 1996, Etuge 3441; Ijim Mountain Forest, 2160m, 19 Nov. 1996, Kamundi 614; Laikom, 2400m, fr., 19 Nov. 1996, Satabie 1075; Ijim plateau above Fon's Palace, Laikom, 1900m, fr., 3 Dec. 1998, Cheek 9741.
Bambui: Bambui, fl., Dec. 1945, Pedder 1; Bambui, 2000m, fl., 21 Dec. 1952, Boughey GC 10752 & 10784.
Kumbo: Kumbo to Lake Oku, mile 3, montane grassland, 2000m, fl., Feb. 1958, Charter FHI 38004.

Hyparrhenia umbrosa (Hochst.) T.Anderson ex Clayton

Local name: Shambawal (Bororo at Bambui) – Pedder 4.
Culms stout, 1.2–1.8 m, supported by stilt roots. Fallow; roadside.
Kilum: Roadside above Djottin, 1660m, fl., 21 Jun. 1962, Brunt 591.
Bambui: Bambui, fl., Dec. 1945, Pedder 4.

Hyparrhenia sp. nov. ?
RED DATA CANDIDATE
Kilum: Bui, Elak, near maternity hospital, 2000m, fl., 10 Jun. 1996, Asonganyi 1327 & 1328.
Ijim: Swamp E of Ardo's compound, Laikom, fr., 5 Dec. 1998, Cheek 9792.

Imperata cylindrica (L.) Raeusch.

Perennial; culms to 1.3 m, from extensively creeping rhizomes; leaf-blades usually erect, broader and flatter than the type variety, sword-like; spikelets 3–5.7 mm. Fallow.
Kilum: KD, fl., 9 Jun. 1996, Etuge 2204; Near maternity hospital, 2000m, 10 Jun. 1996, Asonganyi 1325.
Bambili: Bambili, 2000m, fl., 28 Feb. 1970, Bauer 24.

Leersia hexandra Sw.

Perennial; culms weak, usually long decumbent, arising from a rhizomatous base, forming matted carpets 30–90 cm in shallow water; the retrorse spinulose midrib of the leaf can inflict the most painful lacerations; spikelets 3.5–4 mm. Swamp.
Mbi Crater: 1950m, fr., 9 Dec. 1998, Cheek 9894.
Babungo: Babungo, Ndop Plain, 1330m, fl., 12 Feb. 1963, Brunt 963.
Jakiri: Jakiri-Kumbo road, fr., Feb. 1958, Charter FHI 38006; Near 'Journey's End' House, 1660m, fl., 18 Jun. 1962, Brunt 532.

Loudetia arundinacea (A.Rich.) Steud.

Tufted perennial to 3 m; leaf-blades 30–70 cm; panicle robust, 20–60 cm. Savanna; rocky slopes; swampy soils.
Kumbo: Near Banso, 1930m, fl., 19 Jun. 1962, Brunt 586.

Loudetia phragmitoides (Peter) C.E.Hubb.

A pampas-like perennial to 4.5 m, densely caespitose; panicle 30–60 cm, contracted and dense. Swamp.
Ndop: Baba, Brunt 122 (fide F.W.T.A.).

Loudetia simplex (Nees) C.E.Hubb.

Local name: Sufuel (Bororo at Bambui) – Pedder 6.
Tufted perennial 0.3–1.5 m; basal leaf-sheaths usually woolly tomentose. Grassland; savanna.
Ijim: Above Fon's Palace, Laikom, about 30 minutes walk along the path to Fulani settlement, 2050m, 21 Nov. 1996, Cheek 8719; Ijim plateau above Fon's Palace, Laikom, 1900m, fl., 3 Dec. 1998, Cheek 9738.
Bambui: Bambui, fl., Dec. 1945, Pedder 6.

Melinis effusa (Rendle) Stapf

Perennial; culms 60–90 cm; panicle moderately dense, the branches ascending. Savanna.
Ndop: Near Ndop, 1160m, fl., 20 Dec. 1952, Boughey GC 11129.

Melinis minutiflora P.Beauv. var. *minutiflora*

Perennial, erect or ascending from a prostrate base; culms to 2 m; leaves covered in sticky hairs and smelling strongly of molasses or linseed oil. Forest edge.
Kilum: KD, fl., 9 Jun. 1996, Etuge 2200; Near maternity hospital, 2000m, 10 Jun. 1996, Asonganyi 1331; Towards KJ, 2100m, fl., 30 Oct. 1996, Pollard 24; Lake Oku to Jikijem, 2200m, 6.1214N, 10.2733E, fl., 5 Nov. 1996, Cheek 8604.
Ijim: From Aboh, path towards Tum, 1950m, 6.1116N, 10.2543E, fl., 22 Nov. 1996, Etuge 3581; Isebu, montane rain forest along the ridge, 1650m, fl., 26 November 1996, Kamundi 707; Ijim plateau above Fon's Palace, Laikom, 1900m, fr., 3 Dec. 1998, Cheek 9746; About 2 km E of Laikom, near plot 146 of Kilum-Ijim Forest Project, 1900m, fr., 11 Dec. 1998, Cheek 9914.

Melinis repens (Willd.) Zizka

Annual 6–90 cm; panicles fluffy, silvery white to pink. Forest edge; fallow weed.
Syn. *Rhychelytrum repens* (Willd.) C.E.Hubb., F.W.T.A. 3:

454 (1972).

Kilum: KD, 9 Jun. 1996, Etuge 2249; Near maternity hospital, 2000m, 10 Jun. 1996, Asonganyi 1329; Junction of KJ and KN above Elak, 2200m, 10 Jun. 1996, Cable 3021; Path from BirdLife HQ to KA, 2200m, 6.1349N, 10.3112E, fr., 30 Oct. 1996, Cheek 8495; Oku-Elak, lower parts of KA, 2200m, 6.1349N, 10.3112E, fl. fr., 28 Nov. 1996, Cheek 8479.

Ijim: Anyajua Forest, along the ridge, 1800m, fl., 27 Nov. 1996, Kamundi 719.

Olyra latifolia L.
A cane-like grass to 3 m; leaf-blades 10–20 × 3–6 cm, with about 8 primary nerves either side of the midrib; fascicles of lateral shoots bearing very small leaves occasionally occur. Forest edge.

Bambui: 2 miles N of Bambui on road to Njinikom, 1330m, fr., 7 Aug. 1963, Brunt 1207.

Oplismenus hirtellus (L.) P.Beauv.
Straggling perennial, often with long aerial roots from the nodes; rather variable leaf and panicle characters. Forest.

Kilum: Oku Shrine, 2140m, 3 Nov. 1996, Munyenyembe 858.

Ijim: Zitum road, 1860m, 6.1116N, 10.2543E, fl., 21 Nov. 1996, Etuge 3550; Chuhuku River near Anyajua Forest, riparian forest, 1580m, 28 Nov. 1996, Kamundi 729.

Bambui: Bambui Experimental Station, 1660m, fl., 12 Jul. 1963, Brunt 1188.

Oxyrhachis gracillima (Baker) C.E.Hubb.
Caespitose perennial about 30 cm; leaves filiform. Swamp.

Jakiri: Jakiri, 2000m, fl., 18 Feb. 1958, Hepper 2069.

Panicum acrotrichum Hook.f.
A straggling grass, rooting from the nodes; leaf-blades 3–8 × 0.8–2.3 cm, with transverse veins. Forest.

Kilum: Oku, in partial shade beside path near village, 2000m, fl., 17 Feb. 1958, Hepper 2028; Path from BirdLife HQ to KA, 2600m, 6.1349N, 10.3112E, 30 Oct. 1996, Cheek 8523; KJ, 2400m, fl., 28 Oct. 1996, Buzgo 613; KC-8, 2500m, fl., 30 Oct. 1996, Pollard 12.

Ijim: Above Aboh Village, 1.5 hours walk along Gikwang road to Lake Oku, 2350m, 6.1112N, 10.2529E, 18 Nov. 1996, Cheek 8644.

Panicum calvum Stapf
Perennial, rambling or creeping; leaf-blades lanceolate or narrowly lanceolate; spikelets 2–2.5 mm. Forest edge.

Kilum: KD, 2800m, fl., 10 Jun. 1996, Etuge 2262.

Ijim: Gikwang road towards Nyasosso forest, 2300m, fl., 19 Nov. 1996, Etuge 3491; Ijim Mountain Forest Reserve, along the ridge towards a point where Lake Oku can be viewed, 2500m, fl., 22 Nov. 1996, Kamundi 691; Anyajua, montane rain forest along the ridge, 1900m, fl., 27 Nov. 1996, Kamundi 713.

Panicum ecklonii Nees
Inflorescence an open panicle; spikelets symmetrical in profile. Grassland.

Bambili: Bambili, 1660m, fl., 3 Mar. 1970, Bauer 40.
Bambui: Bambui Experimental Station, 2000m, fl., 4 Mar. 1962, Brunt 77.

Panicum hochstetteri Steud.
Local name: Kekwel (Oku) – Cheek 8480; Ndzifisih (Bekom) – Kamundi 692; Ekfuleh Finnjanngeluh Finjangl (Kom) – Cheek 9742. Loosely tufted perennial; culms to 1 m, weak. Forest edge; grassland.

Kilum: Oku-Elak, lower parts of KA, 2200m, 6.1349N, 10.3112E, fl. fr., 28 Oct. 1996, Cheek 8480.

Ijim: Gikwang road towards Oku, 2260m, fl., 15 Nov. 1996, Etuge 3402; Gikwang road towards Nyasosso forest, 2300m, fr., 19 Nov. 1996, Etuge 3491; Ijim Mountain Forest, 2160m, fl. fr., 19 Nov. 1996, Kamundi 611; Ijim Mountain Forest, along the ridge towards a point where Lake Oku can be viewed, 2500m, fl., 22 Nov. 1996, Kamundi 692; Ijim plateau above Fon's Palace, Laikom, 1900m, fr., 3 Dec. 1998, Cheek 9742; About 2 km E of Laikom near plot 146 of Kilum-Ijim, 1900m, fr., 11 Dec. 1998, Cheek 9913.
EM Plot 146: No. 7.

Panicum monticola Hook.f.
Spikelets 2.5–3 mm. Forest edge; fallow.

Kilum: KC, 2350m, fl., 30 Oct. 1996, Pollard 11.

Ijim: Anyajua, montane forest along the ridge and disturbed farmland and abandoned garden on lower slopes, 1800m, fl., 27 Nov. 1996, Kamundi 721; Akwo-Mufu, 1760m, fr., 5 Dec. 1998, Etuge 4555.

Panicum phragmitoides Stapf
Robust reed-like grass to 2 m; leaf-blades dark green, glabrous or pubescent, 40–70 cm. Savanna.

Ndop: Rest house, Ndop, 1260m, fl., 20 Dec. 1952, Boughey GC 11148.

Panicum pusillum Hook.f.
A weak prostrate annual, forming loose mats; leaf-blades 1–2 cm, occasionally more. Grassland; forest edge.

Kilum: Ewook Etele Mbae, 10 km from Oku, 2350m, 6 Nov. 1996, Zapfack 1217; Oku-Elak, lower parts of KA, 2200m, 6.1349N, 10.3112E, fl. fr., 28 Oct. 1996, Cheek 8485.

Ijim: Gikwang road towards Nyasosso forest, 2300m, fl., 19 Nov. 1996, Etuge 3478; Near Chufekhe, 2010m, 6.1124N, 10.2525E, 20 Nov. 1996, Kamundi 654; Forest along the ridge towards a point where Lake Oku can be viewed, 2500m, fl., 22 Nov. 1996, Kamundi 693; Gikwang road, Aboh to Akeh, about 1 km after first track, 2400m, 25 Nov. 1996, Cheek 8756 & 8758; Ijim plateau above Fon's Palace, Laikom, 1900m, fr., 3 Dec. 1998, Cheek 9744 & 9755.

Panicum walense Mez
Annual 30–60 cm, often forming a dense cover over shallow or disturbed sites; inflorescence delicate, bushy. Wet places.

Kilum: Ewook Etele Mbae, 10 km from Oku, 2350m, fl., 6 Nov. 1996, Zapfack 1223.

Paspalum conjugatum P.J.Bergius
Creeping stoloniferous perennial to 60 cm; spikelets orbicular, greenish yellow. Forest edge.
Bambui: Bambui Experimental Station, 1660m, fr., 12 Jul. 1963, Brunt 1189.

Paspalum scrobiculatum L. var. *scrobiculatum*
Perennial, ± caespitose or rarely stoloniferous, to about 60 cm. Wet places.
Syn. *Paspalum orbiculare* G.Forst., F.W.T.A. 3: 446 (1972).
Ijim: Chuhuku River near Anyajua Forest, 1600m, 28 Nov. 1996, Kamundi 730.

Pennisetum clandestinum Hochst. ex Chiov.
A creeping perennial with stout stolons forming a dense mat; stamens with silvery filaments conspicuously exserted from the leaf-sheath when in flower. Grassland.
Kilum: Summit, 2900m, 10 Jun. 1996, Cable 3061.
Ijim: Gikwang road, Aboh to Akeh, about 1 km after first track, 2400m, 6.15N, 10.26E, 25 Nov. 1996, Cheek 8754.

Pennisetum giganteum A.Rich.
Stout perennial to 5 m. Streambanks.
Ijim: Fundong, Touristic Hotel, waterfall on Chumni river, 1400m, fl., 22 Nov. 1996, Cheek 8739.

Pennisetum glaucocladum Stapf ex C.E.Hubb.
Stout perennial 1.2–2.4 m. Streambanks.
Kilum: KJ, above water tank from Elak, 2 km into forest, 2400m, 10 Jun. 1996, Cable 3013.

Pennisetum hordeoides (Lam.) Steud.
Annual, to 1.2 m. Fallow.
Ndop: Grassland, near Ndop rest house, 1200m, fl., 20 Dec. 1952, Boughey GC 11105.

Pennisetum monostigma Pilg.
Densely caespitose perennial to 1 m; culms straggling, geniculate; leaf-blades to 10 mm broad, but usually narrower; rhachis hairy, with conspicuous peduncle stumps. Grassland.
Kilum: Cliff-face, 2800m, 31 Oct. 1996, Munyenyembe 827; Summit, 2800m, fl. 31 Oct. 1996 Zapfack 1117; Above Mboh, path above village to forest through Wambeng's farm, 2000m, 6.1256N, 10.2911E, 1 Nov. 1996, Cheek 8572.

Pennisetum purpureum Schumach.
A robust perennial to 8 m and 2.5 cm diameter at the base; leaf-blades to 40 mm broad; rhachis densely pubescent. Streambanks..
Kilum: Lake Oku, 2200m, fl., 12 Jun. 1996, Etuge 2326; Towards KJ, 2100m, 29 Oct. 1996, Pollard 7.

Pennisetum thunbergii Kunth.
Perennial with slender culms 30–90 cm; rhachis glabrous. Wet places.
Syn. *Pennisetum glabrum* Steud., F.W.T.A. 3: 463 (1972).
Kilum: Ewook Etele Mbae, 10 km from Oku, 2350m, 6 Nov. 1996, Zapfack 1221.

Pennisetum trachyphyllum Pilg.
Local name: Yieh (Oku) – Cheek 8587.
Perennial to 2 m; spikelets 6–7 mm. Forest edge.
Kilum: KJ, 2500m, fl. 29 Oct. 1996, Zapfack 1079; Lake Oku, near Baptist rest house, 2300m, 6.1214N, 10.2733E, fl., 4 Nov. 1996, Cheek 8587.
Ijim: Isebu Mountain Forest, along the ridge, 1800m, fl., 26 Nov. 1996, Kamundi 705.

Pennisetum unisetum (Nees.) Benth.
Local name: Boos (Oku) – Cheek 8487.
A perennial to 3.5 m; leaf-blades firm, to 60 × 2 cm. Forest edge.
Syn. *Beckeropsis uniseta* (Nees) K.Schum., F.W.T.A. 3: 459 (1968).
Kilum: Oku-Elak, lower parts of KA, 2200m, 6.1349N, 10.3112E, fl., 28 Oct. 1996, Cheek 8487.
Ijim: Ijim Mountain Forest, 2160m, fl. 19 Nov. 1996, Kamundi 615; About 2 km E of Laikom near plot 146 of Kilum-Ijim Forest Project, 1900m, fr., 11 Dec. 1998, Cheek 9912.

Poa annua L.
An annual weed of temperate regions, extending to highland tropical areas. Fallow; roadsides.
Kilum: KA, 2760m, fl., 10 Jun. 1996, Asonganyi 1305; Summit, 2900m, 10 Jun. 1996, Cable 3064.
Ijim: Gikwang road, Aboh to Akeh, about 1 km after first track, 2400m, 25 Nov. 1996, Cheek 8757.

Poa leptoclada Hochst. ex A.Rich.
Straggling or tufted perennial, 15–60 cm; panicle contracted, almost linear; spikeletes densely clustered. Forest edge; fallow.
Kilum: KA, 2760m, fr., 10 Jun. 1996, Asonganyi 1306; Path from BirdLife HQ to KA, 2500m, 6.1349N, 10.3112E, fl., 30 Oct. 1996, Cheek 8511.

Poecilostachys oplismenoides (Hack.) Clayton
A trailing grass with long aerial roots from the lower nodes; leaf-blades lanceolate, 6–12 × 0.5–1.5 cm. Forest.
Syn. *Chloachne oplismenoides* (Hack.) Stapf ex Robyns, F.W.T.A. 3: 436 (1972).
Kilum: Lumeto Forest near Oku-Elak, 2200m, fl., 7 Nov. 1996, Cheek 8612.
Ijim: Laikom, 2000m, fl., 3 Dec. 1998, Etuge 4528.

Pseudechinolaena polystachya (Kunth) Stapf
Culms slender, ascending, the prostrate portion rooting from the nodes and often forming a dense cluster. Forest.
Ndop: Boughey GC 11112.

Rhytachne rottboellioides Desv.
Densely caespitose perennial, 30–60 cm. Swamp.
Mbi Crater: 1950m, fr., 9 Dec. 1998, Cheek 9891.

Sacciolepis chevalieri Stapf
Erect perennial growing from a short rhizome; culms soft below, to 60 cm. Swamp.
Mbi Crater: 1950m, fl., 9 Dec. 1998, Cheek 9895.

Setaria longiseta P.Beauv.
Perennial, usually 60–90 cm; leaf-blades linear, firm. Forest edge.
Ijim: Above Aboh, along the ridge towards a point where the lake can be viewed, 2250m, fl., 22 Nov. 1996, Kamundi 683.

Setaria poiretiana (Schult.) Kunth.
Local name: Keehawos (Oku) Cheek 8481.
Tufted perennial 1.2–1.8 m; leaf-blades broad, plicate; spikelets 3–4 mm. Forest.
Syn. *Setaria caudula* Stapf, F.W.T.A. 3: 424 (1972).
Kilum: Oku evergreen forest, 2000m, fl., 21 Jun. 1962, Brunt 612; Oku-Elak, lower parts of KA, 6.1349N, 10.3112E, fl., 28 Oct. 1996, Cheek 8481.
Ijim: Ijim Mountain Forest, along the ridge towards a point where Lake Oku can be viewed, 2160m, fl., 22 Nov. 1996, Kamundi 682.

Setaria pumila (Poir.) Roem. & Schult.
Annual grass, ascending to about 60 cm; spikes 2–8 cm. Fallow.
Syn. *Setaria pallide-fusca* (Schumach.) Stapf & C.E.Hubb. , F.W.T.A. 3: 423 (1972).
Ijim: Path from Ijim to Tum, 2100m, 6.1116N, 10.2543E, fl., 25 Nov. 1996, Etuge 3624.

Setaria sphacelata (Schumach.) Stapf & C.E.Hubb. ex M.B.Moss var. *sphacelata*
Local name: Kegiy-Yio (Oku) – Cheek 8488.
Caespitose perennial about 1 m, arising from a short rhizome. Forest; swamp; fallow; streamside.
Kilum: Near maternity hospital, 2000m, fl., 10 Jun. 1996, Asonganyi 1315; Oku-Elak, lower parts of KA, 2200m, 6.1349N, 10.3112E, fl. fr., 28 Oct. 1996, Cheek 8488; Towards KJ, 2100m, fl., 30 Oct. 1996, Pollard 23; Elak-Oku, secondary forest near the village, fr., 6 Nov. 1996, Onana 509.
Ijim: Ijim Mountain Forest, 2160m, 19 Nov. 1996, Kamundi 607; Above Aboh village, 1.5 hours walk along Gikwang road to lake Oku, 2000m, 6.1112N, 10.2529E, fl. fr., 18 Nov. 1996, Cheek 8655; Gikwang road, Aboh to Akeh, about 1 km after first track, 2400m, fl., 25 Nov. 1996, Cheek 8760; Main path from Ijim to Tum, 2100m, fr., 25 Nov. 1996, Etuge 3624; Grassland plateau above Fon's Palace, Laikom, 2020m, fr., 3 Dec. 1998, Pollard 288 & 290; Sappel, below Ardo's uncle's compound, 2300m, fr., 5 Dec. 1998, Cheek 9813; Afua Swamp, 1950m, fr., 7 Dec. 1998, Etuge 4563.

Sporobolus africanus (Poir.) Robyns & Tournay
Local name: Ntung-Gee (Oku) – Cheek 8477; Chwunighi (Bekom) – Kamundi 613.
Perennial to about 60 cm; basal leaf-sheaths broad and membranous. Grassland.
Kilum: Near maternity hospital, 2000m, 10 Jun. 1996, Asonganyi 1323; KA, 2200m, fl., 11 Jun. 1996, Etuge 2310; Oku-Elak, lower parts of KA, 2200m, 6.1349N, 10.3112E, fl. 28 Oct. 1996, Cheek 8477.
Ijim: Ijim Mountain Forest, 2160m, fl., 19 Nov. 1996, Kamundi 613; Ijim plateau above Fon's Palace, Laikom, 1900m, fr., 3 Dec. 1998, Cheek 9724.

Sporobolus infirmus Mez
Slender annual 10–30 cm, rather variable; leaf-blades mostly 1–5 cm; spikelets 1.3–1.7 mm (exceptionally 0.8–2 mm). Grassland.
Bambui: Bambui Experimental Station, grassland near top of escarpment above station, 2000m, fl., 4 Mar. 1962, Brunt 72.

Sporobolus subulatus Hack. ex Scott-Elliot
Densely caespitose perennial 30–60 cm; basal sheaths broad, papery, horny and yellowish; leaf-blades linear, involute, 15–30 × 0.3 cm; spikelets greyish-green, 3–4 mm. Grassland.
Syn. *Sporobolus mauritianus* (Steud.) T.Durand & Schinz, F.W.T.A. 3: 407 (1972).
Bambui: Jackson 2505.
Kumbo: Kumbo, 1600m, fl., 1 Feb. 1970, C.N.A.D. 1796.

Sporobolus paniculatus (Trin.) T.Durand & Schinz
Annual to 60 cm; leaf-blades conspicuously pectinate-ciliate on the margins; spikelets dark red. Fallow.
Syn. *Sporobolus micranthus* (Steud.) T.Durand & Schinz. F.W.T.A 3: 407 (1972).
Kilum: Oku-Elak, lower parts of KA, 2200m, 6.1349N, 10.3112E, 28 Oct. 1996, Cheek 8483.
Ijim: Fundong to Belo road, 2050m, fl., 21 Nov. 1996, Cheek 8728; Ijim plateau above Fon's Palace, Laikom, 1900m, fr., 3 Dec. 1998, Cheek 9731.

Sporobolus pyramidalis P.Beauv.
Densely tufted perennial 0.9–1.6 cm; leaf-blades 3–10 mm broad. Fallow.
Ndop: Baba, 1260m, fl., 17 May 1962, Brunt 428.

Streblochaete longiaristata (A.Rich.) Pilg.
A loosely tufted perennial to 1 m; leaf-blades narrowly lanceolate-linear, to 1 cm broad; leaf-sheath entire. Forest edge.
Ijim: From Aboh path towards Tum, 2100m, 6.1116N, 10.2543E, fl., 22 Nov. 1996, Etuge 3574; Forest patch towards foot of waterfall above Akwamofu sacred forest, 1860m, fr., 11 Dec. 1998, Etuge 4607.

Trichopteryx elegantula (Hook.f.) Stapf
A delicate annual; culms erect or geniculately ascending, 5–17 cm. Forest edge; grassland.
Ijim: Gikwang road towards Nyasosso forest, 2300m, fl., 19 Nov. 1996, Etuge 3477; Above Fon's Palace, Laikom, about 30 minutes walk along path to Fulani settlement, 2050m, 21 Nov. 1996, Cheek 8718; Along path from Fon's Palace to grassland plateau above Laikom, 2020m, fr., 2 Dec. 1998, Pollard 281; Ijim plateau above Fon's Palace, Laikom, 1900m, fr., 3 Dec. 1998, Cheek 9723.

Trichopteryx marungensis Chiov.
Perennial; culms wiry, trailing, 30–60 cm. Wet places.
Kumbo: Kumbo to Oku, 3 miles on Oku road, 1850m, fl., 15 Feb. 1958, Hepper & Charter 2019.

Tripogon major Hook.f.

Local name: Owush (Oku) – Munyenyembe 799.
Densely caespitose perennial, about 30 cm; leaf-blades narrow, involute, to 15 cm. Grassland.
Kilum: Summit, 3000m, 29 Oct. 1996, Munyenyembe 799.

Triticum aestivum L.

The common bread wheat of temperate regions. Spikelets 10–15 mm, 5–9-flowered, square in section. Cultivated.
Kilum: Near maternity hospital, 2000m, 10 Jun. 1996, Asonganyi 1335.

Urelytrum digitatum K.Schum.

Caespitose perennial to 2 m; racemes 4–9; spikelets sessile, 6–7 mm, glabrous. Savanna.
Syn. *Urelytrum fasciculatum* Stapf ex C.E.Hubb., F.W.T.A. 3: 504 (1972).
Kilum: Tan or Ntan, Jakiri area, AlHadji Gey's land, 1600m, fl., 6 Nov. 1998, Fokom 172.
Jakiri: Jakiri, 1330m, fl., 23 Jun. 1962, Brunt 689.

HYACINTHACEAE
Det. B.J.Pollard (K)

Albuca nigritana (Baker) Troupin

Single inflorescence to about 80 cm, arising from a single leafless bulb (or with young leaves) during the dry season; flowers yellowish green, 1–2 cm, perianth segments with a median green stripe and pale green or whitish margins. Grassland.
Ijim: Ijim mountain forest above Aboh, 2250m, fl., 19 Nov. 1996, Pollard 59; Grassland plateau above Fon's Palace, Laikom, top of path at grassland and gallery forest boundary, 6.17N, 10.19E, fl., 2 Dec. 1998, Pollard 268;

HYDROCHARITACEAE

Ottelia ulvifolia (Planch.) Walp.

Local name: Mbasjioh (Oku), all plants in lake known by this name – Cheek 8581.
A submerged aquatic; roots numerous, slender; leaves averaging about 30 cm (often much smaller), glabrous, very thin, often purple-tinged; flowers borne just above water level, yellow. Muddy pools.
Kilum: Lake Oku, near Baptist Rest House, 2300m, fr., 4 Nov. 1996, Cheek 8581;

HYPOXIDACEAE
Det. B.J.Pollard (K)

Hypoxis angustifolia Lam.

A small herb; rhizome fusiform; leaves from a few to 30 × 0.5 cm; perianth segments 4–5 mm, yellow, inner whorl glabrous or pilose in the middle. Grassland.
Ijim: 5 km E of Laikom, along grassland ridge, 1900m, 6.1630N, 10.1951E, fl. fr., 12 Dec 1998, Pollard 383.

Hypoxis camerooniana Baker

Subterranean rhizome giving rise to several massive erect rhizomes bearing succulent white roots and masses of old leaf fibres; inflorescences appearing during the dry season before the leaves, which develop before flowering is completed; flowers yellow, frequented by bees. Grassland.
Syn. *Hypoxis recurva* Nel, F.W.T.A. 3: 172 (1968).
Ijim: 5 km E of Laikom, along grassland ridge, 1900m, 6.1630N, 10.1951E, fl., 12 Dec 1998, Pollard 384.

IRIDACEAE
Det. M.Cheek (K) & B.J.Pollard (K)

Aristea abyssinica Pax

Perennial herb; leaves linear, to 20 cm; inflorescence markedly winged, 10–60 cm; flowers borne in solitary or paired clusters, bright blue or purple. On rocky hills.
Syn. *Aristea alata* Baker subsp. *abyssinica* (Pax) Weim. , F.W.T.A. 3: 139 (1968).
Ijim: Bum to Nchan, 1330–1660m, fl., Jun. 1931, Maitland 1600.

Aristea angolensis (N.E.Br.) Weim. subsp. angolensis var. angolensis

Slender herb, 0.3–1.0 m; basal leaves to 30 cm, stem leaves 4–6 , reducing in size apically; inflorescence sometimes branched; flowers bright or pale bluish. Streamside.
Kilum: Tadu stream, 2000m, fl., 11 Jul. 1998, Maisels 153.

Gladiolus aequinoctialis Herb. var. aequinoctialis

Perennial herb; stems about 30 cm from a subterranean corm; leaves very long and rather weak, 1–2 cm broad; flowers showy, 5 cm across, white, the limb blotched with purple in the lower part. Savanna.
Syn. *Acidanthera aequinoctialis* (Herb.) Baker, F.W.T.A. 3: 139 (1968).
Bambili: Ujor FHI 29970.

Gladiolus sp. nov.

Local name: Ndong kwin (Kom at Laikom – Yama Peter) = 'Potato of the hill' – Pollard 334.
RED DATA LISTED
Perennial herb, 0.6–1.5 m; inflorescence of 1–2 (–3) flowers; flowers appearing apricot orange (actually yellow, heavily speckled red), outer tepals closely appressed to laterals, not splayed; flowering at the end of the rainy season. Swamp.
Ijim: Swamp, half a mile W of Ardo's Fulani compound, Ijim ridge grassland plateau above Laikom, 2060m, 6.1609N, 10.2051E, fl. fr., 7 Dec. 1998, Pollard 334.
Mbi Crater: Ndawara area, Mbi swamp, 1900m, 6.05N, 10.21E, fl. fr., 9 Dec. 1998, Etuge 4586.

Gladiolus cf. gregarius Welw. ex Baker
Local name: Ndong Kwin (Kom) – Cheek 9752.
Flowering material needed. Grassland.
Ijim: Laikom, Ijim plateau above Fon's Palace, 1900m,
6.16N, 10.19E, fr., 3 Dec. 1998, Cheek 9752.

Hesperantha petitiana (A.Rich.) Baker
Erect herb about 30 cm; corm very small, subglobose; leaves
few, cauline, narrowly linear, to 17 cm; inflorescence about
3-flowered; flowers pink. Rocky outcrops; grassland.
Syn. *Hesperantha alpina* (Hook.f.) Pax ex Engl., F.W.T.A.
3: 141 (1968).
Kilum: Mount Oku, cliff-face, 2800m, fl., 31 Oct. 1996,
Munyenyembe 828.

Moraea schimperi (Hochst.) Pic.Serm.
Erect herb; corm 2 cm diameter; leaves linear-lanceolate,
about 1 m, appearing after flowering; inflorescence of 1–3
flowers; peduncle about 30 cm; perianth 4–4.5 cm, mauve.
Marshes, streamsides.
Kilum: Tadu, 11 km WNW of Kumbo, 2430m, fl., 26 Jun.
1973, Mbenkum 349.

JUNCACEAE
Det. K.Lye (NLH), J.Kirschner (PR) & M.Cheek
(K)

Juncus dregeanus Kunth subsp. *bachitii* (Hochst.
ex Steud.) Hedberg
Tufted perennial; rhizome short; stems 10–40 cm, rounded
or slightly angular, glabrous; inflorescence terminal,
consisting of a dense solitary head of flowers, or more
commonly of one sessile head and 1–4 stalked heads, each
head 5–12 mm diameter, with 5–20 flowers; capsule
subglobose, 2.2–2.5 mm, light to dark brown. Swamp.
Kilum: Ntogemtuo swamp, Oku summit, on edges of
swamp, 2900–2950m, fl., 14 May 1997, Maisels 49; Mount
Oku, Afro-alpine marsh, 2900m, fl., 25 Jun. 1998, Maisels
141; Kinkolong swamp, on tussocks in the 'pond' and beside
bog, 2900m, fl., 8 Jul. 1998, Maisels 152.

Juncus oxycarpus E.Mey. ex Kunth
Tufted perennial; stems 30–60 cm, rounded, sometimes
decumbent and rooting and branching at the nodes; leaves 3–
5 per stem, 5–25 cm, cylindric, transversely septate;
inflorescence usually of one sessile and 3–5 stalked capitula;
capitula subspherical, 8–15 mm in diameter, consisting of 20
or more flowers; capsule 2.5–3.5 mm, triangular, surface
rather shiny, usually light brown below and dark reddish-
brown to almost black above. Swamp.
Ijim: Laikom, swamp E of Ardo's compound, 2200m, 6.16N,
10.23E, fl., 5 Dec. 1998, Cheek 9797; Sappel, below Ardo's
uncle's compound, 2300m, 6.16N, 10.24E, fl., 5 Dec. 1998,
Cheek 9818.

MUSACEAE
Det. B.J.Pollard (K)

Ensete gilletii (De Wild.) Cheesman
Local name: Ingomajya (whole plant); Atengla (peduncle);
Fubu fingom (flower-head),which is also the same word for
baboon; Atu ingom (bunch of fruits); Molo olo' o (individual
flowers), which is also the same word used for Juju-dancers
anklet shakers; Ban ingom (banana).
All this information from Yama Peter, based on Pollard 329.
NB: It is quite possible, but perhaps unlikely, that some of
these terms apply to locally cultivated *Musa spp*.
Monocarpic herb, 1.5–3 m; leaves spread across the stem
and not aggregated towards the apex, lower leaves to 1.5 m,
reducing in size upwards so that the upper leaves become
bracteate; male bracts 4.5–9 × 17–25 cm; stamens 6–12 mm;
female flower with rudimentary stamens; fruit squat,
angular, rather obconic, about 5 cm; seeds 7–9 mm diameter,
brown or black, hard. A wild banana of hilly grassland
savanna.
Ijim: Near Akwamofu sacred forest, on slope towards
waterfall, Laikom, 1820m, 6.1633N, 10.1940E, fl. fr., 5 Dec.
1998, Pollard 329.

ORCHIDACEAE
Det. P.Cribb (K), Zapfack Louis (YA) &
J.DeMarco (Birdlife International)
The species names listed below are derived from three
sources:
1. Names based on specimens collected in 1996 that were
    determined at K by Cribb;
2. those based on pre-1996 collections that are taken from
    Summerhayes' F.W.T.A. account;
3. those based on the post-1996 collections of both
    Zapfack and DeMarco that are provisional
    determinations made in Cameroon with F.W.T.A. and
    yet to be confirmed.

Aërangis biloba (Lindl.) Schltr.
Distance between bracts subtending flowers less than 2.5
cm.; flowers white, sometimes flushed pink; ovary pink,
fragrant at night. Forest; woodland.
Kilum: Elak, KJ, 2400m, fr., 29 Oct. 1996, Zapfack 1070.

Aërangis gravenreuthii (Kraenzl.) Schltr.
Distance between bracts subtending flowers more than 2.5
cm.; flowers white; spur reddish. Forest; woodland.
Kilum: Elak, 2000m, fl., 9 Jun. 1996, Zapfack 820; KJ,
2400m, fr., 29 Oct. 1996, Zapfack 1070.
Ijim: Ijim Mountain Forest, 2250m, fr., 19 Nov. 1996,
Pollard 58; Ijim forest, near Aboh road, 2250m, fl., 29 Mar.
2000, DeMarco 61.

Ancistrorhynchus cephalotes (Rchb.f.) Summerh.
Epiphyte; leaf apex bilobed; margins quite entire; flowers
white, with green or yellow blotches on the labellum,
fragrant. Forest; woodland.
Ijim: Aboh, 2230m, 6.1116N, 10.2511E, 27 Nov. 1998,
Zapfack 1593; Afua, 3 Jun. 1999, Zapfack 1647.

Ancistrorhynchus serratus Summerh.

Epiphyte; leaf apex serrate or with a few sharp teeth on each side just below; flowers white. Forest; woodland.
Kilum: Path above water-tower, KJ, about 1 km into forest, 2200m, 9 Jun. 1996, Cable 2961; Elak, 2080m, 9 Jun. 1996, Zapfack 821; KA, 2200m, fl., 11 Jun. 1996, Etuge 2300; Lake Oku, 2200m, 11 Jun. 1996, Zapfack 862; Elak, 2400m, fr., 29 Oct. 1996, Zapfack 1068.

Angraecopsis ischnopus (Schltr.) Schltr.

Epiphyte, inflorescence 3–9 cm; spur longer than the labellum, 0.6–3.8 cm, very slightly or not at all swollen in the apical part. Grassland; woodland.
Ijim: Fl., 29 Sep. 1998, DeMarco 37.

Angraecum moandense De Wild.

Epiphyte; leaf apex unequally bilobed with the two lobes rounded, 3 –7 cm; flowers yellow, yellow-green or greenish-white; labellum with a recurved pointed apex, nearly 1 cm; spur gently curved, slightly swollen in apical part, 15–20 mm. Grassland; woodland.
Syn. *Angraecum chevalieri* Summerh., F.W.T.A. 3: 257 (1968).
Ijim: Aboh, 2230m, 27 Nov. 1998, 6.1102N, 10.2510E, Zapfack 1592; Lake Oku, Tawatoh, 2350m, 28 Nov. 1998, 6.1133N, 10.2557E, Zapfack 1596.

Brachycorythis macrantha (Lindl.) Summerh.

A stout herb to 40 cm, the upper half bearing to 20 green and mauve flowers, all but lowest flowers considerably longer than the bracts; petals quite free from column. Forest; riversides.
Ijim: Anyajua, near waterworks, 1500m, fl., 27 Jun. 1999, Toh 100.
Bambui: Agricultural Farm, 1600m, fl., 20 May 1963, Brunt 1136.

Brachycorythis ovata Lindl. subsp. schweinfurthii (Rchb.f.) Summerh.

A slender leafy herb, terrestrial, 0.4–1 m; leaves numerous, closely overlapping, stiff and rigid; flowers numerous in a long spike, rather fleshy, purple-spotted. Grassland.
Ijim: Laikom, 2000m, fl., Apr. 1931, Maitland 1787.
Bambili: 1660m, fl., 18 May 1970, Bauer 72.

Brachycorythis pubescens Harv.

A slender terrestrial herb 40–80 cm; leaves softly velvety; flowers pink, bluish, reddish purple, or rarely white; labellum without obvious spur. Grassland; woodland.
Ijim: Laikom, 2000m, fl., Apr. 1931, Maitland 1780.
Bambili: College of Arts, 1500m, fl., 5 May 1971, Bauer 188.
Jakiri: Grassland, fl., 8 Jun. 1947, Gregory 144.

Bulbophyllum calvum Summerh.

Epiphytic herb; pseudobulbs 2–3 cm diameter; sepals pale yellow-green; labellum red with yellow base and white flecks above. Forest; woodland.
Ijim: Ijim forest, fl., 18 Oct. 1998, DeMarco 41.

Bulbophyllum cochleatum Lindl. var. bequaertii (De Wild.) Vermeulen

Epiphytic herb; pseudobulbs very narrowly conical or cylindrical, 4–12 × 3–8 mm; flowers red or reddish-brown. Forest; woodland.
Syn. *Bulbophyllum bequaertii* De Wild., F.W.T.A. 3: 236 (1968).
Ijim: Ijim area, DeMarco 22.

Bulbophyllum cochleatum Lindl. var. cochleatum

Epiphytic herb; pseudobulbs and leaves often purplish; flower purple or red, or sepals yellow-green with purple labellum. Forest; woodland.
Ijim: Aboh, 2100m, 6.1107N, 10.2440E, 27 Nov. 1998, Zapfack 1583 & 1587.

Bulbophyllum cochleatum Lindl. var. gravidum (Lindl.) Vermeulen

Epiphytic herb; pseudobulbs elongate-ovoid, 4-angled, 2–3.5 cm; labellum 5.5 ×1.5 mm, base very thick and fleshy. Forest; woodland.
Syn. *Bulbophyllum gravidum* Lindl., F.W.T.A. 3: 236 (1968).
Ijim: Ijim forest, fl., 8 Nov. 1998, DeMarco 39.

Bulbophyllum cochleatum Lindl. var. tenuicaule (Lindl.) Vermeulen

Epiphytic herb; pseudobulbs narrowly conical or cylindrical, 2–4 × 0.3–1.3 cm; flowers red-purple, whitish at base. Forest; woodland.
Syn. *Bulbophyllum tenuicaule* Lindl., F.W.T.A. 3: 236 (1968).
Ijim: Ijim forest, fl., 18 Oct. 1998, DeMarco 40.

Bulbophyllum falcatum (Lindl.) Rchb.f. var. velutinum (Lindl.) Vermeulen

Epiphytic herb; rhachis flattened, 4–15 mm broad, green variously tinged purple, or entirely purple; pseudobulbs ellipsoid or conical-ovoid, 1.5–4.5 cm; flowers 2.5–4 mm apart, red or purple. Forest; woodland.
Syn. *Bulbophyllum velutinum* (Lindl.) Rchb.f., F.W.T.A. 3: 239 (1968).
sin. loc., DeMarco s.n.

Bulbophyllum josephii (Kuntze) Summerh. var. josephii

Epiphytic herb; pseudobulbs narrowly ovoid or conical-ovoid, 2–3 cm; flowers orange or orange-red and whitish green; sepals orange in apical part; labellum pale orange. Forest; woodland.
Kilum: Ethiale, near the stream, 2060m, fl., 4 Nov. 1996, Zapfack 1178.
Ijim: Gikwang towards Nyasosso, 2270m, fl., 18 Nov. 1996, Etuge 3424.

Bulbophyllum lupulinum Lindl.

Epiphytic herb; rhachis and bracts with numerous short blackish or purplish scaly hairs; bracts pale brown with purplish tinge; flowers yellow with red spots, red or dark purple. Forest; woodland. sin. loc., DeMarco s.n.

Bulbophyllum oreonastes Rchb.f.
Epiphytic herb; rhachis angled or narrowly winged, the wings quite entire; flowers somewhat incurved, yellow, orange, brown or reddish-brown. Forest; woodland.
Ijim: Anyajua 26 May 2000, DeMarco 77.

Bulbophyllum pumilum (Sw.) Lindl.
Epiphytic herb; pseudobulbs ovoid, reddish with wrinkled surface; leaves often purplish beneath; flowers very densely placed; sepals pale, white or greenish at base, upper part red or purple, apices sometimes yellow or brownish. Forest; woodland.
Syn. *Bulbophyllum winkleri* Schltr. 1906, F.W.T.A. 3: 234 (1968).
Ijim: Ijim area, fl., 26 Jun. 1998, DeMarco 45.

Bulbophyllum saltatorium Lindl. var. albociliatum (Finet) Vermeulen
Epiphytic herb; sepals yellowish green or variously tinged maroon or purplish; labellum maroon or purplish, longer hairs arising from margins both above and beneath and also from the surfaces. Forest; woodland.
Syn. *Bulbophyllum distans* Lindl., F.W.T.A. 3: 236 (1968).
Ijim: Aboh, 1750m, 6.1128N, 10.2143E, 26 Nov. 1998, Zapfack 1576.

Bulbophyllum scaberulum (Rolfe) Bolus var. fuerstenbergianum (De Wild.) Vermeulen
Epiphytic herb; rhachis flattened, 7–20 mm broad, sinuate at the edge, green, mottled brown; flowers green mottled brown, petals crimson, purple or chocolate-coloured; dorsal sepal lanceolate, acute, sides inrolled, 6–8 mm. Forest; woodland.
Syn. *Bulbophyllum fuerstenbergianum* (De Wild.) De Wild., F.W.T.A., 3: 239 (1968).
Ijim: Ijim area, fl., 2 Dec. 1999, DeMarco 19.

Bulbophyllum scaberulum (Rolfe) Bolus var. scaberulum
Epiphytic herb; pseudobulbs elongate, ovoid or conical, 3–5 angled, 2–7.5 cm; rhachis ± flattened, to 10 mm broad, green, variously mottled red or crimson, or entirely crimson; flowers yellow, greenish or crimson. Forest; woodland.
Syn. *Bulbophyllum congolanum* Schltr., F.W.T.A. 3: 239 (1968).
Belo: Mbam hills, 1500m, 6.1397N, 10.1549E, 29 Nov. 1998, Zapfack 1614.

Calyptrochilum christyanum (Rchb.f.) Summerh.
Epiphyte; stem woody, to 50 cm, pendulous or almost horizontal; inflorescence usually 6–9 or sometimes to 12-flowered, rather lax, to 4 cm; flowers white or cream, the base of the labellum often greenish, yellow or orange. Forest; woodland.
Kilum: Lake Oku area, fl., 14 Apr. 1998, DeMarco 44.
Belo: Mbam hills, 1500m, 6.1359N, 10.1628E, 29 Nov. 1998, Zapfack 1611.

Chamaeangis vesicata (Lindl.) Schltr.
Epiphyte, pendulous; flowers single in lower part of inflorescence, in twos or threes in upper part, lime green, yellowish or orange, fragrant at night. Forest; woodland.
Ijim: Aboh, 2230m, 6.1102N, 10.2510E, fl., 27 Nov. 1998, Zapfack 1594.

Cribbia confusa P.J.Cribb
Epiphyte; stem to 10 cm, sometimes branched at the base; leaves distichous; inflorescence several, erect, arising below the leaves, very slender, rather laxly to 12-flowered; flowers translucent, pale yellow-green, sometimes tinged with orange towards the tips of the floral parts. Forest; woodland.
Kilum: Elak, 2300m, fl., 9 Jun. 1996, Zapfack 823.
Ijim: Ijim area, fl., 29 Sep. 1998, DeMarco 38.

Cynorkis anacamptoides Kraenzl. var. anacamptoides
A slender terrestrial herb 15–60 cm with a dense spike of small pink or purple flowers. Grassland; swamp; streamsides; forest edge.
Ijim: Afua swamp, fl., 17 Jun. 2000, DeMarco 73.
Jakiri: 1780m, fl., 13 Feb. 1958, Hepper 1958.

Cyrtorchis arcuata (Lindl.) Schltr. subsp. arcuata
Robust herb; stems woody to about 30 cm; leaves 8–24 cm × 1.5–4 cm; inflorescence 6–20 cm, fairly densely 5–14-flowered; flowers waxy white, turning apricot as they fade; fragrant; spur almost straight, or ± curved, but apex not hooked or rolled up. Forest; woodland.
Ijim: Lake Oku, Tawatoh, 2350m, 6.1133N, 10.2557E, 28 Nov. 1998, Zapfack 1595.

Cyrtorchis aschersonii (Kraenzl.) Schltr.
Epiphyte; stem to 30 cm; leaves set closely together, 8–22 × 0.5–1.5 cm, with almost parallel sides, often very fleshy and apparently terete; flowers white, fragrant; spur greenish or brownish. Forest.
Kilum: Joh, Kwifon traditional forest, 1950m, fr. 3 Nov. 1996, Zapfack 1168.
Bambui: Experimental Station, rest house garden, 1660m, fl., 15 Jun. 1962, Brunt 494.

Cyrtorchis chailluana (Hook.f.) Schltr.
Epiphyte, the largest of the species of *Cyrtorchis*; stems long, becoming pendulous; leaves to 25 × 3 cm, ligulate, oblanceolate; margin sometimes undulate; inflorescence to 25 cm, 7–10-flowered; flowers waxy white or cream, turning apricot; spur 9–16 cm, very slender; sepals 3–5 cm; bracts broad and sheathing. Forest.
sin. loc., DeMarco s.n.

Cyrtorchis ringens (Rchb.f.) Summerh.
Local name: Nken-feka–Cheek 8646.
Epiphyte, stems woody to 30 cm in old plants; roots 2 mm diameter; leaves usually 6–7, thick and leathery; inflorescence 6–7(–16) cm, to 12-flowered; flowers closely placed, creamy white, weakly scented. Forest.
Kilum: Path above water-tower, KJ, about 1 km into forest, 2250m, fl., 9 Jun. 1996, Cable 2987; On the road, 2300m,

fl., 11 Jun. 1996, Zapfack 844.

Ijim: Above Aboh village, 1.5 hours walk along Gikwang road, 18 Nov. 1996, Cheek 8646; Gikwang, towards Nyasosso bush, 2280m, 18 Nov. 1996, Etuge 3456; Ijim basal road, 2500m, 6.1406N, 10.2539E, 1 Jun. 1999, Zapfack 1638.

Diaphananthe bueae (Schltr.) Schltr.
RED DATA LISTED
Epiphyte; flowers white and green; sepals 6–8.5 mm. Forest.
Kilum: N side of Lake Oku, 2350m, fl., 22 May 2000, DeMarco 69.

Diaphananthe cf. fragrantissima (Rchb.f.) Summerh.
Epiphyte. Forest.
Ndop: Baba, Ndop plain, DeMarco 35.

Diaphananthe kamerunensis (Schltr.) Schltr.
Robust epiphyte; roots stout, 4–5 mm diameter; inflorescence 10–30 cm; flowers pale green or creamy yellow; spur not quite so long as the labellum. Forest; woodland.
Ijim: Ijim area, fl., 21 Sep. 1998, DeMarco 36.

Diaphananthe pellucida (Lindl.) Schltr.
Robust epiphyte; roots fine, 2 mm diameter; inflorescence many-flowered, pendulous, 15–55 cm; flowers white, translucent creamy yellow, pale green or pinkish; sepals 9–11 mm. Forest.
Belo: Mbam hills, 1500m, 6.1359N, 10.1633E, 29 Nov. 1998, Zapfack 1613.

Diaphananthe plehniana (Schltr.) Schltr.
Epiphyte; inflorescence to 13 cm, many-flowered; flowers pink to apricot-coloured; sepals 4.5–6.5 mm. Forest.
Kilum: Mbidjiame Traditional forest, 2020m, fr., 4 Nov. 1996, Zapfack 1185;

Diaphananthe polyantha (Kraenzl.) Rasm.
Local name: Mbolevekak (Oku) – Munyenyembe 834.
Epiphyte; inflorescence 4–22 cm, many-flowered; flowers greenish-white or very pale yellow; sepals 2.5–4 mm; labellum 2.7–3.6 mm; spur shorter than the labellum, swollen in the apical part. Forest.
Syn. *Sarcorhynchus polyanthus* (Kraenzl.) Schltr., F.W.T.A. 3: 263 (1968).
Kilum: Elak, 2600m, fl., 31 Oct. 1996, Munyenyembe 834; Summit, 2800m, fl., 31 Oct. 1996, Zapfack 1141; KDH, 2300m, fl., 1 Nov. 1996, Zapfack 1153.
Ijim: Above Aboh village, 1.5 hours walk along Gikwang road, 2350m, fl., 18 Nov. 1996, Cheek 8648; Gikwang road towards Nyasosso forest, 2300m, fl., 19 Nov. 1996, Etuge 3479; Ijim Mountain Forest, 2020m, fl., 19 Nov. 1996, Kamundi 635; Above Fon's palace, Laikom, about 30 minutes walk on path to Fulani settlement, 2050m, fl., 21 Nov. 1996, Cheek 8713.
Kumbo: Ledermann 5741 & 5763.

Diaphananthe polydactyla (Kraenzl.) Summerh.
Epiphyte; inflorescence 6–11 cm, 5–9-flowered; flowers greenish-white; labellum 3-lobed, lateral lobes longly pectinate, front lobe much smaller, tooth-like, entire; sepals and petals 4–4.5 mm. Forest.
Kumbo: Kufum, 1750m, fl., 17 Dec. 1909, Ledermann 5716A.

Diaphananthe rohrii (Rchb.f.) Summerh.
Epiphyte; inflorescence 6–28 cm, 6–20-flowered; flowers greenish white or greenish-yellow; sepals 3.5–5.5 mm. Forest.
Syn. *Diaphananthe quintasii* (Rolfe) Schltr., F.W.T.A. 3: 261 (1968).
Ijim: Above Fon's palace, Laikom, about 30 minutes walk on path to Fulani settlement, 2050m, fr., 21 Nov. 1996, Cheek 8711.

Disa erubescens Rendle subsp. *erubescens*
A terrestrial herb, 0.3–1 m, with the lower sheaths spotted; leaves on separate barren shoots, linear or narrowly lanceolate, 10–40 cm; spike about 5–15 cm, rather lax, 3–10-flowered; flowers orange, flame-coloured or deep red, spotted in parts. Grassland; swamp.
Ijim: Mbizenaku, 2000m, Maitland 1503.
Bambili: 1660m, fl., 5 Jun. 1970, Bauer 84.

Disa hircicornis Rchb.f.
A slender to rather stout terrestrial herb, 0.3–1 m, with a dense cylindrical flower spike 5–20 cm; flowers purple with darker markings. Grassland; swamp.
Ijim: Laikom, fl., Jun. 1931, Maitland 1781.

Disa nigerica Rolfe
A slender terrestrial herb 20–30 cm; inflorescence rather lax, about 10 cm; flowers small, purple. Grassland.
Ijim: Laikom plateau, 2000m, fl., 15 Apr. 2000, DeMarco 83.

Disa ochrostachya Rchb.f.
An erect terrestrial herb, 0.3–1 m, with sheathing lanceolate leaves on flowering stems; foliage leaves on separate shoot, narrow, to 45 cm; flower-spike long and slender; flowers yellow with orange markings. Grassland.
Ijim: Nchan, 1660–2000m, fl., Apr. 1931, Maitland 1779.

Disa welwitschii Rchb.f. subsp. *occultans* (Schltr.) Linder
An erect terrestrial herb, 0.3–1 m; leaves linear, to 50 cm on sterile shoots; scape bearing sheaths; flowers pink or purple. Grassland; swamps.
Ijim: Mbizenaku-Laikom, 1660–2000m, fl., Jun. 1931, Maitland 1616bis; Bum to Nchan, 1330–1660m, Maitland 1788; Ijim grassland, DeMarco 57.
Bambili: 1660m, fl., 5 Jun. 1970, Bauer 84.

Disperis nitida Summerh.
RED DATA LISTED
An epiphytic or terrestrial herb 12–30 cm; leaves dark green, satiny; flowers 1–4 in a short raceme, white. Forest, shade.

Kilum: Forest above Lake Oku, 2000–2330m, fl., 4 Sep. 1952, Savory UCI 451; KDH, 2630m, fr., 30 Oct. 1996, Zapfack 1156; KDH, 2300m, fl., 1 Nov. 1996, Zapfack 1152.

Disperis parvifolia Schltr.
RED DATA LISTED
Leaves to 7 × 4 mm; flowers pink or yellowish; hood formed by the dorsal sepal and petals, helmet-shaped, narrow, 4–6 mm. Grassland.
Ijim: Mbizenaku to Nchan, 1660m, fl., 1931, Maitland s.n.

Eulophia cf. barteri Summerh.
Terrestrial herb. Grassland.
Ijim: Ijim grassland, DeMarco 49.

Eulophia cucullata (Sw.) Steud.
Robust terrestrial herb to 1 m; leaves 3–4, appearing after the flowers, eventually 20–70 × 0.5–1.5 cm; flowers few, large, pink or purplish with yellow or white inside the broad spur. Grassland; savanna.
Bamessi: 1260m, fl., 30 Mar. 1962, Brunt 271.

Eulophia horsfallii (Batem.) Summerh.
Robust terrestrial herb, 1–3 m; leaves in a tuft, 0.3–2.5 m × up to 0.15 m, ribbed, dark green; scapes to 3 m; inflorescence a long raceme; flowers white and pink or purple. Forest edge; swamp; fallow.
Kilum: Kumbo to Oku, fl., 1958, Hepper 1999.
Ijim: Main path from Ijim to Tum, 2100m, fl., 25 Nov. 1996, Etuge 3629; Anyajua, streamside near waterworks, 1500m, fl., 12 Jun. 1999, DeMarco 48.
Bambui: Hill above and to E of Agricultural Station, 1660m, fl., 1 Jun. 1971, Bauer 189.

Eulophia odontoglossa Rchb.f.
Terrestrial herb; flowers in a dense head, white, yellow with brown or purple markings, or entirely brownish or purplish. Rough grassland; roadsides.
Syn. *Eulophia shupangae* (Rchb.f.) Kraenzl., F.W.T.A. 3: 247 (1968).
Ijim: Bum to Nchan, 1330–1860m, fl., Jun. 1931, Maitland 1638; Fundianse hill above Juambum, fl., 7 May 1999, Nkwah 99.
Ndop: Ndop village rest house, 1260m, fl., 13 Apr. 1962, Brunt 363; Between Ndop and Sabgo, 1500m, fl., 27 Jun. 1962, Brunt 769.
Kumbo: Near Banso, in grass by roadside, 1930m, fl., 19 Jun. 1962, Brunt 582.

Eulophia stachyodes Rchb.f.
Herb to 80 cm with broad leaves and a rather close spike of brown and white flowers. Swamp; grassland.
Mbi Crater: Crater rim, 2050m, fl., 20 May 2000, DeMarco 79.

Genyorchis macrantha Summerh.
RED DATA LISTED
Epiphyte; pseudobulb 2-leaved, scarcely flattened; scape to 6.5 cm, to 7-flowered; sepals white or very pale pink;

labellum pink with yellow apex. Forest.
Ijim: Ijim forest near Mbesa, fl., 2 Mar. 2000, DeMarco 56.

Graphorkis lurida (Sw.) Kuntze
Robust epiphyte; pseudobulbs cylindrical-fusiform or conical ovoid, 3–9 × 1–3 cm, 4–6-leaved, yellowish and ribbed; inflorescence appearing before the leaves, 15–50 cm, branches spreading; flowers yellowish flushed with brown; sepals about 5–6 mm; spur sharply bent forward, nearly as long as the labellum. Forest.
Ijim: Tuantoh, Ijim forest, 28 Dec. 1999, DeMarco 80.
Belo: Mbam hills, 1360m, 6.1330N, 10.1622E, 29 Nov. 1998, Zapfack 1605.

Habenaria bracteosa A.Rich.
An erect, leafy terrestrial herb 0.15–1 m; inflorescence a long, dense spike; flowers green or greenish-yellow. Grassy glades in forest, especially by streams, damp places.
Ijim: Laikom, 2000m, fl., 1931, Maitland s.n.

Habenaria maitlandii Summerh.
RED DATA LISTED
A terrestrial herb to 35 cm; flowers white in short racemes. Grassland.
Ijim: Nchan, on mountain slope among grass, stones, 1860m, fl., Jun. 1931, Maitland 1386.

Habenaria malacophylla Rchb.f. var. *malacophylla*
A slender terrestrial herb, 0.3–1 m, leafy in the middle of the stem; inflorescence a long loose raceme; flowers numerous, small, green. Forest, occasionally grassland above forest.
Kilum: JOH, Kwifon traditional forest, 1950m, fl., 3 Nov. 1996, Zapfack 1171.
Ijim: Zitum road, 1900m, fl., 21 Nov. 1996, Etuge 3532.

Habenaria mannii Hook.f.
An erect herb to 70 cm; stem leafy; inflorescence a dense raceme of a few to 25 flowers; flowers large, green or whitish green. Grassland.
Ijim: Mbizenaku to Laikom, fl., Jun. 1931, Maitland 1617; Nchan, fl., Jun. 1931, Maitland 1643b.

Habenaria microceras Hook.f.
Local name: Klandikikie (Oku) – Munyenyembe 775.
RED DATA LISTED
A leafy terrestrial or epiphytic herb to 75 cm; inflorescence a narrow spike, 5–25 cm; flowers small, green. Forest; grassland.
Kilum: Elak, 2800m, fl., 29 Oct. 1996, Munyenyembe 775; KJ, 2400m, fl., 29 Oct. 1996, Zapfack 1069; Summit, 2800m, fl., 31 Oct. 1996, Zapfack 1139.

Habenaria obovata Summerh.
RED DATA LISTED
A terrestrial herb to 45 cm; inflorescence a dense spike; flowers small, green. Grassland.
Kilum: Summit, 2800m, fl., 31 Oct. 1996, Zapfack 1140.

Habenaria zambesina Rchb.f.
A ground orchid to 1.2 m; stem leafy; inflorescence a dense spike; flowers white. Grassland; swamp.
Ijim: Nchan, 1660m, fl., Jun. 1931, Maitland 1387.
Jakiri: Grassland, fl., 8 Jun. 1947, Gregory 145.

Holothrix aphylla (Forssk.) Rchb.f.
A dwarf herb to 12.5 cm; leaves 2, basal, flattened; flowers mauve-white. Grassland.
Ijim: Ijim grassland near Mbesa, fl., 29 Feb. 2000, DeMarco 55.
Kumbo: Lake Babessi, 1200m, fl., 27 Dec. 1908, Ledermann 1988; Kufum, 2000m, fl., 29 Dec. 1908, Ledermann 2005.

Liparis suborbicularis Summerh.
Epilith; scape longer than the leaves; flowers greenish-yellow; labellum suborbicular. Rock crevices.
Kilum: KC, 2320m, fl., 30 Oct. 1996, Zapfack 1155; Ewook Etele, Oku-Mbae, 2350m, fr., 6 Nov. 1996, Etuge 3344.

Malaxis maclaudii (Finet) Summerh.
Terrestrial herb; inflorescence apex with flowers close together, forming a false umbel; flowers flesh to deep rose-coloured; labellum bearing 2 parallel ridges at the base, often with a needle-like projecting point just in front of the median vein. Forest.
Ijim: Ntum, Ijim forest, fl., 14 Jun. 2000, Nkwah 101.

Microcoelia microglossa Summerh.
Epiphyte; flowers with a brown line down centre of sepals; limb of labellum very small, 2–4 mm, less than half as long as the straight spur. Woodland.
Mbam: Mbam Nchouo, 1440m, 6.1348N, 10.1635E, 30 Nov. 1998, Zapfack 1615.

Plectrelminthus caudatus (Lindl.) Summerh.
Robust epiphyte; stem leafy, to 15 cm; roots long and stout, sometimes even 10 mm diameter; inflorescence 25–60 cm, 4–10-flowered; tepals yellow-green, sometimes flushed with bronze; labellum white. Forest.
Babanki: Main road, fl., 27 Jul. 1998, DeMarco 43.

Podangis dactyloceras (Rchb.f.) Schltr.
Epiphyte; stem short, sometimes forming clumps; roots very fine; inflorescence to 6 cm; flowers translucent white with a crystalline texture; sepals and petals ± elliptical, obtuse, 3.5–5 mm; anther green. Forest.
Kilum: On the road, 2500m, fr., 11 Jun. 1996, Zapfack 847.

Polystachya albescens Ridl. subsp. albescens
Epiphytic herb; flowers greenish or whitish, sometimes with reddish veins; labellum tinged or veined red. Forest; woodland.
Bambui: Bambui to Ndop, Brunt 450 (fide F.W.T.A.).

Polystachya alpina Lindl.
Epiphytic herb; flowers white or rose. Forest; woodland.
Kilum: Elak, 2400m, fl., 9 Jun. 1996, Zapfack 822; JOH, Kwifon traditional forest, fr., 3 Nov. 1996, Zapfack 1170.

Ijim: Gikwang road towards Nyasosso forest, 2300m, fl., 19 Nov. 1996, Etuge 3483.

Polystachya bennettiana Rchb.f.
Epiphytic herb; flowers pale yellow, often ± tinged with pink, rarely mauve-pink. Forest; woodland.
Syn. **Polystachya stricta** Rolfe, F.W.T.A., 3: 221 (1968).
Kilum: Mbidjiame Traditional forest, 2020m, fr., 4 Nov. 1996, Zapfack 1184.

Polystachya bicalcarata Kraenzl.
RED DATA LISTED
A small densely tufted epiphyte; flowers white and purple or rose. Forest; woodland.
Kilum: Around Project Guest House, 2080m, fr., 4 Nov. 1996, Zapfack 1187.

Polystachya cultriformis (Thouars) Spreng.
Epiphytic herb; inflorescence to 22 cm, usually paniculate in the upper part; flowers white, yellowish, pink or purplish. Forest; woodland.
Ijim: Mbingo Back Valley, 10 Nov. 1999, DeMarco 81.

Polystachya elegans Rchb.f.
Epiphytic herb; flowers greenish white or yellowish white, sometimes tinged mauve. Forest; woodland.
Kumbo: Kumbo to Jakiri, M55, 1860m, fl., 2 Sep. 1952, Savory 385.

Polystachya fusiformis (Thouars) Lindl.
Epiphytic herb; flowers greenish suffused with purple, to deep purple or plum coloured. Forest; woodland.
Kilum: On the road, 2500m, 11 Jun. 1996, Zapfack 845; JOH, Kwifon traditional forest, 1950m, fr., 3 Nov. 1996, Zapfack 1167.

Polystachya galeata (Sw.) Rchb.f.
Epiphytic herb; flowers large, helmet-shaped, white, green, yellow or yellow-green with red or purplish markings. Forest; woodland.
Ijim: Aboh, 1750m, 6.11245N, 10.23844E, 26 Nov. 1998, Zapfack 1581; Mboh, 2000m, 6.1912N, 10.2054E, Zapfack 1618.

Polystachya laxiflora Lindl.
Epiphytic herb; flowers white, yellow or orange-yellow, sometimes with red markings on the sides of the labellum. Forest; woodland.
Belo: Mbam hills, fl., 2 Jun. 1999, Zapfack 1650.
sin. loc., DeMarco s.n.

Polystachya odorata Lindl. var. odorata
Epiphytic herb; flowers white or yellow, tinged rose. Forest; woodland.
Ijim: Anyajua, DeMarco 82.
Belo: Mbam hills, 1500m, 6.1348N, 10.1651E, 29 Nov. 1998, Zapfack 1600; Mbam hills, 1500m, 6.1359N, 10.1633E, 29 Nov. 1998, Zapfack 1612; fl., 2 Jun. 1999, Zapfack 1651.

Polystachya parva Summerh.
Epiphytic herb; flowers white, tinged greenish; labellum sometimes orange or purple; anther purple. Forest; woodland.
Kilum: On the road, 2400m, fl., 11 Jun. 1996, Zapfack 846.

Polystachya steudneri Rchb.f.
Epiphytic herb; flowers yellow or greenish yellow with red stripes or markings. Forest; woodland.
Babungo-Ndawara: Road, fl., 29 Jan. 2000, DeMarco 51.

Polystachya superposita Rchb.f.
RED DATA LISTED
Epiphytic herb; flowers reddish. Forest; woodland.
Kilum: Elak, 2420m, fl., 9 Jun. 1996, Zapfack 824; JOH, Kwifon traditional forest, 1950m, 3 Nov. 1996, Zapfack 1169.

Satyrium breve Rolfe
A terrestrial herb, mostly glabrous; sterile stem to 19 cm; flowers almost spreading, sessile, pink, deep red, mauve or purple, sometimes with darker markings. Grassland; swamp.
Ijim: Ardo's swamp, 2200m, fl., 4 May 2000, Formbui 101.

Satyrium coriophoroides A.Rich.
A terrestrial herb to 1 m. Grassland.
Babanki: 1500m, fl., Apr. 1931, Maitland 1786.
Kumbo: 1660m, fl. fr., 7 Jun. 1947, Gregory 140.

Satyrium crassicaule Rendle
A terrestrial, moderately stout to stout herb to 1 m. Grassland; swamp.
Ijim: Nchan, 1660m, fl., Jun. 1931, Maitland 1385; Ijim basal road, 2500m, 6.1406N, 10.2539E, 1 Jun. 1999, Zapfack 1639; Ardo's swamp, 2200m, fl., 4 May 2000, Formbui 102.

Satyrium volkensii Schltr.
A slender terrestrial herb to 1 m. Grassland.
Ijim: Laikom, grassy plateau, 2000m, fl., May 1931, Maitland 1366.
Mbi Crater: Crater rim, 2050m, fl., 20 May 2000, DeMarco 84.

Stolzia repens (Rolfe) Summerh. var. *repens*
A creeping dwarf epiphytic herb to 1 cm; pseudobulbs prostrate except at apex, to 3 × 0.3 cm, bearing 2 leaves near insertion of next pseudobulb; flowers yellow, brown or reddish, ± striped red or brown. Forest.
Belo: Near Council Office, 1200m, fl., 9 Jun. 2000, Nkwah 100.

Tridactyle tridactylites (Rolfe) Schltr.
Epiphyte; stems to 1 m, usually trailing, sometimes erect; leaves distichous, borne on apical half or third of stem; flowers greenish-cream or yellow-cream, orange or brownish orange, fragrant. Forest.
Kilum: Mbidjiame traditional forest, 2020m, 4 Nov. 1996, Zapfack 1183.

Belo: Mbam hills, 1460m, 6.1341N, 10.1628E, fl., 29 Nov. 1998, Zapfack 1607.

Tridactyle tridentata (Harv.) Schltr.
Epiphyte; stem 10–50 cm, erect or pendulous; flowers whitish, pale ochre-yellow or salmon-pink. Woodland.
Ijim: Anyajua, 1500m, fl., 18 Oct. 1998, DeMarco 30.

PALMAE
Det. W.J.Baker (K) & M.Cheek (K)

Eremospatha macrocarpa (G.Mann & H.Wendl.) H.Wendl.
A climbing palm; stems smooth, polished, of great length, 1.5 cm diameter; leaf segments more than 5 times longer than broad, usually narrow, 15–35 × 2–3 cm; rhachis prolonged into a cirrus with opposite hooks, but not armed with prickles; inflorescence conspicuous, buff-yellow. In swamp-forest.
Ndop: Brunt 207.

Phoenix reclinata Jacq. var. *reclinata*
A tufted palm, often forming clumps; stem occasionally as as 10 m, but commonly much less. Open wet places.
Ijim: Zitum road, 6.1116N 10.2543E, 2000m, fr., 21 Nov. 1996, Etuge 3572; Laikom, forest patch towards foot of waterfall, above Akwamofu sacred forest, 1860m, st., 11 Dec. 1998, Etuge 4596.

Raphia mambillensis Otedoh
RED DATA LISTED
Palm with no clear trunk; stem very short, covered by petioles; fronds clustered at ground level. Common near villages at the edge of water courses.
Syn. *Raphia farinifera* sensu F.W.T.A. 3: 166 (1968).
Ijim: Sans loc. 1999, DeMarco s. n.
Bambui: Brunt 1139 & 1140.

POTAMOGETONACEAE

Potamogeton schweinfurthii A.Benn.
Submerged aquatic; submersed leaves about 20 mm broad, broadly linear, narrowed to each end, 15–25 cm, very thin; upper leaves rarely floating; spikes about 4 cm, many-flowered. Freshwater lakes.
Kilum: N side of Lake Oku, lake shore, water level, 2100m, fl., 5 Nov. 1996, Buzgo 702.

Potamogeton sp. 1
Leaves filiform. Freshwater lakes.
Local name: Mbasijioh (Oku) – Cheek 8579 & 8597.
Kilum: Lake Oku, near Baptist Rest House, 2300m, 4 Nov. 1996, Cheek 8579; Lake Oku, below Baptist Rest House, 2200m, 5 Nov. 1996, Cheek 8597.

Potamogeton sp. 2
Not matched in F.W.T.A. or F.D.C.– Cheek 9884.
Mbi Crater: Approach to Mbi crater by short cut from Belo-Afua, 6.0562N, 10.23E, fl. fr., 9 Dec. 1998, Cheek 9884.

SMILACACEAE

Smilax kraussiana Meisn.

A climbing or scrambling shrub with prickly shoots; leaves alternate, ovate-elliptic to broadly elliptic; flowers greenish white; fruits subglobose, nearly 1 cm diameter, red. Forest..
Kilum: Forest patches in Oku-Elak, below BirdLife HQ, 6.1526N 10.2949E, 1800m, 6 Nov. 1996, Buzgo 708.
Ijim: Aboh village, 6.15N, 10.26E, 25 Nov. 1996, Onana 628; Zitum road, 1900m, 6.1116N, 10.2543E, 21 Nov. 1996, Etuge 3563; Cheek 9827.

XYRIDACEAE
Det. J.M.Lock (K)

Xyris capensis Thunb.

A herb with fan-like tufts of leaves arising from a perennial rhizome; scape 30–60 cm; capitula broader than long; flowers yellow. Swamp.
Ijim: From Aboh path towards Tum, 2200m, 6.1116N, 10.2543E, fl., 22 Nov. 1996, Etuge 3598; Laikom, swamp E of Ardo's compound, 2000m, 6.16N, 10.22E, fl., 5 Dec. 1998, Cheek 9791; Sappel, below Ardo's uncle's compound, 2300m, 6.16N, 10.24E, fl., 5 Dec. 1998, Cheek 9808; Afua swamp, 1950m, 6.08N, 10.24E, fl., 7 Dec. 1998, Cheek 9844.
Mbi Crater: Approach to Mbi Crater by short cut from Belo-Afua, 1750m, 6.0562N, 10.23E, fr., 9 Dec. 1998, Cheek 9887; Mbi Crater, 1950m, 6.0536N, 10.21E, fl., 9 Dec. 1998, Cheek 9901.

Xyris congensis Büttner

Herb growing in dense clumps; scapes to 1.2 m; capitula about 15 mm, somewhat cylindrical, longer than broad and tapering apically. Swamp.
Mbi Crater: 1950m, 6.0536N, 10.21E, fr., 9 Dec. 1998, Cheek 9898.

Xyris cf. filiformis Lam.
RED DATA CANDIDATE
A slender terrestrial annual herb, c. 10–20 cm; stems and leaves reddish-brown; capitula about 5 mm, as broad as long. Grassland.
Ijim: Laikom, Ijim ridge between Ardo's compound and Fon's Palace, 2000m, 6.16N, 10.19E, fl. fr., 3 Dec. 1998, Cheek 9784.

Xyris rehmannii L.A.Nilsson

A stout, clump forming herb to 1.5 m; leaves leathery, glossy, ribbon-like; capitula shortly ovoid, 10–20 mm, loose with chaffy scales; flowers yellow. Swamp.
Mbi Crater: 1950m, 6.0536N, 10.21E, fl. fr., 9 Dec. 1998, Cheek 9899.

Xyris welwitschii sensu Lewis non Rendle
(≡ de Wilde 4259 & 8635).
RED DATA CANDIDATE
Local name: Fuee Fichou (Kom–Yama Peter).
A slender herb to 50 cm; capitula c. 5mm when young, 10 mm when mature, as broad as long; flowers yellow. Swamp.

Xyris welwitschii sens. strict. is known from Angola and Zambia.
Ijim: Laikom, in bog below Ardo's compound, 2100m, 6.16N, 10.21E, fl., 7 Dec. 1998, Gosline 201.

Xyris sp. A
RED DATA CANDIDATE
Herb, 20–40 cm; capitula glossy, ovoid, 5–7mm, as broad as long; flowers yellow. Swamp.
Kilum: Ntogemtuo swamp, Oku summit, 2900m, fl., 14 May 1997, Maisels 51.

ZINGIBERACEAE

Aframomum sp. A
Local name: Achoch (Kom) – Etuge 3453; Achoh (Kom at Laikom) – Etuge 4505; Acho (Kom at Laikom) – Cheek 9874.
Inflorescence radical; fruits red, smooth. Forest edge.
Kilum: KA, 2200m, fl., 11 Jun. 1996, Etuge 2317; Oku, 2010m, fr., 2 Nov. 1996, Etuge 3336; Mbidjiame traditional forest, 2020m, fr., 4 Nov. 1996, Zapfack 1182.
Ijim: Zitum road, 1860m, fr., 21 Nov. 1996, Etuge 3543; Laikom, grassland and gallery forest on volcanic plateau, beside a stream called Jrua Akuamufi, 2000m, fl., 2 Dec. 1998, Etuge 4505; Laikom, Plot between Fon's Palace and Ardo's compound, 2000m, st., 8 Dec. 1998, Cheek 9874.

Aframomum sp. B
Local name: Achoch–Etuge 3542; Acho Afooboo (Kom at Laikom) –Cheek 9870.
Inflorescence radical; fruits red, ribbed. Forest edge.
Ijim: Zitum road, 1920m, fr., 21 Nov. 1996, Etuge 3542; Isebu Mountain Forest, 1700m, fr., 26 Nov. 1996, Kamundi 702; Afua swamp, wetland along valley bottom, 1950m, 6.08N, 10.24E, st., 7 Dec. 1998, Cheek 9848; Plot between Fon's Palace and Ardo's compound, 2000m, 6.16N, 10.21E, st., 8 Dec. 1998, Cheek 9870.

Aframomum sp.
Poor material.
Kilum: Emoghwo traditional forest, 1900m, st., 3 Nov. 1996, Zapfack 1159;

GYMNOSPERMAE
PINOPSIDA

CUPRESSACEAE
Det. A.Farjon (K)

Cupressus lusitanica Mill.
Tree to 30–35 m, evergreen, monoecious; trunk monopodial, large trees buttressed, up to 2 m d.b.h; leaves scale-like; seed cones solitary or in groups near the upper ends of lateral branches, terminal on short leafy branchlets, maturing in 2 growing seasons, persistent. From cultivation.
Ijim: Path from Fon's Palace downhill towards Akwamofu sacred forest, Laikom, 1900m, 6.1630N, 10.1951E, fr., 12 Dec. 1998, Pollard 376.

PODOCARPACEAE

Podocarpus latifolius (Thunb.) Mirb.
Local name: Ebshieko (Oku) – Munyenyembe 761.
Dioecious shrub or tree to 35 m; stems much-branched, sympodial; leaves alternate; male cones flesh-pink, about 2.5 cm; receptacle large, fleshy, scarlet; seeds glaucous. Forest.
Kilum: Lake Oku, Johnstone 233/31; Elak, 2550m, fr., 9 Jun. 1996, Zapfack 799; Summit, 2850m, fr., 10 Jun. 1996, Cable 3080 & Cable 3081; Elak, KD, 3000m, fl. fr., 10 Jun. 1996, Etuge 2285; Summit, 2900m, 29 Oct. 1996, Munyenyembe 761; KC, 2500m, fr., 30 Oct. 1996, Zapfack 1102; Summit, 2800m, fr., 31 Oct. 1996, Zapfack 1143 & 1144.

PTERIDOPHYTA
LYCOPSIDA

ISOETACEAE

Isoetes biafrana Alston
RED DATA LISTED
Submerged (45–150 cm deep in November) aquatic, stemless and bottom rooting; leaf rosettes 7–20 × 7–20 cm; leaves linear, 15–20, dark green, the outermost each with a swollen white base containing numerous black megaspores (reproductive bodies); reproductive in November. Lake.
Kilum: Lake Oku, near Baptist Rest House, 2300m, 4 Nov. 1996, Cheek 8577.

LYCOPODIACEAE
P.J.Edwards (K) & J-M.Onana (YA)

Huperzia mildbraedii (Herter) Pic.Serm.
Local name: Ijimfekah (Kom) – Cheek 8659.
Epiphyte; foliage leaves narrow, subulate, c. 15 mm long; branches to 0.5 m long, subpendulous, appressed to subappressed; sporophylls similar but shorter. Forest.
Syn. *Lycopodium mildbraedii* Herter, F.W.T.A. Suppl.: 11 (1959).
Kilum: Elak, 2300m, 8 Jun. 1996, Zapfack 791; Lake Oku, 2300m, 11 Jun. 1996, Zapfack 849; KJ, 2400m, 29 Oct. 1996, Zapfack 1063; KA, 2200m, 11 Jun. 1996, Etuge 2322.
Ijim: Along the top of the ridge, second gate from Aboh village, 2300m, 6.11N, 10.26E, 21 Nov. 1996, Kamundi 670; Gikwang road towards Nyasosso forest, Ijim Mountain Forest Reserve, 2300m, 19 Nov. 1996, Etuge 3480; Above Aboh village, 1.5 hours on Gikwang road to Lake Oku, 2450m, 6.1112N, 10.2529E, 19 Nov. 1996, Cheek 8659.

Huperzia ophioglossoides (Lam.) Rothm.
Local name: Njienenmevakak (Oku) – Munyenyembe 782.
Epiphyte; branches to 0.5 m long; pendulous; leaves narrowly lanceolate; foliage leaves c. 15 cm long, spreading at about 45°, sporophylls much smaller and broad lanceolate, in long and narrow 'tassles' well differentiated from the foliage leaves. Forest.
Syn. *Lycopodium ophioglossoides* Lam., F.W.T.A. Suppl.: 12 (1959).
Kilum: KA, 2500m, 6.15N, 10.26E, 29 Oct. 1996, Onana 461; Elak, 2300m, 8 Jun. 1996, Zapfack 753; 2500m, 29 Oct. 1996, Munyenyembe 782; KJ, 2400m, 29 Oct. 1996, Zapfack 1064; Path above water-tower, KJ, about 1 km into forest, 2250m, 9 Jun. 1996, Cable 3000; Shore of Lake Oku, 2300m, 12 Jun. 1996, Cable 3125; Oku-Elak, lower parts of KA, 2200m, 6.1349N, 10.3112E, 28 Oct. 1996, Cheek 8463.
Ijim: Above Aboh village, 1.5 hours on Gikwang road to Lake Oku, 2380m, 6.1112N, 10.2529E, 18 Nov. 1996, Cheek 8634.

Lycopodiella cernua (L.) Pic.Serm.
Terrestrial; lower parts of stems (stolons) long-creeping, the rest erect, much-branched; foliage leaves c. 4 mm long, very narrow; strobili very compact, c. 1 cm long, drooping. Fallow.
Syn. *Lycopodium cernuum* L., F.W.T.A. Suppl.: 12 (1959).
Kilum: Lake Oku to Jikijem, 2200m, 6.1214N, 10.2733E, 5 Nov. 1996, Cheek 8601.
Ijim: From Aboh, path towards Tum, 2100m, 6.1116N, 10.2543E, 22 Nov. 1996, Etuge 3595.

SELAGINELLACEAE
P.J.Edwards (K) & J-M.Onana (YA)

Selaginella abyssinica Spring
Local name: Eghcok (Oku) – Munyenyembe 829.
Small, (< 30 cm), terrestrial or epiphytic; median leaves strongly heeled with an apical bristle; lateral leaves oblong with toothed hyaline margin. Cliffs and banks.
Kilum: Elak, 2500m, 31 Oct. 1996, Munyenyembe 829; KC, 2500m, 30 Oct. 1996, Zapfack 1106.
Ijim: Zitum road, 1900m, 6.1116N, 10.2543E, 21 Nov. 1996, Etuge 3553; Above Fon's palace, Laikom, about 30 minutes along path to Fulani settlement, 2050m, 6.1642N, 10.1952E, 21 Nov. 1996, Cheek 8714.

FILICOPSIDA

P.J.Edwards (K), J-M.Onana (YA) & B.J.Pollard (K)

ADIANTACEAE

Adiantum poiretii Wikstr.
Terrestrial fern to 1 m; rhizome slender, wide-creeping; fronds closely-spaced; stipe and rachis very dark brown or black, conspicuously glossy; frond 3-pinnate; pinnules broadly obcuneate, articulated to long stalks; sori crescent shaped. Forest.
Kilum: Summit, 2800m, 31 Oct. 1996, Zapfack 1136.
Ijim: Gikwang road towards Nyasosso forest, Ijim Mountain Forest Reserve, 2250m, 19 Nov. 1996, Etuge 3494; Above Aboh village, 1.5 hours on Gikwang road to Lake Oku, 2450m, 6.1112N, 10.2529E, 19 Nov. 1996, Cheek 8663.

Cheilanthes farinosa (Forssk.) Kaulf. *sens. lat.*
Terrestrial fern to 50 cm; rhizome short, ascending; stipe dark red-brown, glossy; lamina deltoid, 1-pinnate pinnatisect, lower surface covered in a white, mealy powder. Fallow.
Kilum: Summit, 2800m, 31 Oct. 1996, Zapfack 1124.
Ijim: Aboh village, 6.15N, 10.26E, 18 Nov. 1996, Onana 535; Main path from Ijim to Tum, 2100m, 6.1116N, 10.2543E, 25 Nov. 1996, Etuge 3615; Laikom to Fundong, 1700m, 6.17N, 10.20E, 10 Dec. 1998, Etuge 4595.

Coniogramme africana Hieron.
Large terrestrial fern, 1–2 m; lamina 2-pinnate; pinnules lanceolate, large (10–25 cm long); veins 1–2 forked with naked sori along many of them. Forest edge.
Kilum: KA, 2200m, 11 Jun. 1996, Etuge 2308; Above Mboh, path above village to forest through Wambeng's farm, 2200m, 6.1146N, 10.2856E, 1 Nov. 1996, Cheek 8557.
Ijim: Along the top of the ridge, second gate from Aboh village, 2250m, 21 Nov. 1996, Kamundi 678; Zitum road, 1900m, 6.1116N, 10.2543E, 21 Nov. 1996, Etuge 3559; Ijim ridge, about 30 minutes walk above Fon of Kom's Palace, Laikom, 2050m, 6.16N, 10.19E, 2 Dec. 1998, Cheek 9773.
Plot voucher: Forest above Oku-Elak, Jun. 1996, Oku 33.

Pellaea quadripinnata (Forssk.) Prantl
Terrestrial fern; rhizome short-creeping; fronds tufted to 1 m, erect to arching; stipe dark red-brown, glossy, 0.3–0.7 m long, being 1–2 times as long as the lamina; lamina 3–4-pinnate, leathery. Forest.
Kilum: Elak, 2520m, 9 Jun. 1996, Zapfack 811; KD, 3000m, 10 Jun. 1996, Etuge 2293.
Ijim: Ijim Mountain Forest Reserve, 1800m, 6.1116N, 10.2543E, 21 Nov. 1996, Pollard 74.

ASPLENIACEAE

Asplenium abyssinicum Fée
Epilithic or epiphytic fern to 40 cm; rhizome short; stipe and rachis red-brown, very shiny; lamina narrowly lanceolate, tripinnate, very delicate, the base of many of the rounded lobes bearing one rounded or slightly elongated sorus. Forest.
Kilum: Elak, 2360m, 9 Jun. 1996, Zapfack 807; KJ, 2500m, 29 Oct. 1996, Zapfack 1084; Junction of KJ and KN above Elak, 2550m, 10 Jun. 1996, Cable 3030; Path from BirdLife HQ to KA, 2500m, 6.1349N, 10.3112E, 30 Oct. 1996, Cheek 8507.
Ijim: Main path from Ijim to Tum, 2100m, 6.1116N, 10.2543E, 25 Nov. 1996, Etuge 3625.

Asplenium aethiopicum (Burm.f.) Bech.
Terrestrial, epilithic or epiphytic fern; rhizome erect; fronds oblong-lanceolate; stipe black or very dark brown, glossy; lamina 1–2-pinnate pinnatisect; scales deciduous; pinnae very acute to long caudate; sori extending along pinna-midrib (costa) and for c. 2/3 pinna length. Forest.
Kilum: KD, 2800m, 9 Jun. 1996, Etuge 2214; Elak, 2360m, 9 Jun. 1996, Zapfack 810; KD, 2800m, 10 Jun. 1996, Etuge 2268 & 2275; KJ, above water tank from Elak, 2 km into forest, 2400m, 10 Jun. 1996, Cable 3015; KD, 2900m, 10 Jun. 1996, Cable 3055; Lake Oku, 2300m, 11 Jun. 1996, Zapfack 879; Shore of Lake Oku, 2300m, 12 Jun. 1996, Cable 3129; KJ to KD, 3000m, 13 Jun. 1996, Etuge 2352; KJ, 2500m, 29 Oct. 1996, Zapfack 1082; KA path, 2600–2800m, 6.15N, 10.26E, 30 Oct. 1996, Onana 472 & 473.
Ijim: Aboh village, 6.15N, 10.26E, 18 Nov. 1996, Onana 525; Gikwang road towards Nyasosso forest, Ijim Mountain Forest Reserve, 2300m, 19 Nov. 1996, Etuge 3476.
Plot voucher: Forest above Oku-Elak, Jun. 1996, Oku 81.

Asplenium anisophyllum Kunze
Asplenium elliottii C.H.Wright is included here. According to C.H.Wright the base of the lower side of the pinnae in *A. elliottii* is much more obtuse than in *A. anisophyllum*, and is sometimes almost parallel to the rachis. The specimens cannot be clearly divided on the basis of this single character, further study of specimens from this complex (including *A.boltonii*) is required to clarify the specific delimitations.
Terrestrial or epilithic fern to 1 m; rhizome short, erect; scales brown, broad; lamina linear-lanceolate, pinnate; larger fronds 60 × 18 cm; pinnae linear-oblong, apex acuminate, base unequally cuneate, bluntly serrate; sori elongate, short, at c. 45°; bud/plantlet at base of terminal pinna. Forest.
Syn. *Asplenium geppii* Carruth. , F.W.T.A. Suppl.: 56 (1959).
Kilum: Path above water-tower, KJ, about 1 km into forest, 2200m, 9 Jun. 1996, Cable 2955; KD, 2800m, 9 Jun. 1996, Etuge 2217; Elak, 2400m, 9 Jun. 1996, Zapfack 818; KD, 2600m, 10 Jun. 1996, Cable 3043; KA, 2200m, 11 Jun. 1996, Etuge 2307; Lake Oku, 2300m, 11 Jun. 1996 Zapfack 853; Elak, lower parts of KA, 2200m, 6.1349N, 10.3112E, 28 Oct. 1996, Cheek 8467; KC, 2300m, 30 Oct. 1996, Zapfack 1096.
Ijim: Gikwang main road towards Nyasosso forest, 2260m, 6.1116N, 10.2543E, 20 Nov. 1996, Etuge 3516; Near Aboh-Anyajua, 2500m, 20 Nov. 1996, Satabie 1080; Along the top of the ridge, second gate from Aboh village, 2360m, 6.11N, 10.26 E, 21 Nov. 1996, Kamundi 667; Ijim ridge, about 30 minutes walk above Fon of Kom's Palace, Laikom, 2050m, 6.16N, 10.19E, 2 Dec. 1998, Cheek 9774 & 9776;

Plot between Fon's Palace and Ardo's compound, 2000m,
6.16N, 10.21E, 8 Dec. 1998, Cheek 9865.
Plot voucher: Forest above Oku-Elak, Jun. 1996, Oku 12; 43;
76 & 87.

Asplenium sp. aff. biafranum Alston & Ballard
(Asplenium friesiorum group*)*
Very similar to *Asplenium friesiorum*, but: pinnae more
shallowly serrate; sori very oblique, c. 30° to the costa.
Forest.
Kilum: Elak, KJ, 2300m, 29 Oct. 1996, Zapfack 1047.

Asplenium dregeanum Kunze
Epiphyte to 25 cm; rhizome short-creeping; fronds densely
tufted; lamina ovate-lanceolate in outline; dimidiate pinnae
deeply dissected, one linear sorus on many lobes. Forest.
Ijim: Akwo-mofu, Laikom, 1760m, 6.17N, 10.20E, 5 Dec.
1998, Etuge 4561; Plot between Fon's Palace and Ardo's
compound, 2000m, 6.16N, 10.21E, 8 Dec. 1998, Cheek
9872.

Asplenium erectum Bory ex Willd. var. usambarense (Hieron.) Schelpe
Local name: Nk 'um (Oku), all ferns known by this name
but used differently for medicine – Cheek 8585.
Terrestrial, epilithic or epiphytic fern to 40 cm; rhizome
short, suberect; lamina 1-pinnate, linear in outline; pinnae
deeply crenate, auricled at base; lower pinnae gradually
decrescent. Forest.
Syn. *Asplenium quintasii* Gand., F.W.T.A. Suppl.: 57
(1959).
Kilum: Elak, 2420m, 9 Jun. 1996, Zapfack 808; KA,
2200m, 11 Jun.1996, Etuge 2306; Shore of Lake Oku,
2300m, 12 Jun. 1996, Cable 3092; Oku-Elak, lower parts of
KA, 2200m, 6.1349N, 10.3112E, Cheek 8465; KC, 2300m,
30 Oct.1996, Zapfack 1099; Lake Oku, near Baptist Rest
House, 2300m, 6.1214N, 10.2733E, 4 Nov. 1996, Cheek
8585.
Ijim: Aboh village, 6.15N, 10.26E, 25 Nov. 1996, Onana
623; Plot between Fon's Palace and Ardo's compound,
2000m, 6.16N, 10.21E, 8 Dec. 1998, Cheek 9878.

Asplenium friesiorum C.Chr.
Similar to *Asplenium anisophyllum*, but: rhizome creeping;
scales narrow, glossy, very dark brown; pinnae long-caudate,
deeply and doubly serrate; sori more elongate, lying ±
parallel to the costa (pinna midrib). Forest.
Kilum: KD, 2800m, 10 Jun. 1996, Etuge 2273; Shore of
Lake Oku, 2300m, 12 Jun. 1996, Cable 3103 & 3128; Elak,
KJ, 2400m, 29 Oct. 1996, Zapfack 1066; KJ, 2500m, 29 Oct.
1996, Zapfack 1083.

Asplenium mannii Hook.
Very small epiphyte to 6 cm; rooting runners present; lamina
2-pinnate, one linear sorus per lobe. Forest.
Kilum: Elak, 2400m, 9 Jun. 1996, Zapfack 816; Lake Oku,
2300m, 11 Jun. 1996, Zapfack 850; Shore of Lake Oku,
2300m, 12 Jun. 1996, Cable 3100B; Oku-Elak, lower parts
of KA, 2200m, 6.1349N, 10. 3112E, 28 Oct. 1996, Cheek
8460; KC, 2600m, 30 Oct. 1996, Zapfack 1107; KD, 2300m,

1 Nov. 1996, Zapfack 1151.
Ijim: Plot between Fon's Palace and Ardo's compound,
2000m, 6.16N, 10.21E, 8 Dec. 1998, Cheek 9864.

Asplenium preussii Hieron.
Terrestrial or epilithic fern to 60 cm ; rhizome short; lamina
linear-lanceolate in outline, 1-pinnate pinnatisect, larger
fronds c. 40 × 12 cm; pinnae ± pectinate; basal acroscopic
lobe very enlarged and almost stipitate, most lobes with a
single elongated sorus along 1/2 to 3/4 their length. Forest.
Ijim: Zitum road, 1860m, 6.1116N, 10.2543E, 21 Nov.
1996, Etuge 3551; Aboh village, 6.15N, 10.26E, 25 Nov.
1996, Onana 622; Ijim ridge, about 30 minutes walk above
Fon of Kom's Palace, Laikom, 2050m, 6.16N, 10.19E, 2 Dec.
1998, Cheek 9778; Akwo-mofu, Laikom, 1760m, 6.17N,
10.20E, 5 Dec. 1998, Etuge 4550.

Asplenium protensum Schrad.
Like a larger form of *Asplenium erectum* var. *usambarense*,
but: pinnae more deeply lobed; rachis pubescent; lamina
with bud/plantlet near apex. Forest.
Kilum: Above Mboh, path above village to forest through
Peter Wambeng's farm, 2200m, 6.1146N, 10.2856E, 1 Nov.
1996, Cheek 8563.
Ijim: Plot between Fon's Palace and Ardo's compound,
2000m, 6.16N, 10.21E, 8 Dec. 1998, Cheek 9873.

Asplenium theciferum (Kunth) Mett. var. cornutum (Alston) Benl
Small epiphyte to 15 cm tall; rhizome short-creeping; lamina
2-pinnate, with a rounded sorus embedded at the tip of many
of the lobes. (Smaller plants are similar to *Asplenium
mannii*, but runners are absent). Forest.
Syn. *Asplenium cornutum* Alston, F.W.T.A. Suppl.: 60
(1959).
Kilum: Path above water-tower, KJ, about 1 km into forest,
2200m, 9 Jun. 1996, Cable 2957; KD, 2800m, 9 Jun. 1996,
Etuge 2210; Elak, 2460m, 9 Jun. 1996, Zapfack 812; Elak,
2200m, 9 Jun. 1996, Zapfack 813; Shore of Lake Oku,
2300m, 12 Jun. 1996, Cable 3100B & Cable 3127; Elak,
KDH, 1 Nov. 1996, Zapfack 1150; Elak, JOH, Kwifon
traditional forest, 1950m, 3 Nov. 1996, Zapfack 1166.
Ijim: Main path from Ijim to Tum, 2100m, 6.1116N,
10.2543E, 25 Nov. 1996, Etuge 3627.
Plot voucher: Forest above Oku-Elak, Jun. 1996, Oku 37; 82.

Asplenium sp. A
Very variable, certainly more than one taxon here, but for
now all included under this name, **sens. lat.** Forest.
Kilum: Path above water-tower, KJ, about 1 km into forest,
2200m, 9 Jun. 1996, Cable 2958; Elak, KJ, 2500m, 29 Oct.
1996, Zapfack 1081; KC, 2300m, 30 Oct. 1996, Zapfack
1097.
Ijim: From Aboh, path towards Tum, 2200m, 6.1116N,
10.2543E, 22 Nov. 1996, Etuge 3599.
Plot voucher: Forest above Oku-Elak, Jun. 1996, Oku 59.

CYATHEACEAE

Cyathea dregei Kunze

Terrestrial tree-fern to 2 m or more; fronds to 2.5 m long, 2 pinnate-pinnatisect; old fronds persistent (pendulous/hanging down trunk). Forest edge.
Local name: Lenga Lengha (Kom) – Etuge 4508.
Ijim: Above Fon's palace, Laikom, about 30 minutes on path to Fulani settlement, 2050m, 6.1642N, 10.1952E, 21 Nov. 1996, Cheek 8700; Zitum road, 1900m, 6.1116N, 10.2543E, 21 Nov. 1996, Etuge 3570; Laikom, 5 km E of Fundong, 2100m, 21 Nov. 1996, Satabie 1084; By stream, Laikom, 2000m, 6.17N, 10.20E, 2 Dec. 1998, Etuge 4508; Plot between Fon's Palace and Ardo's compound, 2000m, 6.16N, 10.21E, 8 Dec. 1998, Cheek 9879.

DENNSTAEDTIACEAE

Blotiella glabra (Bory) A.F.Tryon

Large terrestrial to 2.5 m; rhizome erect to suberect; stipe densely covered in very long red-brown hairs; lamina desnely and coarsely short-hairy, 2-pinnate pinnatisect, to 2 × 0.8 m; sori U-shaped in the rounded sinuses. Forest.
Syn. *Lonchitis gracilis* Alston, F.W.T.A. Suppl.: 34 (1959).
Kilum: KJ, 2400m, 29 Oct. 1996, Zapfack 1065.

Blotiella mannii (Baker) Pic.Serm.

Large terrestrial to 2.5 m; rhizome erect to suberect; stipe with long dark brown hairs near the base; lamina densely and coarsely short hairy, to 2 × 0.5 m; pinnae of larger fronds deeply lobed near the base, progressiveyl shallower so that the apex has an entire margin; sori running along most of the margins, very elongated to continuous. Forest.
Syn. *Lonchitis mannii* (Baker) Alston, F.W.T.A. Suppl.: 34 (1959).
It is quite possible this species is conspecific with *Blotiella currorii* (Hook.) A.F.Tryon.
Ijim: Plot between Fon's Palace and Ardo's compound, 2000m, 6.16N, 10.21E, 8 Dec. 1998, Cheek 9871.

Pteridium aquilinum (L.) Kuhn subsp. *aquilinum*

Terrestrial fern; thicket forming; rhizome long-creeping, subterranean; fronds to 1.5m tall; stipe erect, the base black (remainder brown); lamina 3-pinnate pinnatisect; sori marginal with fimbriate indusia on both sides. Forest.
Kilum: Lake Oku, 2300m, 12 Jun. 1996, Etuge 2338.

DRYOPTERIDACEAE

Didymochlaena truncatula (Sw.) J.Sm.

Terrestrial erect tufted fern to 2m tall; stipe and rachis with many, broad, dark-brown scales; lamina 2-pinnate; pinnules dimidiate, trapeziform. Forest.
Ijim: Zitum road, 1900m, 6.1116N, 10.2543E, 21 Nov. 1996, Etuge 3564; Plot between Fon's Palace and Ardo's compound, 2000m, 6.16N, 10.21E, 8 Dec. 1998, Cheek 9875.

Dryopteris athamantica (Kunze) Kuntze

Terrestrial fern; rhizome prostrate; frond coriaceous; pinnules obliquely cuneate at base; pinnule segments untoothed. Forest.
Kilum: Secondary forest near the village, 6.15N, 10.26E, 6 Nov. 1996, Onana 512.

Dryopteris pentheri (Krasser) C.Chr.

Terrestrial fern; rhizome suberect; frond herbaceous; pinnules truncate at base; pinnule segments with acute teeth. Forest.
Kilum: Elak, 2420m, 9 Jun. 1996, Zapfack 809; Elak, 2600m, 10 Jun. 1996, Zapfack 837; KD, 2800m, 10 Jun. 1996, Etuge 2263; Lake Oku, 2300m, 11 Jun. 1996, Zapfack 854; KA to summit, 2800m, 6.1349N, 10.3112E, 30 Oct. 1996, Cheek 8536.

Polystichum transvaalense N.C.Anthony

Terrestrial fern; rhizome suberect with many scales and matted roots; stipe, rachis and pinna-rachis densely brown-scaly; lamina 2-pinnate; pinnules ovate to obliquely transversely rhomboid, c. 12–15 mm long, serrate, aristate.
Kilum: Path above water-tower, KJ, about 1 km into forest, 2200m, 9 Jun. 1996, Cable 2956 & 2962; Shore of Lake Oku, 2300m, 12 Jun. 1996, Cable 3130; KJ, 2620m, 29 Oct. 1996, Zapfack 1086; KC, 2500m, 30 Oct. 1996, Pollard 14.

Polystichum wilsonii H.Christ

Similar to *Polystichum transvaalense*, but: stipe shorter; fronds narrower and with the basal pinnae more reduced and deflexed. Grassland.
Syn. *Polystichum fuscopaleaceum* Alston, F.W.T.A. Suppl.: 70 (1959).
Kilum: Summit, 2950m, 10 Jun. 1996, Cable 3079; KD, 2800m, 10 Jun. 1996, Etuge 2280.

Tectaria fernandensis (Baker) C.Chr.

Local name: Nkon (Oku) – Etuge 3351.
Terrestrial fern; rhizome erect; fronds tufted to 1 m; lamina mostly 1-pinnate pinnatisect with 3–4 pairs of pinnae, but the basal pair with much enlarged basal pinnules. Forest edge.
Kilum: Main road from Jikijem to Oku, 2350m, 7 Nov. 1996, Etuge 3351.
Ijim: Zitum road, 1925m, 6.1116N, 10.2543E, 21 Nov. 1996, Etuge 3540.

GLEICHENIACEAE

Dicranopteris linearis (Burm.f.) Underw. var. *linearis*

Terrestrial fern; thicket forming, scrambling to 2 m; rhizome subterranean and surface-rooting; stipes erect, very long; lamina dichotomous forming a complex of divaricating 'branches' bearing deeply dissected pectinate pinnules.
Syn. *Gleichenia linearis* (Burm.) C.B.Clarke, F.W.T.A. Suppl.: 22 (1959).
Kilum: Lowland forest patch at lower altitude than the BirdLife Headquarters, 1800m, 6.15N, 10.29E, 6 Nov. 1996, Buzgo 709.

GRAMMITIDACEAE

Xiphopteris villosissima (Hook.) Alston subsp. *villosissima* var. *villosissima*
Small epiphyte to 15 cm; rhizome short-creeping; stipe wiry; lamina entire, deply-lobed (pinnatisect). Forest.
Kilum: KJ, 2300m, 29 Oct. 1996, Zapfack 1054.

HYMENOPHYLLACEAE

Microgonium ballardianum (Alston) Pic.Serm.
Very small epiphyte on tree trunks; rhizome long-creeping, wiry, dark brown; liverwort-like hepatic with undivided translucent fronds, ± circular, broadly ovate to obovate; lamina c. 1 × 0.5–1 cm. Forest.
Kilum: Oku-Elak, lower parts of KA, 2200m, 6.1349N, 10.3112E, 28 Oct. 1996, Cheek 8459B.

Vandenboschia melanotricha (Schltdl.) Pic.Serm.
Very small trunk-epiphyte; rhizome long-creeping, black with glossy black hairs; lamina deeply and unevenly lobed to near the rachis; rachis winged, indusium on basal lobe. Forest.
Kilum: Elak, 2360m, 9 Jun. 1996, Zapfack 817; Lake Oku, 2300m, 11 Jun. 1996, Zapfack 875; Oku-Elak, lower parts of KA, 2200m, 6.1349N, 10.3112E, 28 Oct. 1996, Cheek 8459A; KJ, 2300m, 29 Oct. 1996, Zapfack 1050; KJ, 2620m, 29 Oct. 1996, Zapfack 1087; Path from Project HQ to KA, 2500m, 6.1349N, 10.31112E, 30 Oct. 1996, Cheek 8510; KC, 2500m, 30 Oct. 1996, Zapfack 1103.

LOMARIOPSIDACEAE

Elaphoglossum chevalieri H.Christ
Small epiphyte to 30 cm; stipe c. half to three-quartesrs length of the narrowly oblong-elliptic lamina; both surfaces with well-spaced hair-like appressed red-brown scales.
Ijim: From Aboh, path towards Tum, 2100m, 6.1116N, 10.2543E, 22 Nov. 1996, Etuge 3588.

Elaphoglossum kuhnii Hieron.
Similar to *Elaphoglossum chevalieri*, but fronds to 50 cm or more long; stipe and sterile lamina densely covered on both surfaces with lanceolate, appressed, red-brown ciliate scales.
Kilum: Shore of Lake Oku, 2300m, 12 Jun. 1996, Cable 3095; Lometo Traditional Forest, 2050m, 7 Nov. 1996, Zapfack 1239.

MARATTIACEAE

Marattia fraxinea J.Sm.
Very large terrestrial fern; rhizome erect to 40 × 30 cm; fronds tufted to 4 m, stiff, fleshy; stipe with brown flushing and long white or green streaks; swollen base with a pair of green to dark brown, thick, fleshy stipules; lamina ovate in outline, 2-pinnate, to 2 × 1 m. Forest.
Kilum: KA, 2200m, 11 Jun. 1996, Etuge 2318; Lake Oku, 2800m, 4 Nov. 1996, Pollard 37.

OLEANDRACEAE

Arthropteris monocarpa (Cordem.) C.Chr.
Epilithic or epiphytic fern; rhizome long-creeping, ± black below, with brown scales and woolly hairs when shed; stipe 1/5 to 1/3 length of the lamina, articulated near base; lamina 1-pinnate-pinnatisect, 15–30 × 8–10 cm; 1 sorus near base of each pinna-lobe in fertile fronds. Forest.
Kilum: Ethiale, near the stream, 2060m, fl., 4 Nov. 1996, Zapfack 1180.

Nephrolepis undulata (Afzel. ex Sw.) J.Sm. var. *undulata*
Local name: Nken (Oku) – Cheek 8605.
Rhizome vestigial, < 1 cm; stolons numerous, long and wiry (many bearing tubers); fronds tufted, erect; lamina 1-pinnate, with 30–100 pairs of slightly crenate, elliptic-lanceolate pinnae; bases auricled; sori semi-circular. Forest.
Kilum: Lake Oku to Jikijem, 2200m, 6.1214N, 10.2733E, 5 Nov. 1996, Cheek 8605; Iwook Etele Mbae, 2500m, 5 Nov. 1996, Zapfack 1191; Manchok, lowland areas, path leading towards Kumbo road, 2100m, 9 Nov. 1996, Etuge 3361.
Ijim: N facing, seasonally damp rockface amongst dense vegetation, in 'gallery forest' along path from Fon's Palace to grassland plateau above Laikom, 2040m, 6.1642N, 10.1950E, 2 Dec. 1998, Pollard 278.

Oleandra distenta Kunze var. *distenta*
Epilithic or epiphytic fern; rhizome 3–5 mm diam., long-creeping, scandent to 10 m, with short side branches, densely covered in closely appressed dark-brown scales; stipe 5–30 mm long, articulated near the base and this lower part persistent after the shedding of the rest of the frond; lamina entire, linear-lanceolate, 15–30 × 2–6 cm. Forest.
Ijim: From Aboh, path towards Tum, 6.1116N, 10.2543E, 2200m, 22 Nov. 1996, Etuge 3601.

OSMUNDACEAE

Osmunda regalis L. var. *regalis*
Local name: Lenga lengha (Kom at Laikom) – Etuge 4512.
Terrestrial fern; rootstock erect becoming very large, embedded in matted, fibrous roots; fronds erect to 2 m; lamina 2-pinnate to 1 m long; 2–5 fertile pinnae borne in the apical portion of some fronds. Near streams.
Ijim: Above Fon's palace, Laikom, about 30 minutes on path to Fulani settlement, 2050m, 6.1642N, 10.1952E, 21 Nov. 1996, Cheek 8702; Beside a stream, Laikom, 2000m, 6.17N, 10.20E, 2 Dec. 1998, Etuge 4512.

POLYPODIACEAE

Lepisorus excavatus (Bory ex Willd.) Ching
Epiphytic fern; rhizome long-creeping; stipe short; lamina very thin, glabrous, 15–30 × 2–3.5 cm; sori circular, large, in a single series each side of the midrib. Forest.
Syn. *Pleopeltis excavata* (Bory ex Willd.) T.Moore
Kilum: Lake Oku, 2300m, 11 Jun. 1996, Zapfack 852.

Lepisorus preussii (Hieron.) Pic.Serm.

Very similar to *Pleopeltis macrocarpa* var. *macrocarpa* (and probably synonymous with), but rhizome scales not glossy; fronds generally smaller. Forest.

Syn. *Pleopeltis preussii* (Hieron.) Tardieu

Kilum: Path above water-tower, KJ, about 1 km into forest, 2200m, 9 Jun. 1996, Cable 2960A; KD, 2800m, 9 Jun. 1996, Etuge 2212; Elak, 2100m, 9 Jun. 1996, Zapfack 814; Elak, 2260m, 9 Jun. 1996, Zapfack 815; Shore of Lake Oku, 2300m, 12 Jun. 1996, Cable 3124; KJ, 2300m, 29 Oct. 1996, Zapfack 1053; Summit, 2800m, 31 Oct. 1996, Zapfack 1137; Near to the stream in savanna zone, 2500m, 5 Nov. 1996, Zapfack 1207.

Plot voucher: Forest above Oku-Elak, Jun. 1996, Oku 74.

Loxogramme abyssinica (Baker) M.G.Price

Epiphytic fern; rhizome wide-creeping, 1–2 mm diam.; fronds mostly with short stripes; lamina entire, glabrous, very leathery, 5–30 × 1–2.5 cm; sori very elongated, forming oblique parallel lines near midrib. Forest.

Syn. *Loxogramme lanceolata* (Sw.) C.Presl., F.W.T.A. Suppl.: 48 (1959).

Kilum: Path above water-tower, KJ, about 1 km into forest, 2200m, 9 Jun. 1996, Cable 2959; Lake Oku, 2300m, 11 Jun. 1996, Zapfack 851; Shore of Lake Oku, 2300m, 12 Jun. 1996, Cable 3110 & 3126.

Ijim: Plot between Fon's Palace and Ardo's compound, 2000m, 6.16N, 10.21E, 8 Dec. 1998, Cheek 9863.

Plot voucher: Forest above Oku-Elak, Jun. 1996, Oku 19.

Pleopeltis macrocarpa (Bory ex Willd.) Kaulf. var. *macrocarpa*

Small leathery epiphytic fern similar to *Loxogramme abyssinica*, but lower surface with many scattered small brown scales; sori as a single series of large oval spots each side of the midrib. Forest.

Syn. *Pleopeltis lanceolata* (L.) Kaulf., F.W.T.A. Suppl.: 49 (1959).

Kilum: Path above water-tower, KJ, about 1 km into forest, 2200m, 9 Jun. 1996, Cable 2960B; Summit, 2800m, 31 Oct. 1996, Zapfack 1138.

PTERIDACEAE

Pteris pteridioides (Hook.) Ballard

Terrestrial fern to 1.5 m; rhizome erect to 10 cm diam.; fronds tripartite, deltoid in outline, 40–80 × 30–70 cm. Forest.

Kilum: Oku-Elak, lower parts of KA, 2200m, 6.1349 N, 10.3112 E, 28 Oct. 1996, Cheek 8473; Elak, KJ, 2500m, 29 Oct. 1996, Zapfack 1085.

Ijim: Zitum road, 1900m, 6.1116N, 10.2543E, 21 Nov. 1996, Etuge 3554.

Plot voucher: Forest above Oku-Elak, Jun. 1996, Oku 5; 17.

Pteris quadriaurita Retz. subsp. *togoensis* (Hieron.) Schelpe

Probably a part of the very ill-defined *Pteris catoptera* complex.

Terrestrial to 1 m tall; rhizome erect to 6 cm diam.; stipe and basal part of rachis pink to red; lamina ovate in outline, 50–70 × 30–40 cm, the lowest pair of pinnae bifid; costa and costules with small spines on upperside. Forest.

Syn. *Pteris togoensis* Hieron., F.W.T.A. Suppl.: 40 (1959).

Kilum: Path above water-tower, KJ, about 1 km into forest, 2250m, 9 Jun. 1996, Cable 2998; Path from BirdLife HQ to KA, 2600m, 6.1349N, 10.3112E, 30 Oct. 1996, Cheek 8525; KD, 2400m, 1 Nov. 1996, Zapfack 1154.

Ijim: Gikwang main road towards Nyasosso forest, 2250m, 6.1116N, 10.2543E, 20 Nov. 1996, Etuge 3515; Aboh village, 2450m, 6.15N, 10.26E, 20 Nov. 1996, Onana 553; Plot between Fon's Palace and Ardo's compound, 2000m, 6.16N, 10.21E, 8 Dec. 1998, Cheek 9877.

Plot voucher: Forest above Oku-Elak, Jun. 1996, Oku 29; 42.

THELYPTERIDACEAE

Christella guineensis (H.Christ) Holttum

Terrestrial fern to 90 cm; rhizome stout, suberect or short-creeping; larger fronds softly hairy on both surfaces; lamina 1-pinnate pinnatisect, c. 50 × 15 cm, excluding the 34 pairs of very reduced basal pinnae running down the stipe. Streamsides in forest.

Syn. *Thelypteris guineensis* (H.Christ) Alston, F.W.T.A. Suppl.: 61 (1959).

Ijim: From Aboh, path towards Tum, 2100m, 6.1116N, 10.2543E, 22 Nov. 1996, Etuge 3591.

Pneumatopteris unita (Kunze) Holttum

Terrestrial fern to 2.5 m; rhizome erect to suberect; lamina 1-pinnate pinnatifid, glabrous; larger fronds c. 1 × 0.3 m with a terminal bud/plantlet; sori neatly dotted along main vein of each lobe. Forest.

Syn. *Cyclosorus patens* (Fée) Copel., F.W.T.A. Suppl.: 62 (1959).

Ijim: Plot between Fon's Palace and Ardo's compound, 2000m, 6.16N, 10.21E, 8 Dec. 1998, Cheek 9868.

Thelypteris confluens (Thunb.) Morton

Small fern; fronds erect to 80 cm; rhizome long-creeping, black; lamina 1-pinnate pinnatisect, lanceolate-elliptic in outline; pinna midribs (costae) underside bearing conpicuous circular scales. Swamp tussocks.

Ijim: Afua swamp, 1950m, 6.08N, 10.24E, 7 Dec. 1998, Cheek 9834.

WOODSIACEAE

Athyrium ammifolium (Mett.) C.Chr.

Local name: Fghuin f-aku (Kom at Laikom) – Etuge 4507.

Terrestrial fern; rhizome erect; lamina broadly lanceolate to deltoid in outline, 20–40 × 12–30 cm, 2-pinnate pinnatisect; sori mostly J-shaped. Forest.

Ijim: Grassland and gallery forest on volcanic plateau, Laikom, 2000m, 6.17N, 10.20E, 2 Dec. 1998, Etuge 4507; Plot between Fon's Palace and Ardo's compound, 2000m, 6.16N, 10.21E, 8 Dec. 1998, Cheek 9876;

Athyrium schimperi Moug. ex Fée

Terrestrial fern to 80 cm; rhizome short-creeping; lamina narrowly lanceolate in outline, 15–40 × 8–16 cm; sori mostly J-shaped. Forest.

Ijim: Gikwang road towards Nyasosso forest, Ijim Mountain Forest Reserve, 2260m, 19 Nov. 1996, Etuge 3495.

Cystopteris fragilis (L.) Bernh.

Small terrestrial or epilithic fern; stipe very brittle; sori dark brown at base, yellowish above; lamina lanceolate in outline; 1-pinnate pinnatisect, c. 15 × 4 cm. Grassland.

Kilum: Summit, 2950m, 10 Jun. 1996, Cable 3067.

Lunathyrium boryanum (Willd.) H.Ohba

Large terrestrial fern to 2 m; rhizome erect; lamina 2-pinnate pinnatisect; larger fronds c. 1 × 0.6 m; sori circular, elongate and J-shaped. Forest.

Syn. *Athyrium glabratum* (Mett.) Alston, F.W.T.A. Suppl.: 64 (1959).

Kilum: KA, 2200m, 11 Jun. 1996, Etuge 2321.

INDEX OF CHECKLIST TAXA

Didymochlaena truncatula (Sw.) J.Sm., 196
Digitaria abyssinica (Hochst. ex A.Rich.) Stapf, 178
Digitaria debilis (Desf.) Willd., 178
Digitaria diagonalis (Nees) Stapf var. *diagonalis*, 178
Digitaria diagonalis (Nees) Stapf var. *hirsuta* (De Wild. & T.Durand) Troupin, 178
Digitaria ternata (A.Rich.) Stapf, 178
DILLENIACEAE, 130
Dioscorea schimperiana Hochst. ex Kunth, 176
DIOSCOREACEAE, 176
DIPSACACEAE, 130
Dipsacus narcisseanus Lawalree, 130
Dipsacus pinnatifidus Steud. ex A.Rich., 130
Disa erubescens Rendle subsp. *erubescens*, 188
Disa hircicornis Rchb.f., 188
Disa nigerica Rolfe, 188
Disa ochrostachya Rchb.f., 188
Disa welwitschii Rchb.f. subsp. *occultans* (Schltr.) Linder, 188
Discopodium penninervium Hochst., 164
Disperis nitida Summerh., 188
Disperis parvifolia Schltr., 189
Dissotis bambutorum Gilg & Ledermann ex Engl., 148
Dissotis bamendae Brenan & Keay, 147
Dissotis brazzae Cogn., 147
Dissotis elliotii Gilg var. *elliotii*, 147
Dissotis irvingiana Hook., 147
Dissotis perkinsiae Gilg, 147
Dissotis princeps (Kunth) Triana var. *princeps*, 147
Dissotis senegambiensis (Guill. & Perr.) Triana var. *senegambiensis*, 147
Dissotis thollonii Cogn. ex Büttner var. *elliotii* (Gilg) Jacq.-Fél., 147
Dolichos sericeus E.Mey., 142
Dombeya ledermannii Engl., 165
Dovyalis sp. nov., 133
Dracaena deisteliana Engl., 176
Dracaena fragrans (L.) Ker-Gawl, 176
DRACAENACEAE, 176
Dregea schimperi (Decne.) Bullock, 115
Droguetia iners (Forssk.) Schweinf. subsp. *iners*, 167
Drosera madagascariensis DC., 130
DROSERACEAE, 130
Drymaria cordata (L.) Willd., 119
Drymaria villosa Cham. & Schltdl. subsp. *villosa*, 119
DRYOPTERIDACEAE, 196
Dryopteris athamantica (Kunze) Kuntze, 196
Dryopteris pentheri (Krasser) C.Chr., 196
Dyschoriste nagchana (Nees) Bennet, 111
Echinochloa crus-pavonis (Kunth) Schult., 178
Echinops gracilis O.Hoffm., 124
Echinops mildbraedii Mattf., 124
Elaphoglossum chevalieri H.Christ, 197
Elaphoglossum kuhnii Hieron., 197
Elatostema monticola Hook.f., 167
Elatostema paiveanum Wedd., 168
Elatostema welwitschii Engl., 168
Eleusine indica (L.) Gaertn., 178
Elymandra androphila (Stapf) Stapf, 178
Embelia mildbraedii Gilg & G.Schellenb., 150

Embelia schimperi Vatke, 150
Emilia abyssinica (Sch.Bip. ex A.Rich.) C.Jeffrey var. *abyssinica*, 124
Emilia coccinea (Sims) G.Don, 124
Englerastrum gracillimum T.C.E.Fr., 137
Englerina gabonensis (Engl.) Balle, 145
Ensete gilletii (De Wild.) Cheesman, 185
Entada abyssinica Steud. ex A.Rich., 140
Entandrophragma angolense (Welw.) C.DC., 148
Epilobium salignum Hausskn., 151
Eragrostis atrovirens (Desf.) Trin. ex Steud., 178
Eragrostis camerunensis Clayton, 178
Eragrostis gangetica (Roxb.) Steud., 179
Eragrostis macilenta (A.Rich.) Steud., 179
Eragrostis mokensis Pilg., 179
Eragrostis pobeguinii C.E.Hubb., 179
Eragrostis tenuifolia (A.Rich.) Hochst. ex Steud., 179
Eragrostis volkensii Pilg., 179
Eremomastax polysperma (Benth.) Dandy, 111
Eremomastax speciosa (Hochst.) Cufod., 111
Eremospatha macrocarpa (G.Mann & H.Wendl.) H.Wendl., 191
Erica mannii (Hook.f.) Beentje, 131
Erica tenuipilosa (Engl. ex Alm & T.C.E.Fr.) Cheek subsp. *tenuipilosa*, 131
ERICACEAE, 131
Erigeron bonariensis L., 122
Erigeron floribundus (Kunth.) Sch.Bip., 122
ERIOCAULACEAE, 176
Eriocaulon asteroides S.M.Phillips, 176
Eriocaulon parvulum S.M.Phillips, 177
Eriocaulon zambesiense sensu F.W.T.A., 176
Eriosema montanum Baker f. var. *montanum*, 142
Eriosema parviflorum E.Mey., 142
Eriosema scioanum Avetta subsp. *lejeunei* (Staner & Ronse Decr.) Verdc. var. *lejeunei*, 142
Erythrina sp. aff. excelsa Baker, 142
Erythrococca hispida (Pax) Prain, 132
Eucalyptus sp. 1, 150
Eugenia gilgii Engl. & Brehmer, 150
Eulophia cf. barteri Summerh., 189
Eulophia cucullata (Sw.) Steud., 189
Eulophia horsfallii (Batem.) Summerh., 189
Eulophia odontoglossa Rchb.f., 189
Eulophia shupangae (Rchb.f.) Kraenzl., 189
Eulophia stachyodes Rchb.f., 189
Euphorbia depauperata Hochst. ex A. Rich., 132
Euphorbia hirta L., 132
Euphorbia schimperiana Scheele var. *schimperiana*, 132
EUPHORBIACEAE, 131
Fagara rubescens (Planch. ex Hook.f.) Engl., 161
Fagopyrum snowdenii (Hutch. & Dandy) S.P.Hong, 153
Festuca camusiana subsp. *chodatiana* St.-Yves, 179
Festuca mekiste Clayton, 179
Ficus cf. cyathistipula Warb., 149
Ficus cf. sur Forssk., 149
Ficus chlamydocarpa Mildbr. & Burret, 149
Ficus natalensis Hochst., 149
Ficus ottoniifolia (Miq.) Miq., 149
Ficus sp. aff. lutea Vahl, 149

Passiflora edulis Sims fa. **edulis**, 151
PASSIFLORACEAE, 151
Paullinia pinnata L., 162
Pauridiantha paucinervis (Hiern) Bremek. subsp.
  paucinervis, 159
Pavetta hookeriana Hiern var. **hookeriana**, 159
Pavonia urens Cav. var. **glabrescens** (Ulbr.) Brenan, 146
Pavonia urens Cav. var. **urens**, 146
Pellaea quadripinnata (Forssk.) Prantl, 194
Pennisetum clandestinum Hochst. ex Chiov., 182
Pennisetum giganteum A.Rich., 182
Pennisetum glabrum Steud., 182
Pennisetum glaucocladum Stapf ex C.E.Hubb., 182
Pennisetum hordeoides (Lam.) Steud., 182
Pennisetum monostigma Pilg., 182
Pennisetum purpureum Schumach., 182
Pennisetum thunbergii Kunth., 182
Pennisetum trachyphyllum Pilg., 182
Pennisetum unisetum (Nees.) Benth., 182
Pentarrhinum abyssinicum Decne. subsp. **angolense**
  (N.E.Br.) Liede & Nicholas, 115
Pentarrhinum abyssinicum Decne. subsp. **ijimense** Goyder,
  115
Pentas ledermanii Krause, 159
Pentas pubiflora S.Moore subsp. *bamendensis* Verdc., 159
Pentas pubiflora S.Moore subsp. **pubiflora**, 159
Pentas purpurea Oliv. subsp. **purpurea**, 159
Pentas schimperiana (A.Rich.) Vatke subsp. **occidentalis**
  (Hook.f.) Verdc., 159
Peperomia cf. laeteviridis Engl., 152
Peperomia fernandopoiana C.DC., 152
Peperomia molleri C.DC., 152
Peperomia retusa (L.f.) A.Dietr., 152
Peperomia tetraphylla (G.Forst.) Hook. & Arn., 152
Peperomia thomeana C.DC., 152
Peperomia vulcanica Baker & C.H.Wright, 152
Periploca nigrescens Afzel., 115
Peucedanum angustisectum (Engl.) Norman, 166
Peucedanum camerunensis Jacq.-Fél., 167
Peucedanum cf. angustisectum (Engl.) Norman, 166
Peucedanum winkleri H.Wolff, 167
Phaulopsis angolana S.Moore, 112
Philippia mannii (Hook.f.) Alm & T.C.E.Fr., 131
Phoenix reclinata Jacq. var. **reclinata**, 191
Phragmanthera polycrypta (Didr.) Balle, 146
Physalis peruviana L., 164
Phytolacca dodecandra L'Hér., 152
PHYTOLACCACEAE, 152
Pilea ceratomera Wedd., 168
Pilea rivularis Wedd., 168
Pilea tetraphylla (Steud.) Blume, 169
Piliostigma thonningii (Schum.) Milne-Redh., 140
Pimpinella oreophila Hook.f., 167
Pimpinella sp. aff. **praeventa** Norman, 167
PINOPSIDA, 193
Piper capense L.f., 152
PIPERACEAE, 152
PITTOSPORACEAE, 152
Pittosporum viridiflorum Sims *"mannii"*, 152
Platostoma rotundifolium (Briq.) A.J.Paton, 137

Plectranthus alpinus (Vatke) Ryding, 137
Plectranthus assurgens (Baker) J.K.Morton, 137
Plectranthus cyaneus Gürke, 137
Plectranthus esculentus N.E.Br., 137
Plectranthus glandulosus Hook.f., 137
Plectranthus gracillimus T.C.E.Fr. ex Hutch. & Dandy, 137
Plectranthus hadiensis Forssk., 137
Plectranthus insignis Hook.f., 137
Plectranthus punctatus L'Hér. subsp. **lanatus** J.K.Morton,
  137
Plectranthus sp. nov. ?, 138
Plectranthus sylvestris Gürke, 138
Plectrelminthus caudatus (Lindl.) Summerh., 190
Pleopeltis excavata (Bory ex Willd.) T.Moore, 197
Pleopeltis lanceolata (L.) Kaulf., 198
Pleopeltis macrocarpa (Bory ex Willd.) Kaulf. var.
  macrocarpa, 198
Pleopeltis preussii (Hieron.) Tardieu, 198
PLUMBAGINACEAE, 153
Plumbago zeylanica L., 153
Pneumatopteris unita (Kunze) Holttum, 198
Poa annua L., 182
Poa leptoclada Hochst. ex A.Rich., 182
Podangis dactyloceras (Rchb.f.) Schltr., 190
PODOCARPACEAE, 193
Podocarpus latifolius (Thunb.) Mirb., 193
PODOSTEMACEAE, 153
Poecilostachys oplismenoides (Hack.) Clayton, 182
Polygala albida Schinz subsp. **stanleyana** (Chodat) Paiva,
  153
Polygala tenuicaulis Hook.f. subsp. **tayloriana** Paiva, 153
POLYGALACEAE, 153
POLYGONACEAE, 153
Polygonum cf glomeratum Dammer, 153
Polygonum cf salicifolium Brouss. ex Willd., 154
Polygonum cf strigosum R.Br., 154
Polygonum nepalense Meisn., 154
Polygonum salicifolium Brouss. ex Willd., 154
POLYPODIACEAE, 197
Polyscias fulva (Hiern) Harms, 114
Polystachya albescens Ridl. subsp. **albescens**, 190
Polystachya alpina Lindl., 190
Polystachya bennettiana Rchb.f., 190
Polystachya bicalcarata Kraenzl., 190
Polystachya cultriformis (Thouars) Spreng., 190
Polystachya elegans Rchb.f., 190
Polystachya fusiformis (Thouars) Lindl., 190
Polystachya galeata (Sw.) Rchb.f., 190
Polystachya laxiflora Lindl., 190
Polystachya odorata Lindl. var. **odorata**, 190
Polystachya parva Summerh., 191
Polystachya steudneri Rchb.f., 191
Polystachya superposita Rchb.f., 191
Polystichum fuscopaleaceum Alston, 196
Polystichum transvaalense N.C.Anthony, 196
Polystichum wilsonii H.Christ, 196
Popowia littoralis Bagsh. & Baker.f., 114
Potamogeton schweinfurthii A.Benn., 191
Potamogeton sp. 1, 191
Potamogeton sp. 2, 191

Sericostachys scandens Gilg & Lopr., 113
Sesbania macrantha Welw. ex E.Phillips & Hutch., 143
Setaria caudula Stapf, 183
Setaria longiseta P.Beauv., 183
Setaria pallide-fusca (Schumach.) Stapf & C.E.Hubb., 183
Setaria poiretiana (Schult.) Kunth., 183
Setaria pumila (Poir.) Roem. & Schult., 183
Setaria sphacelata (Schumach.) Stapf & C.E.Hubb. ex
    M.B.Moss var. *sphacelata*, 183
Sibthorpia europaea L., 163
Sida acuta Burm.f. subsp. *carpinifolia* (L.f.) Borss.Waalk.,
    146
Sida rhombifolia L., 147
Sigesbeckia abyssinica (Sch.Bip.) Oliv. & Hiern, 126
Sigesbeckia orientalis L., 126
SIMAROUBACEAE, 163
SMILACACEAE, 192
Smilax kraussiana Meisn., 192
Smithia elliotii Baker f. var. *elliotii*, 144
SOLANACEAE, 163
Solanecio mannii (Hook.f.) C.Jeffrey, 126
Solanum aculeastrum Dunal var. *albifolium* (C.H.Wright)
    Bitter, 164
Solanum aculeatissimum Jacq., 164
Solanum distichum Thonn., 164
Solanum indicum L. subsp. *distichum* (Thonn.) Bitter var.
    grandemunitum Bitter, 164
Solanum indicum L. subsp. *distichum* (Thonn.) Bitter var.
    distichum Bitter, 164
Solanum indicum L. subsp. *distichum* (Thonn.) Bitter var.
    modicearmatum Bitter, 164
Solanum nigrum L. *sens. lat.*, 164
Solanum terminale Forssk., 165
Solenostemon decumbens (Hook.f.) Baker, 139
Solenostemon mannii (Hook.f.) Baker, 139
Solenostemon repens (Gürke) J.K.Morton, 139
Sonchus angustissimus Hook.f., 126
Sonchus elliotianus Hiern, 125
Sonchus rarifolius Oliv. & Hiern, 126
Sonchus schweinfurthii Oliv. & Hiern, 126
Sopubia mannii Skan var. *mannii*, 163
Sorindeia grandifolia Engl., 114
Spermacoce natalensis (Hochst.) K.Schum. ex S.Moore,
    161
Spermacoce princeae K.Schum. var. *princeae*, 161
Spermacoce pusilla Wall., 161
Spermacoce sphaerostigma (A.Rich.) Vatke, 161
Sphaerocodon caffrum (Meisn.) Schltr., 115
Spilanthes africana DC., 121
Sporobolus africanus (Poir.) Robyns & Tournay, 183
Sporobolus infirmus Mez, 183
Sporobolus mauritianus (Steud.) T.Durand & Schinz, 183
Sporobolus micranthus (Steud.) T.Durand & Schinz, 183
Sporobolus paniculatus (Trin.) T.Durand & Schinz, 183
Sporobolus pyramidalis P.Beauv., 183
Sporobolus subulatus Hack ex Scott-Elliot, 183
Stachys aculeolata Hook.f. var. *aculeolata*, 139
Stachys pseudohumifusa Sebsebe subsp. *saxeri* Y.B.Harv.,
    139
Stellaria mannii Hook.f., 119

Stellaria media (L.) Vill., 119
Stephania abyssinica (Quart.-Dill. & A.Rich.) Walp. var.
    abyssinica, 148
STERCULIACEAE, 165
Stereospermum acuminatissimum K.Schum., 117
Stolzia repens (Rolfe) Summerh. var. *repens*, 191
Streblochaete longiaristata (A.Rich.) Pilg., 183
Streptocarpus elongatus Engl., 134
Strombosia scheffleri Engl., 150
Strophanthus sp., 114
Succisa trichotocephala Baksay, 130
Swertia abyssinica Hochst., 133
Swertia mannii Hook.f., 134
Swertia usambarensis Engl., 134
Symphonia globulifera L.f., 135
Syzygium staudtii (Engl.) Mildbr., 150
Tabernaemontana cf. ventricosa Hochst. ex A.DC., 114
Tapinanthus globiferus (A.Rich.) Tieghem, 146
Tapinanthus letouzeyi (Balle) Polhill & Wiens, 146
Tarenna pavettoides sensu F.W.T.A., non (Harv.) Sim, 161
Tectaria fernandensis (Baker) C.Chr., 196
Tephrosia preussii Taub., 144
Tephrosia vogelii Hook.f., 144
Terminalia avicennioides Guill. & Perr., 120
Ternstroemia polypetala Melch., 165
Tetracera alnifolia Willd., 130
Thalictrum rhynchocarpum Quart.-Dill. & A.Rich. subsp.
    rhynchocarpum, 155
THEACEAE, 165
THELYPTERIDACEAE, 198
Thelypteris confluens (Thunb.) Morton, 198
Thelypteris guineensis (H.Christ) Alston, 198
Thunbergia cf. fasciculata Lindau, 113
THYMELAEACEAE, 165
TILIACEAE, 165
Tithonia diversifolia (Hemsl.) A.Gray, 126
Torilis arvensis (Huds.) Link, 167
Tragia benthamii Baker, 133
Trema orientalis (L.) Blume, 165
Trichopteryx elegantula (Hook.f.) Stapf, 183
Trichopteryx marungensis Chiov., 183
Tridactyle tridactylites (Rolfe) Schltr., 191
Tridactyle tridentata (Harv.) Schltr., 191
Trifolium baccarini Chiov., 144
Trifolium simense Fresen., 144
Trifolium usambarense Taub., 144
Tripogon major Hook.f., 184
Triticum aestivum L., 184
Triumfetta annua L., 165
Triumfetta cordifolia A.Rich., 165
Triumfetta rhomboidea Jacq., 165
Triumfetta tomentosa Bojer var. *tomentosa*, 165
Tylophora cf. oblonga N.E.Br., 115
Uebelinia abyssinica Hochst., 119
Uebelinia hispida Pax, 120
ULMACEAE, 165
UMBELLIFERAE, 166
Umbilicus botryoides Hochst. ex A.Rich., 129
Urelytrum digitatum K.Schum., 184
Urelytrum fasciculatum Stapf ex C.E.Hubb., 184